MEMBRANE FLUIDITY

Biophysical Techniques and Cellular Regulation

EXPERIMENTAL BIOLOGY AND MEDICINE

MEMBRANE FLUIDITY: Biophysical Techniques and Cellular Regulation, Edited by **Morris Kates** and **Arnis Kuksis**, 1980

MEMBRANE FLUIDITY

Biophysical Techniques and Cellular Regulation

Edited by

Morris Kates

University of Ottawa

and

Arnis Kuksis

University of Toronto

The HUMANA Press, Inc. • Clifton, New Jersey

Library of Congress Cataloging in Publication Data

Main entry under title:

Membrane Fluidity.

(Experimental biology and medicine (Clifton, N.J.))
1. Cell membranes. 2. Fluids. 3. Lipids.
4. Cytology—Technique. I. Kates, Morris.
II. Kuksis, Arnis. III. Series. [DNLM: 1. Cell
membrane—Physiology. 2. Membrane Fluidity. QH601
M5323]
QH601.M4663 574.87'5 79-93347
ISBN 0-89603-020-2

Crescent Manor
P. O. Box 2148
Clifton, NJ 07015

Printed in the United States of America

PREFACE

This book represents the proceedings of a Satellite Symposium of the XIth International Congress of Biochemistry on "Control of Membrane Fluidity" which was held on July 7, 1979 at the Charles H. Best Institute, University of Toronto, Toronto, Canada. The meeting was organized by M. Kates and A. Kuksis and was supported by the International Congress of Biochemistry.

The purpose of the meeting was to review recent progress in many different areas of investigation bearing on the role of lipids in the structural and functional property of the cell membrane commonly referred to as fluidity. The aim was to emphasize the factors controlling membrane fluidity as studied in appropriate in vitro and in vivo experiments. The Symposium included invited review papers and short papers offered by discussants. In assembling the book no distinction has been made between the two types of presentations, nor has any significance been attached to the chronological order of their presentation in the Symposium. As a result it was possible to provide a much more coherent and continuous presentation than that available at the meeting.

The articles are arranged in five sections, the first of which deals with recent methodological advances in the chromatographic and mass spectrometric techniques that are used in the determination of the composition of membrane lipids, and includes a brief description of the Enhorning pulsating-bubble surface tension meter as means of assessing the physical properties of the lipid phase. This section concludes with a brief examination of the physicochemical meaning of the term fluidity.

The second section contains papers describing various correlations of changes in general fluidity of the membrane with changes in the physiological activity of the system or vice versa. It includes papers describing both normal and inverse correlations. In many of these studies thermal changes constitute a major part of the experimental manipulation of the system.

The third section describes those changes in the fatty acid composition of cell membranes that accompany specific physiological events, or vice versa. The fatty acid changes include those characterized by desaturation, and by shortening or extension of chain length, as well as the selective utilization of acids from a supply source.

The fourth section describes alterations in the composition of the phospholipids that accompany specific physiological events, or vice versa. In addition to alterations in the phospholipid head groups, there are usually changes in the composition of the component fatty acids, along with an intermediate involvement of the lysophospholipids. Although the relationship between the head group composition and membrane fluidity has not been as extensively investigated as that between fatty acid composition and membrane fluidity, there is good evidence that certain reciprocal changes requiring experimental attention may be involved.

The last section includes papers that emphasize the homeostatic regulation of membrane fluidity. This regulatory process can be considered to be the basic overall transformation underlying all responses of the cell to internal or external perturbation of the membrane. In addition to adjustments in fatty acid and head group composition, homeostatic regulation probably also involves changes in the free cholesterol content and in the lipid/protein ratio of the membrane.

Among the highlights of the proceedings may be mentioned the following:

1. The recent development of rapid analyses of the molecular species of the lipid phase by capillary chromatography and selected ion mass spectrometry.

2. Correlation of changes in phospholipid composition with certain physiological events.

3. The concept of inverse correlation between membrane fluidity and certain enzymic activities.

4. The discontinuous lateral distribution (or patching) of lipids in cell membranes.

5. The general recognition that the idea of membrane fluidity requires a revision in its working definition, or perhaps a total replacement by some other more suitable concept.

Investigation of membrane fluidity in vivo is complicated by its homeostatic regulation. The problem of lipid patching is extremely important since it complicates the interpretation of the results obtained with randomly introduced reporter molecules. Future work will require detailed knowledge of the distribution of reporter molecules within the membrane monolayer. The concept of lipid patching likewise may put an end to any efforts to calculate membrane fluidity from physicochemical first principles without prior knowledge of the submembrane distribution of the component lipid classes and molecular species. As a result of this Symposium, it is clear that new efforts and approaches will be required for the design of more meaningful experiments, and for the interpretation of their results in terms of the structure and function of cell membranes and the role of the lipid phase therein. It is our hope that this volume will stimulate increased research activity in this area.

We wish to express our thanks to Drs. K. Keough, F. Possmayer, and J. N. Hawthorne for presiding at the sessions, and to the authors for preparing their manuscripts in such short order. We wish also to express our gratitude to Mrs. Anne Brown without whose secretarial assistance the Symposium and this volume would not have been possible.

February, 1980

Morris Kates
Arnis Kuksis

CONTENTS

PART I
MEASUREMENT OF MEMBRANE LIPID COMPOSITION AND FLUIDITY

PART II
CORRELATION OF MEMBRANE FLUIDITY WITH PHYSIOLOGICAL ACTIVITY

PART III
FATTY ACID CHANGES ACCOMPANYING PHYSIOLOGICAL EVENTS

PART IV
PHOSPHOLIPID CHANGES ACCOMPANYING PHYSIOLOGICAL EVENTS

PART V
HOMEOSTATIC REGULATION OF MEMBRANE FLUIDITY

PART I

MEASUREMENT OF MEMBRANE LIPID COMPOSITION AND FLUIDITY

NEW APPROACHES TO LIPID ANALYSES OF LIPOPROTEINS AND CELL MEMBRANES

A. Kuksis and J.J. Myher

Banting and Best Dept. of Medical Research
University of Toronto
Toronto, Ontario M5G 1L6
Canada

ABSTRACT

Conventional and capillary gas chromatography-mass spectrometry (GC/MS) was used to determine the molecular association and positional distribution of fatty acids in diacylglycerols and of molecular association of nitrogenous bases and fatty acids in ceramides following their conversion to the tert-butyl dimethyl silyl (t-BDMS) ethers. The diacylglycerols and ceramides were generated from the corresponding glycerophospholipids and sphingomyelins by hydrolysis with phospholipase C of the original lipoproteins and cell membranes or appropriate total lipid extracts thereof. The abundant M-57 ion provided the molecular weights for both diacylglycerols and ceramides. The abundance ratio of the ions due to losses of the acyloxy radical (M-RCOO) from position 1 and from position 2 indicated the proportion of the reverse isomers of the saturated diacylglycerols. The ratio of the reverse isomers in unsaturated diacylglycerols could be similarly determined after reducing the sample with hydrogen (unsaturated species having higher carbon numbers) or with deuterium in the presence of Wilkinson's catalyst. The alkylacyl and alkenylacyl glycerols were similarly identified by abundant ions at m/e values corresponding to [M-57], [acyl + 74], [RO + 114] and [R-CH=CHO + 114]. The long chain bases in the ceramides were identified via the intense ions $CH_3(CH_2)_n$-$(CH=CH)_m$-CH_2O-t-BDMS and the fatty acids via the intense ions

$CH_3(CH_2)_n-(CH=CH)_m-CONHCH-CH_2O-$ t-BDMS. These methods are suitable for work with stable isotope-labelled phospholipids provided the isotopes are present in sufficient excess at least in one of the neutral lipid moieties of the molecular species. The assays usually do not require silver nitrate prefractionation of the sample and can be executed in the microgram range of material.

The above methods have been applied in analyses of molecular species of phosphatidylcholines and sphingomyelins of high density lipoproteins (HDL) of human plasma, and of the free diacylglycerol fraction obtained from an isolated rat liver during perfusion with deuterium oxide.

INTRODUCTION

There is good evidence to indicate that many structural and functional properties of lipoproteins (Scanu, 1969; Morrisett et al., 1976) and cell membranes (Chapman, 1973; Gregoriadis et al., 1977) are influenced or controlled by the physico-chemical characteristics of the lipid phase. In view of the relatively specific composition of the molecular species of natural glycerophospholipids (Kuksis, 1972a; Holub and Kuksis, 1978) and sphingomyelins (Samuelsson and Samuelsson, 1969), it is obvious that the positional placement and molecular association of the fatty chains in the lipid molecules constitute the determinant factors of both micellar and monolayer or bilayer properties of the lipid phase of lipoproteins and cell membranes. This recognition along with the knowledge that diet (Farias et al., 1975) and disease (Wood, 1973; Wood, 1975) may alter the fatty acid composition of these lipids have led to an expenditure of much ingenuity and effort into identifying and quantitating the composition of the molecular species of the various lipid classes making up specific lipoproteins and cell membranes (Holub and Kuksis, 1978). In the course of the development of appropriate experimental techniques the analytical approaches have ranged from detailed fractionation of large samples of material to superficial appraisal of miniature samples. In the following a brief review is presented of these efforts along with an attempt at an utopian solution of the problem by providing detailed composition of molecular species of minute samples by means of capillary GC/MS with selected ion monitoring of improved chemical derivatives.

OLD APPROACHES

The early methods of analysis of lipids of lipoproteins and cell membranes date back to the first preparations of pure lipoproteins and subcellular fractions and were largely limited to the determination of total fatty acids. In a few instances the fatty acid determinations were preceded by a separation of the polar and non-polar lipid classes by adsorption column chromatography. In rare instances the lipoprotein (Skipski, 1972) and cell membrane (Keenan and Morre, 1970; Colbeau *et al.*, 1971) fatty acids were determined following a preliminary resolution of the phospholipid classes. A major improvement upon these analyses consisted of the inclusion of the earlier demonstrated (Tattrie, 1959; Hanahan *et al.*, 1960; De Haas *et al.*, 1962) separate release of the fatty acids from the sn-1- and 2-positions of the phospholipid molecules by means of phospholipase A_2 digestion. However, these analyses did not allow the identification and quantitation of any specific structures for the molecular species of natural glycerophospholipids or sphingomyelins.

True analyses of the molecular species of glycerophospholipids and sphingomyelins became possible with the advent of thin-layer chromatography (TLC) with silver nitrate and with high temperature gas chromatography (GLC). These methods yielded estimates for either individual molecular species of the phospholipids or small groups thereof, the complete structure of which could be determined by a fatty acid analysis, especially when combined with an enzymic positional analysis. The first successful analyses along these lines were carried out by Renkonen (1965), who employed $AgNO_3$-TLC for the fractionation of the diacylglycerol moieties of the phosphatidylcholines and of the ceramides of sphingomyelins following either a chemical or enzymatic dephosphorylation of the parent compounds. Particularly effective proved the removal of the phosphates of the nitrogenous bases with phospholipase C, which could also be applied to other glycerophospholipids. The enzymes from *Clostridium perfringens* (Stahl, 1973) and *Bacillus cereus* (Mavis *et al.*, 1972), however, have different specificities for the different phospholipid classes. These enzymes attack the plasmalogens and their alkyl ether homologues at markedly slower rates than the corresponding diacyl esters (Renkonen, 1966). Furthermore, there are differences in susceptibility to phospholipase C of free and membrane-bound phospholipids (Rottem et al., 1973).

Although the AgNO3-TLC did not yield individual species, the resulting mixtures were sufficiently simple to provide nearly complete compositions of the fatty acids for the major molecular species (Renkonen, 1965; Renkonen, 1966). A complete account of the molecular species could be obtained by including GLC in the analytical scheme which provided the molecular weight distribution of the diacylglycerols (Kuksis, 1965; Kuksis and Marai, 1967; Renkonen, 1967) or ceramides (Samuelsson and Samuelsson, 1970) from which the quantitative proportions of the fatty chain pairings in each silver nitrate fraction could be readily obtained. A preliminary resolution of the alkylacyl, alkenylacyl and diacyl glycerols by TLC on plain silica gel allowed the separate estimation of the ester and ether subclasses of the glycerophospholipids from those tissues that contained sufficient amounts of both (Renkonen, 1966; Renkonen, 1971). These methods which were originally worked out for the choline phospholipids were eventually adopted in a modified form for the analysis of the molecular species of phosphatidylethanolamine (Wood and Harlow, 1969; Holub and Kuksis, 1971a), phosphatidylinositol (Holub and Kuksis, 1971b) and phosphatidylserine (Wood, 1973), including any plasmalogens (Marai and Kuksis, 1973a&b).

A major drawback to an efficient utilization of these methods of analysis of the molecular species of phospholipids is the necessity of removing the nitrogenous bases and the phosphate moieties of the molecules resulting in a loss of information about the labelling of these parts of the molecules during metabolic transformations. The latter difficulty can be overcome to a considerable extent by utilizing AgNO3-TLC for a resolution of the intact phospholipid molecules either as such (Arvidson, 1965; Arvidson, 1968; Salem, 1976) or after masking (Sundler and Akesson, 1973; Yeung et al., 1977) or destroying (Luthra and Sheltawy 1972a&b) the polar groups. The major unsaturation classes of the glycerophospholipids can then be isolated and subjected to further fractionation by reversed phase partition chromatography (Arvidson, 1967; Shamgar and Collins, 1975a&b), or can be degraded to complete the identification of the species as described. The above methods of analysis of the molecular species of glycerophospholipids have provided the main routines for obtaining the lipid composition of essentially all lipoproteins and cell membranes thus far analyzed in detail (Holub and Kuksis, 1978).

The detailed analyses of the molecular species of the glycerophospholipids and sphingomyelins by the above methods, however, require relatively large amounts of material and are extremely time-consuming, so that only a few complete analyses can be obtained in a reasonable period of time. These analyses usually are not suitable for examination of replicates and multiple samples obtained in time-course studies of reactions. Various attempts have therefore been made to simplify the analyses by increasing either the chromatographic resolution or by employing methods that would allow an assay of the composition of the unfractionated sample, such as infrared and mass spectrometry. A resolution of the diacylglycerols as the trimethylsilyl (TMS) ethers or acetates by GLC using polar liquid phases yields essentially pure molecular species and eliminates the need for AgNO3-TLC (Kuksis, 1971; Kuksis, 1972b). The separations are based on both molecular weight and degree of unsaturation, but reverse isomers and enantiomers are not resolved. The early polar phases did not allow an effective recovery of the polyunsaturated long chain species, but this difficulty has been overcome using the more recently synthesized polar silicone phases (Myher and Kuksis, 1975; Breckenridge _et al_., 1976), which are also suitable for work with triacylglycerols (Tagaki and Itabashi, 1977). However, this approach has thus far found only a limited application in lipoprotein and cell membrane lipid analyses. An adaption of this method to capillary GLC promises to yield improved separation and recovery of all components.

A significant improvement in the identification and quantitation of the molecular species of diacylglycerols has been realized by GC/MS. Both acetyl (Hasegawa and Suzuki, 1975) and TMS (Curstedt and Sjovall, 1974; Satouchi _et al_., 1978) derivatives have been employed with non-polar packed GLC columns. These derivatives, however, yield only low intensities for the higher mass fragments from which the molecular species are being identified. An improved derivative, permitting much more sensitive analyses, is the t-BDMS ether, which yields a prominent M-57 ion for all species of diacylglycerols (Myher _et_ al., 1978;

Satouchi _et al._, 1979) and ceramides (Kuksis _et al._, 1979).

Improved separation and quantitation have recently been obtained by reversed phase HPLC of molecular species of phosphatidylethanolamines and phosphatidylserines after preparation of ultraviolet-absorbing derivatives (Jungalwala _et al._, 1975) and with detection in the region of 200 nm without derivatization (Jungalwala _et al._, 1976). Especially extensive resolution has been achieved for the molecular species of the sphingomyelins (Jungalwala _et al._, 1979). Likewise, effective separations of molecular species by liquid-liquid chromatography have been achieved in the recent past for diacylglycerols (Curstedt and Sjovall, 1974) and triacylglycerols (Lindqvist _et al._, 1974). For effective identification of the resolved components, these methods must await integration with mass spectrometry.

The lipid class composition of plasma lipoproteins without prior separation has been analyzed by infrared spectrometry (Freeman, 1968; Freeman, 1972). The information obtained is essentially that from separate chemical determinations of total cholesterol, total esterified fatty acids and lipid phosphorus. In addition to the determination of the lipid classes, there have been some infrared analyses related to fatty acid characteristics. It is possible to isolate the average chain length in mixed saturated fatty acids from the ratio of C-H to C-O absorption, but this method is of little utility when dealing with a complex mixture of lipid esters. The levels of _trans_-acids may be estimated and conclusions may be reached about the dietary proportion of the corresponding molecular species in the mixture, as they are believed to be exclusively of exogenous origin. Among the old approaches must also be considered the early attempts to identify and quantitate molecular species of glycerophospholipids by direct electron impact mass spectrometry (Klein, 1973). By analysis of the fragment ions observed in the mass spectra of the phosphatidylcholines, it is possible to gain information concerning the fatty acyl chains, the glycerol backbone and the phosphorylcholine moiety. This information is of importance in considering the molecular species present in naturally occurring phosphatides.

Triacylglycerols may be admitted to the mass spectrometer directly (Hites, 1970; Hites, 1975) or by GLC (Murata and Takahashi, 1973). The abundance of the molecular ion in the spectra of long-chain triacylglycerols is generally low and decreases with chain length. There are a number of ion types that can be used to determine the individual fatty acids that make up the triacylglycerol.

NEW APPROACHES

New approaches to analyses of molecular species of lipids in lipoproteins and cell membranes have been directed first of all towards improving the older techniques by taking advantage of newer chromatographic technology and instrumentation. A direct mass spectrometric analysis of molecular species of glycerophospholipids and acylglycerols with considerable future potential is based on field desorp-technique (Evans _et al._, 1974; Wood and Lau, 1974). It allows the estimation of the molecular weight of even the most labile molecular species along with a few characteristic ions sufficient for identification and quantitation of the molecules. At the present time this technique requires much larger quantities of material than any of the other mass spectrometric methods, as well as suffers from poor reproducibility (Main _et al._, 1976; Wood _et al._, 1977). Chemical ionization mass spectrometry yields higher intensities for ions characteristic of the molecular weight of the molecular species and therefore provides a better basis for quantitative analysis of mixed natural acylglycerols by mass spectrometry (Murata, 1977; Murata and Takahashi, 1977). Furthermore, capillary GLC has been substituted for packed column GLC to obtain improved resolution and more sensitive detection of glycerolipids (Grob and Grob, 1969; Schomberg _et al._, 1976) and ceramides (Kuksis _et al._, 1979) on the basis of molecular weight and/or the overall shape of the molecule using nonpolar liquid phases of increased thermal stability. Particularly promising appears to be the combination of capillary GC/MS of the t-BDMS ethers of the diacylglycerols and ceramides. The t-BDMS ethers are stable to moisture and can be isolated from the reaction mixture and purified prior to analysis (Myher, 1978), which is not possible when dealing with the TMS ethers.

Myher et al., 1978). Like the TMS ethers the t-BDMS derivatives of diacylglycerols (Myher et al., 1978; Satouchi and Saito, 1979) and ceramides (Kuksis et al., 1979) possess excellent chromatographic properties and many of the mass spectrometric properties of the TMS ethers. A special advantage of the t-BDMS ethers is the prominent M-57 fragment, which can be utilized for accurate measurement of molecular weight of the diacylglycerols and ceramides.

Figure 1 shows a capillary GC/MS run obtained for the diacylglycerols derived from the phosphatidylcholines of

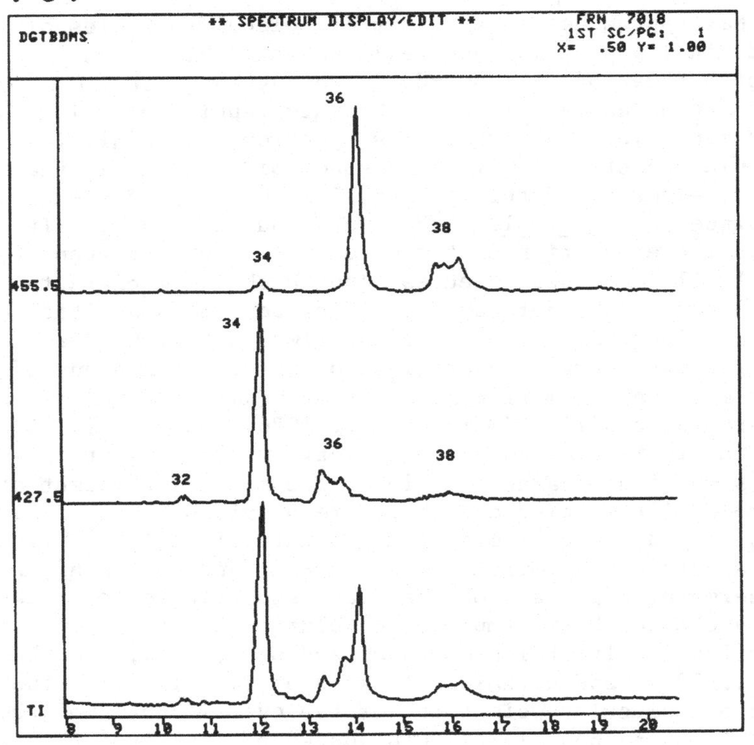

Fig. 1. Capillary GC/MS of t-BDMS ethers of diacylglycerols derived from phosphatidylcholines of plasma HDL. TI, total ion current; 427.5, single ion scan for species containing palmitic acid; 455.5, single ion scan for species containing stearic acid (EI, 70 eV. Hewlett-Packard 5985B quadrupole mass spectrometer).

human plasma HDL, as the t-BDMS ethers. The total ion current gives the complete mass distribution profile of the molecular species based on carbon number and to some extent also on the degree of unsaturation. The unsaturated species are eluted ahead of the saturated species within each carbon number, and for some of the components a complete baseline resolution may be easily achieved. The single ion scans for M-RCOO indicate the distribution of the palmitic and stearic acids among the carbon numbers of the diacylglycerols. It is seen that the palmitic acid is present largely in the peaks with carbon numbers 34 and 36, while the stearic acid is present in the peaks with carbon numbers 36 and 38, as expected.

Figure 2 gives a similar total ion current profile of the diacylglycerols as in Figure 1, but the single ion scans now represent those of the RCO + 74 ions of oleic and eicosatrienoic acids. It is seen that the oleic acid is confined largely to the peaks with carbon numbers 34 and 36 (two peaks), while the eicosatrienoic acid is confined largely to the peaks with carbon numbers 36 and 38. Similar single ion plots have been obtained for all the fatty acids of the various diacylglycerol peaks and appropriate conclusions made about the total carbon number and molecular association of the fatty acids in each molecular species. The quantitative composition of the molecular species can then be calculated provided appropriate calibration factors are used for the response in the mass spectrometer.

The high sensitivity of the GC/MS system is illustrated by the identification of the tiny peak following that of carbon number 32 as a mixture of alkenyl-acylglycerols (see below).

When the diacylglycerols are analyzed as the t-BDMS ethers, it is also possible to differentiate between the reverse isomers of the saturated species (e.g. molecular species differing only in the content of fatty acids in the sn-1- and sn-2-positions, but having the same total composition). Figure 3 shows the mass spectra of the _rac_-1-palmitoyl-2-stearoyl and _rac_-1-stearoyl-2-palmitoylglycerols as the t-BDMS ethers. The major difference is in the abundance ratio of the ions produced due to losses of the acyloxy radical (M-RCOO) from position 1 (or 3) and position 2. About twice as much of the radical is lost from the 2 position. This differentiation, however, is possible only with saturated diacylglycerols, the unsaturated species failing to give sufficiently high intensities for the diagnostic ions. The difficulty can be overcome by reducing the sample with hydrogen, if necessary after a preliminary AgNO3-TLC. An alternative procedure is to reduce each of the fractions from AgNO3-TLC (or the total sample) with deuterium prior to the GC/MS analysis. Not only would this procedure remove the need for calibration factors for unsaturated compounds, but also the original degree of unsaturation for each fatty acid could still be determined from the m/e values of the respective (M-RCOO)

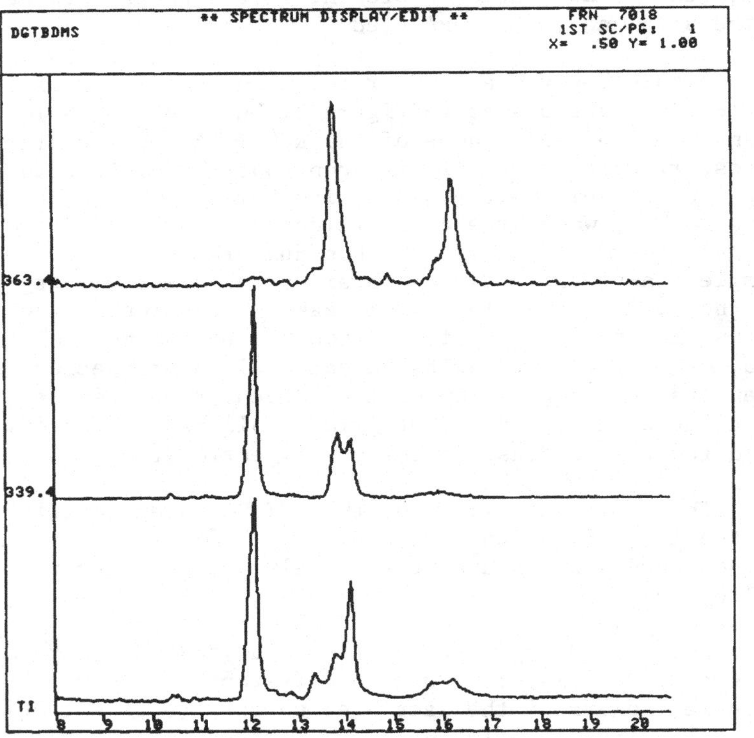

Fig. 2. Capillary GC/MS of t-BDMS ethers of diacylglycerols derived from phosphatidylcholines of plasma HDL. TI, total ion current; 339.4, single ion scan for species containing oleic acid; 363.4, single ion scan for species containing eicosatrienoic acid. GC/MS conditions as in Fig. 1

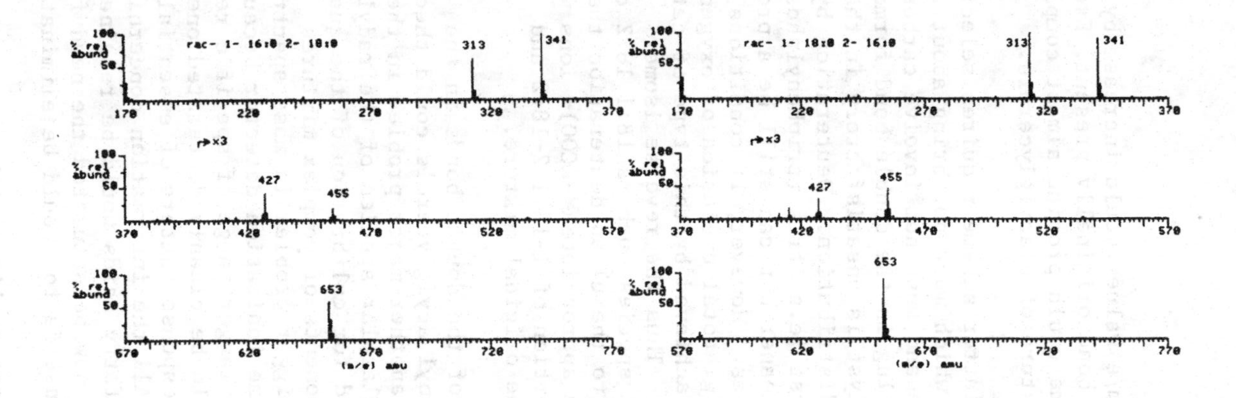

<u>Fig. 3</u>. Mass spectra of t-BDMS ethers of <u>rac</u>-1-palmitoyl 2-stearoyl and <u>rac</u>-1-stearoyl 2-palmitoylglycerols.

or (RCO + 74)+ ions. The m/e values would increase by 2 mass units for each double bond originally present. Except for enantiomers, this scheme could provide almost complete resolution of a natural mixture of diacylglycerols.

To be practical, the latter scheme requires selective catalysts for deuteration, which must not bring about a hydrogen-deuterium exchange and must not involve carbons other than those participating in the double bond formation. Deuteration via Adams catalyst is unsatisfactory in that it produces a wide deuterium distribution. Deuteration by means of a Wilkinson catalyst [e.g. tris (triphenylphosphine) rhodium (1) chloride] is cleaner but can still be a problem for polyunsaturated molecules. However, if conditions are sufficiently controlled (e.g. total exclusion of oxygen) a clean deuteration may be achieved by this type of catalyst (Emken, 1978; Myher, 1978). Thus the reverse isomer content of a diacylglycerol species such as 18:1 18:2 could be determined. Subsequent to the clean deuteration the relative intensities of the appropriate (M-RCOO)+ ions would be used to define the proportion of 1-18:1 2-18:2 and 1-18:2 2-18:1 present in the original mixture.

Catalytic deuteration of the double bonds in the diacyl, alkyl acyl and alkenyl acylglycerols could theoretically be used to overcome another major problem in the mass spectrometry of the molecular species of the radylglycerols, which is the need for calibration of the quantitative response of the components of complex mixtures (Myher, 1978). This is a bigger problem in mass spectrometry than in GLC using flame ionization detector because ion intensities have a much larger range of specific responses. If the samples could be cleanly deuterated one would be dealing with the response factors of essentially saturated molecules only. All the information concerning the original nature of the fatty acids would be retained and since all species would now be saturated the need for a double bond dependent response factor would be eliminated.

Similar methods have been utilized with comparable success for the analysis of the molecular species in the alkylacyl and alkenylacyl subclasses of glycerophospholipids. The molecular weights of these compounds are also recognized by the prominence of the corresponding M-57 ions, along with the characteristic (acyl + 74), (RO + 114) and

(R-CH=CHO + 114) ions. A preliminary TLC on plain silica gel is usually necessary to resolve the alkylacyl, alkenylacyl and diacylglycerol subclasses of the acylglycerols when present in the same sample.

The t-BDMS ethers of the alkenylacylglycerols can be identified by a number of characteristic fragments. The alkenyl portion of the molecule can be identified via ions corresponding to m/e (R"O + 40) and M- (RCOOH + 57), whereas the acyl portion of the molecule is readily identified by ions m/e (RCO + 74) and M-(R"O + 1) or M- (R"O). The molecular weight is derived via the M-57 ion. Figure 4 shows the mass spectra for the tiny peak of the t-BDMS ethers eluting just after the dipalmitoylglycerol species in Figure 1, along with the corresponding standards. It was possible to recognize the main species in the unknown mixture as 16:0" 18:2 with some 16:0" 18:1.

The alkylacylglycerols also give characteristic mass spectra and increased abundance of higher mass ions when analyzed as the t-BDMS derivatives. The alkyl portion of the molecule is identified primarily by means of the (M-(RCOOH + 57))+ ion but other less intense ions include (R'O + 40) and (R')+. The major acyl containing ion is again (RCO + 74)+ but there is also a less intense ion corresponding to RCO+. The molecular weight is derived from the abundant M-57 ion.

The mass spectra of dialkylglycerol t-BDMS ethers contain M-57 ions as well as (M-(R'OH + 57))+ and (RO + t-BDMS-OH) ions.

The t-BDMS ethers also provide improved derivatives for the analysis of the ceramides. The fragmentation pattern for the t-BDMS ethers is very similar to that observed for the TMS ethers (Samuelsson and Samuelsson, 1969) except that the higher mass ions are of greater relative intensities. The t-BDMS ethers are also much more stable than the TMS derivatives and hence can be purified and fractionated by TLC and $AgNO_3$-TLC, if necessary, without fear of hydrolysis. Furthermore, the t-BDMS ethers of the ceramides also do not break down in the molecular separator as do the TMS ethers. The principle cleavage occurs between the carbon atoms containing the secondary O-t-BDMS group and the amide linkage. The

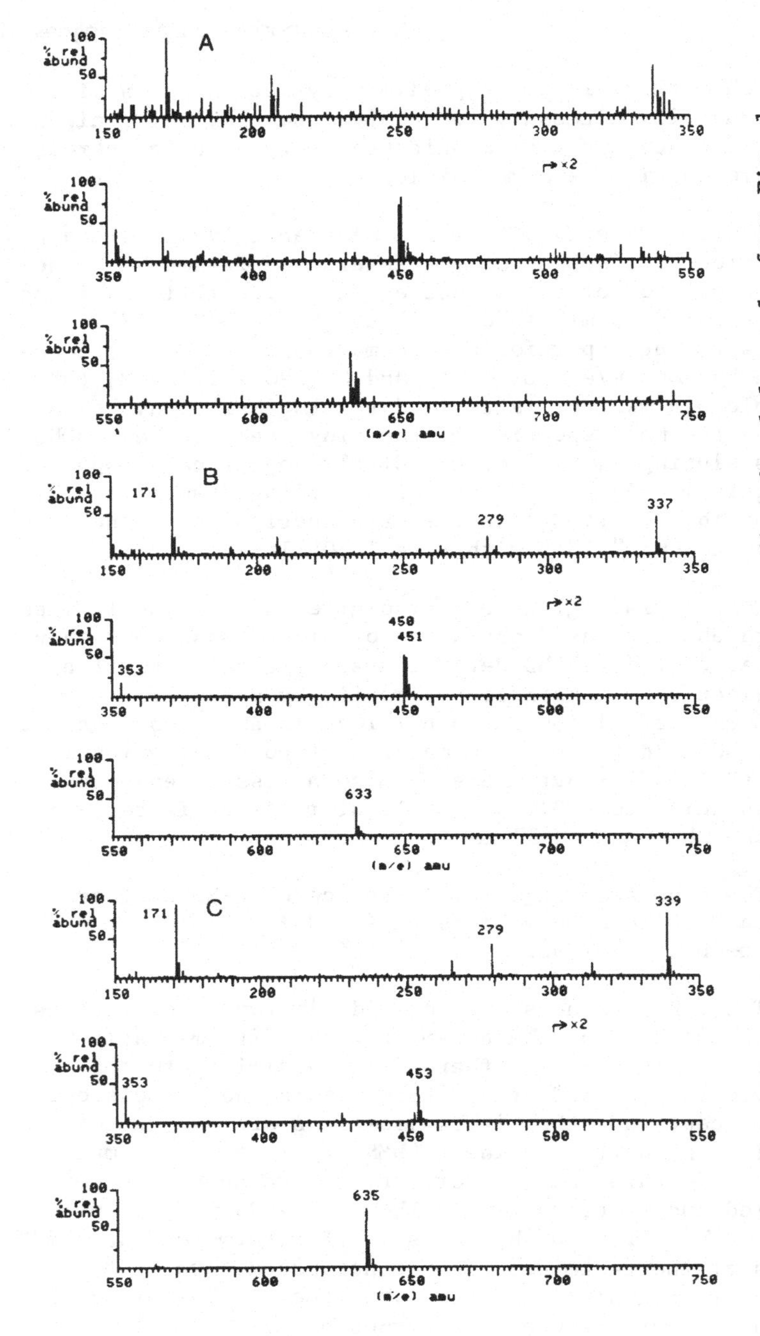

Fig. 4. Mass spectra of t-BDMS ethers of unknown alkenylacylglycerols from Fig. 1 (A), and of standard 1-hexadecenyl 2-linoleoyl (B), and standard 1-hexadecenyl 2-oleoyl (C) glycerols.

ionized form of the two fragments have similar intensities and can be used to identify both the base and fatty acyl portions of the molecule. Unlike the TMS derivative there is also a very intense ion $(M-(RCONH_2 + 57))^+$, due to loss of the fatty amide plus a t-butyl radical. Other ions of lower intensity are also present. The long chain bases in the ceramides are identified via the intense ions $CH_3(CH_2)_n-(CH=CH)_m-CHO$-t-BDMS and the fatty acids via the intense ions $CH_3(CH_2)_n-CH=CH)_m-CONHCH-CH_2O$-t-BDMS.

Figure 5 shows a GC/MS run of the t-BDMS ethers of ceramides derived from the sphingomyelins of human

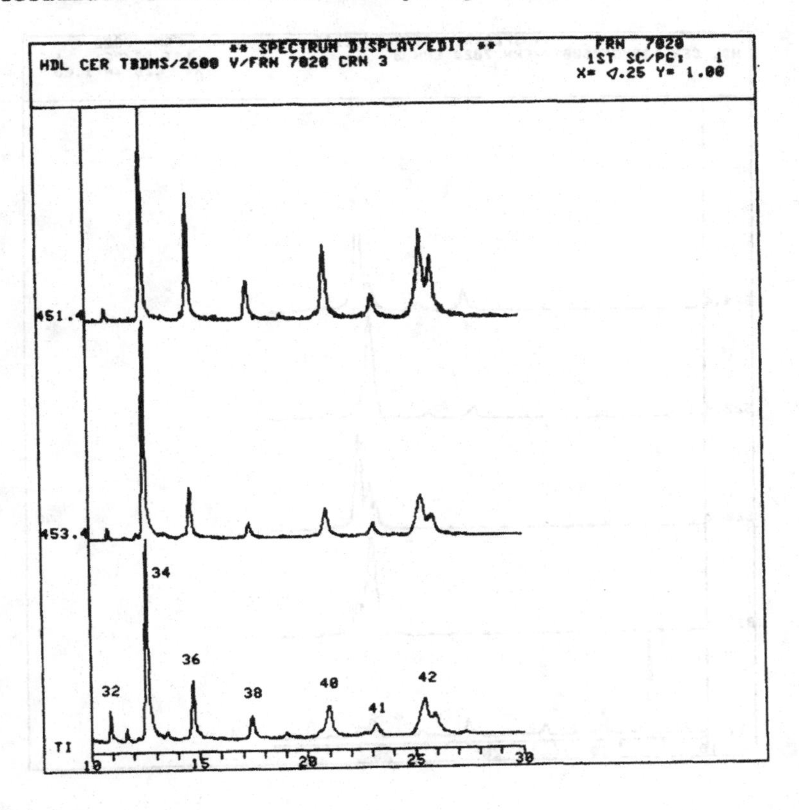

Fig. 5. Capillary GC/MS of t-BDMS ethers of ceramides derived from sphingomyelins of HDL. TI, total ion current; 453.4, single ion scan for species containing sphinga-dienine; 451.4, single ion scan for species containing sphingosine. GC/MS conditions as in Fig. 1.

plasma HDL. In addition to the total ion current scan, which provides the carbon number profile of the total mass of the ceramides, the figure includes single ion scans for the distribution of the nitrogenous bases, sphingosine and sphingadienine, among the different carbon numbers. It is seen that carbon number 36 has a higher sphingadienine/ sphingosine ratio than does the ceramide of carbon number 34. Figure 6 indicates in addition to the total ion current profile of the ceramide mass, the distribution of the lignoceric (24:0) and nervonic (24:1) acids. It is seen that the split in the ceramide peak with carbon number

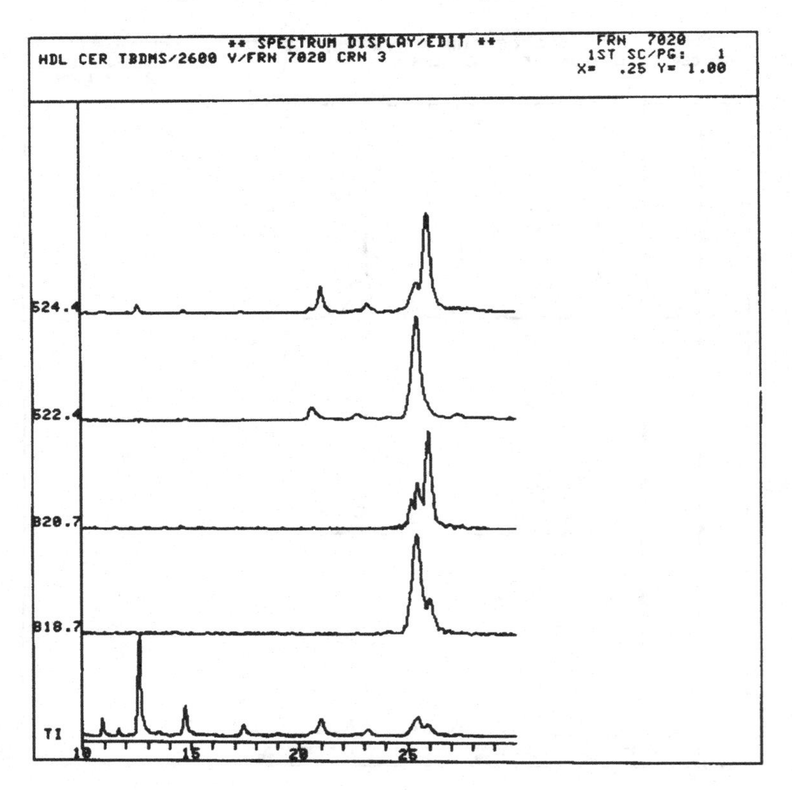

Fig. 6. Capillary GC/MS of t-BDMS ethers of ceramides derived from sphingomyelins of HDL. TI, total ion current; 818.7 and 522.4, single ions scans for species containing nervonic acid; 820.7 and 524.4, single ions scans for species containing lignoceric acid. GC/MS conditions as in Fig. 1.

42 is mainly due to the resolution of the ceramides on the basis of the fatty acid unsaturation rather than the unsaturation of the nitrogenous bases, although some resolution due to the bases may also have taken place. The distribution of the fatty acids is best indicated by the characteristic ions derived from the main cleavage of the molecule as explained above, rather than from the M-57 fragment. Thus, the cleavage products demonstrate that the long chain acids occur in combination not only with the sphingosine and sphingadienine bases but also with bases of 16 and 17 carbon chains. Scans for other fatty acid fragments have revealed a comparable association with both long and short chain bases but the association has not proved to be random. Provided appropriate calibration factors are determined and used, the quantitative composition of the molecular species of the ceramides may be obtained.

The above method is also suitable for work with stable isotope (deuterium) labelled phospholipids provided the isotope is present in sufficient excess at least in one of the neutral lipid moieties of the molecular species. Thus, phosphatidylcholines derived from the perfusion of isolated rat liver with buffers containing deuterium oxide have given palmitoyl species rich in newly synthesized palmitic acid containing 5-24 deuterium atoms per molecule. By this means we have demonstrated differences in the distribution of the old and newly synthesized palmitic acid among the phosphatidylcholines synthesized at different times of the perfusion.

Figure 7 shows two GC/MS scans of the monoenoic diacylglycerol t-BDMS ethers of carbon number 34 from the free diacylglycerol fraction of a rat liver perfused with deuterium oxide. On the left is a scan from the ascending side of the GLC peak showing a substantial number of the 16:0 18:1 molecules that have incorporated palmitic acids having 7 to 22 deuterium atoms per palmitoyl moiety and of glycerol having from 1 to 5 deuterium atoms per molecule. On the right is a scan taken at the center of the peak. Because species with heavily labelled palmitic acid elute slightly earlier than unlabelled molecules the relative number of species having heavily labelled palmitic acid is greatly reduced.

The fragments in the spectra of the t-BDMS ethers of

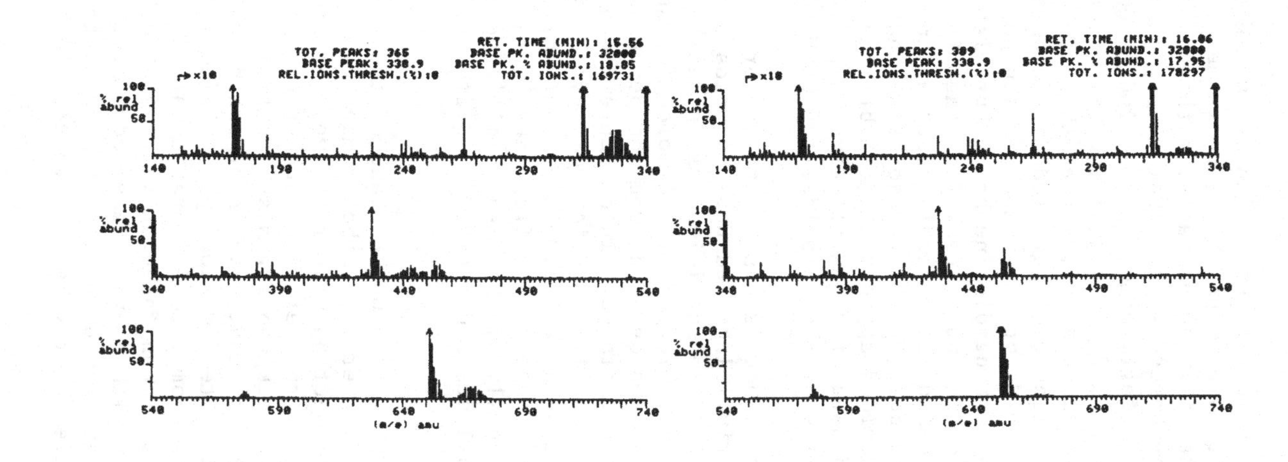

Fig. 7. Mass spectra of t-BDMS ethers of diacylglycerols from the ascending limb (left side) and the center (right side) of the peak with carbon number 34 derived in the free diacylglycerol fraction of a rat liver perfused with deuterium oxide.

the diacylglycerols are well suited for the assessment of the stable isotope enrichment (Figure 8). Since the m/e 171 fragment has been shown (Myher et al., 1978) to contain four glycerol hydrogens and no acyl hydrogens, it can be used to measure total isotope incorporation into glycerol. (Note: at present there is uncertainty as to which glycerol hydrogen is lost). The intensities of m/e 172, 173, 174 and 175 are indeed elevated as shown in the scans in Figure 7. However, quantitation of each ion must be done by integrating the appropriate peaks in the corresponding mass chromatograms. Labelling of the fatty acid portion of the molecule is best done by monitoring the RCO+74 fragments. Here m/e 313 represents ordinary unlabelled palmitic acid and m/e 320-335 are the corresponding ions for species having from 7 to 22 deuterium atoms per palmitoyl residue. Analysis of the M-57 ions completes the assay. In Figure 8 m/e 651.5 is the fragment for the completely unlabelled molecule. An elevation in the intensity of the m/e 652-656 ions indicates diacylglycerol moieties having labelled glycerol but not labelled fatty acid. Ions at m/e 658-675 indicate molecules with labelled fatty acid plus labelled or unlabelled glycerol. For the sample shown in Figure 8 it is possible to demonstrate that virtually all the labelled palmitic acid is coupled with labelled glycerol. This conclusion is derived by subtracting the amount of labelled glycerol that was not found with labelled fatty acid from the total labelled glycerol (m/e 172-175).

A special aspect of analysis of the lipid composition of natural cell membranes, where new approaches have been attempted, concerns the differentiation between the lipid composition of the outer and inner halves of the lipid bilayer (Rothman and Leonard, 1977; Chap et al., 1978; Higgins and Evans, 1978). Various procedures have been proposed and utilized for the selective destruction of one or the other half of the bilayer (Rothman and Leonard, 1977). Analyses of the total and the residual phospholipids has then given the composition of the inner and outer halves of the membrane by subtraction. Similar techniques would appear to be suitable for the determination of the molecular species of the phospholipid classes in each half of the membrane. In this instance effective use could be made of selective reagents, which attack only one phospholipid at a time leaving the other phospholipids untouched and thereby

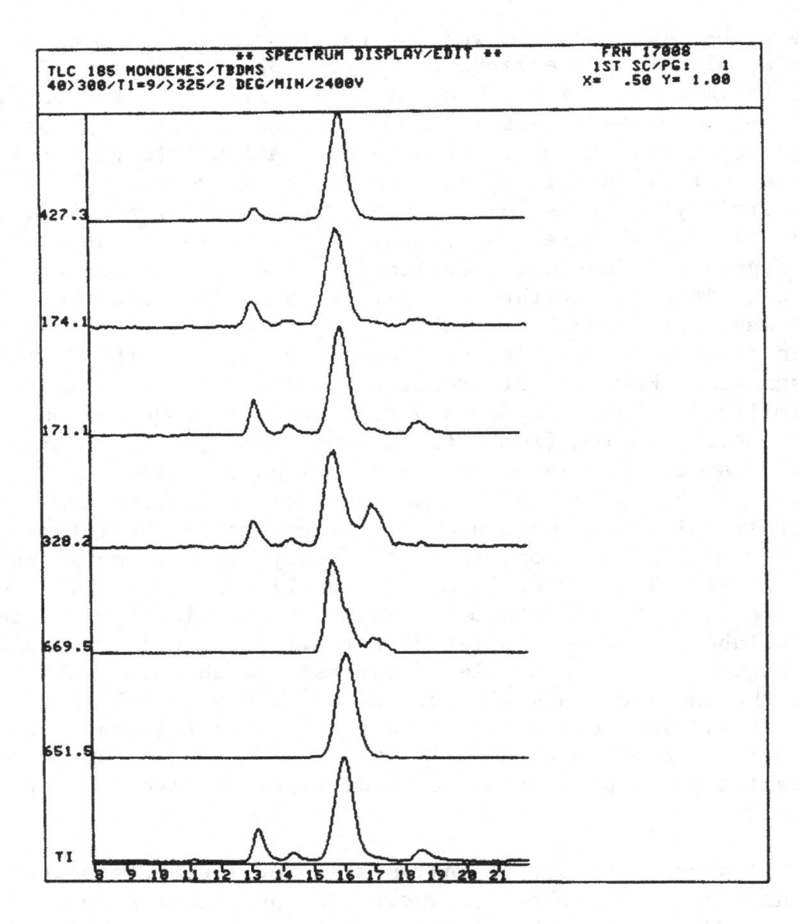

Fig. 8. GC/MS of t-BDMS ethers of free diacylglycerols derived from an isolated rat liver during perfusion with deuterium oxide. TI, total ion current; 651.5, single ion scan for species containing neither labelled palmitate nor labelled glycerol; 669.5, single ion scan for species containing labelled palmitic acid (e.g. 15 deuteriums) plus labelled glycerol (e.g. 3 deuteriums); 328.2, single ion scan for species containing labelled palmitic acid (e.g. 15 deuteriums); 171.1, single ion scan for species containing unlabelled glycerol; 174.1, single ion scan for species containing labelled glycerol (e.g. 3 deuteriums); 427.3, single ion scan for species containing unlabelled palmitic acid. GC/MS conditions as in Fig. 1.

ensuring the integrity of the membrane. However, the possibility of using such general reagents of destruction of the phospholipids of the outer layer as phospholipase A_2 also should not be overlooked. Analyses based on the diacylglycerols released by phospholipase C-type of enzymes even when confined to presumably one class of phospholipids, however, may be unsatisfactory, as there might be other sources of diacylglycerols than the destroyed glycerophospholipid. Furthermore, lipases present in the cell membrane or in the phospholipase C preparation may lead to hydrolysis of diacylglycerols (Mitchell et al., 1973), while in other instances acyltransferases may be activated resulting in a de novo synthesis of diacylglycerols (Waite and Sisson, 1976). Although the triacylglycerols may be eliminated as the source of the diacylglycerols on the basis of the fatty acid composition in some instances, in other cases this difference may be insufficient. In such cases the diacylglycerols released could be assayed for their enantiomer content and an indication obtained of the possibility of triacylglycerol lipolysis, which would presumably yield both sn-1,2- and sn-2,3-isomers, while the hydrolysis of glycerophospholipids would yield only sn-1,2-diacylglycerol isomers. For this purpose effective use may be made of the stereospecific analysis of diacylglycerols proposed by Myher and Kuksis (1979). In this method (Figure 9), the diacylglycerols are first converted into racemic phosphatidylcholines which are then differentially hydrolyzed with phospholipase C to regenerate first the sn-1,2-diacylglycerols and finally the sn-2,3-diacylglycerols, the molecular species composition of which could then be determined as above using the t-butyldimethylsilyl ether derivatives of each. At the present time it is not certain whether or not the asymmetry of phospholipid distribution exists in any other plasma membranes than those derived from the red blood cells (Rothman and Leonard, 1977), the platelets (Chap et al., 1978), and from the plasma membrane of hepatocytes (Higgins and Evans, 1978). There has been no previous work with the sidedness of the plasma membrane of the cells obtained in culture. Microsomal membranes of the liver, however, have been reported to possess symmetrical compositions of the bilayer (Sundler et al., 1977). Furthermore, it has been claimed that the fatty acid composition of the phospholipids in the residual digestion mixture are the same as those in the total mixture, suggesting identical composition of molecular species of phospholipids in both halves of the membrane (Rothman and Leonard, 1977). This claim, however, may be incorrect because there is evidence that gradients may be obtained across the membrane for both dietary fatty acids and for radioactive fatty acids. Unless an absolute exchange of like fatty acids only or like molecular species only takesplace, there must result also a gradient of phospholipids over the two halves of the bilayer. Whether or not this leads to a difference in the fluidity of the two halves of the bilayer has yet to be determined, as it could be subject to a variety of compensating factors, including subtle alterations in cholesterol content.

The new method of stereospecific analysis of diacylglycerols (Myher and Kuksis, 1979) is of greatest value for the determination of the structure of the triacylglycerols of plasma lipoproteins and is presently being employed for this purpose in our laboratory. The earlier methods of stereospecific analysis of triacylglycerols (Brockerhoff, 1965) based on the destruction of one of the diacylglycerol

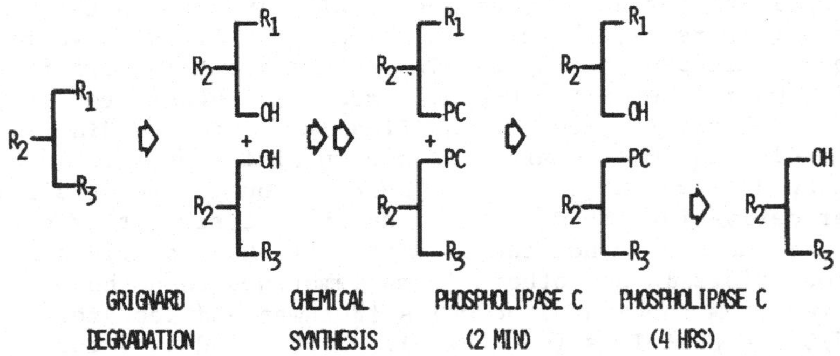

GRIGNARD DEGRADATION
CHEMICAL SYNTHESIS
PHOSPHOLIPASE C (2 MIN)
PHOSPHOLIPASE C (4 HRS)

Fig. 9. A schematic showing stereospecific analysis of triacylglycerols and diacylglycerols via racemic phosphatidylcholines and phospholipase C. R_1, R_2, R_3, different fatty acids at each stereospecifically numbered glycerol position; PC, choline phosphate moiety of phosphatidylcholine.

enantiomers with phospholipase A_2 after formation of the phosphatidylphenols are not suitable for this purpose because they do not permit the regeneration of intact enantiomeric diacylglycerols. The sn-2,3-diacylglycerols can be recovered in the form of the acetates following dephosphorylation by acetolysis (Breckenridge and Kuksis, 1968), but this reaction is known to be accompanied by a partial destruction of the sample and possible isomerization. In any event these methods do not yield free diacylglycerols which could be converted to the t-BDMS ethers that are needed for the detailed gas chromatographic-mass spectrometric assay of the component molecular species. The final triacylglycerol structure is then reconstituted by progressive reiterated fitting of the two diacylglycerol enantiomers with the corresponding fatty acids to approximate the molecular weight and double bond distribution in the original triacylglycerols and in the sn-1,3-diacylglycerols.

CONCLUSION

The capillary GC/MS analysis of the molecular species of diacylglycerols and ceramides, although rapid and applicable to small quantities of sample, is expensive. Therefore, careful thought must be given to the nature of the material used for analysis. New approaches have been made in membrane lipid analyses by differentiating between the inner and outer halves of the bilayer, as well as among various domains of either of the two monolayers. Perhaps these samples will prove suitable and worthwhile for detailed assessment of molecular species. The analyses based on intact molecules avoid errors due to contamination, which frequently plague determination of fatty acid composition by transmethylation and GLC. Hopefully, complete analyses of a few representative samples will allow to establish the rules of the positional placement and molecular association of fatty acids in the glycerolipids so that the molecular composition and physical properties, including the fluidity, of any sample could in the future be calculated from its fatty acid composition.

ACKNOWLEDGMENT

These studies were supported by funds from the Ontario Heart Foundation, the Medical Research Council of Canada, and the Heart and Lung Institute, NIH-NHLI-720-917, MD.

REFERENCES

Arvidson, G.A.E. (1965). J. Lipid Res. 6, 574.

Arvidson, G.A.E. (1967). J. Lipid Res. 8, 155.

Arvidson, G.A.E. (1968). Europ. J. Biochem. 4, 478.

Breckenridge, W.C. and A. Kuksis (1968). J. Lipid Res.

Breckenridge, W.C., S.K.F. Yeung, A. Kuksis, J.J. Myher and M. Chan (1976). Can. J. Biochem. 54, 137.

Brockerhoff, H. (1965). J. Lipid Res.

Chap, H.J., P. Comfurius, L.L.M. Van Deenen, E.J.J. van Zoelen and R.F.A. Zwaal (1978). In R. Dils and J. Knudsen, (eds.), Regulation of Fatty Acid and Glycerolipid Metabolism. FEBS 11th Meeting Copenhagen 1977. Volume 46, p. 83. Pergamon Press, London.

Chapman, D. (1973). Biological Membrane. Academic Press, London. p. 91.

Colbeau, A., J. Nachbaur and P.M. Vignais (1971). Biochim. Biophys. Acta 249, 462.

Curstedt, T. and J. Sjovall (1974). Biochim. Biophys. Acta 360, 24.

De Haas, G.H., F.J.M. Daemen and L.L.M. Van Deenen (1962). Biochim. Biophys. Acta 65, 260.

Emken, E.A. (1978). In A. Kuksis (ed.), Handbook of Lipid Research. Vol. 1. Fatty Acids and Glycerides. Plenum Press, New York. p. 77.

Evans, N., D.E. Games, J.L. Harwood and A.H. Jackson (1974). Biochem. Soc. Trans. 2, 1091.

Farias, R.N., B. Bloj, R.D. Morero, F. Sineriz and R.E. Trucco (1975). Biochim. Biophys. Acta 415, 231.

Freeman, N.K. (1968). J. Am. Oil Chem. Soc. 45, 798.

Freeman, N.K. (1972). In G.J. Nelson (ed.). Blood Lipids and Lipoproteins. Wiley-Interscience, New York. p. 113

Gregoriadis, G., M. Siliprandi and E. Turchetto (1977). Life Sciences 20, 1773.

Grob, K. and G. Grob (1969). J. Chromatogr. Sci. 7, 584.

Hanahan, D.J., H. Brockerhoff and E.J. Barron (1960). J. Biol. Chem. 235, 1917.

Hasegawa, K. and T. Suzuki (1974). Lipids 8, 631.

Hasegawa, K. and T. Suzuki (1975). Lipids 10, 667.

Higgins, J.A. and W.H. Evans (1978). Biochem. J. 174, 563.

Hites, R.A. (1970). Anal. Chem. 42, 1736.

Hites, R.A. (1975). In J.M. Lowenstein (ed.), Methods of Biochemical Analysis. Wiley-Interscience, New York. p. 348.

Holub, B.J. and A. Kuksis (1971a). Can. J. Biochem. 49, 1347.

Holub, B.J. and A. Kuksis (1971b). J. Lipid Res. 12, 699.

Holub, B.J. and A. Kuksis (1978). Adv. Lipid Res. 16, 1.

Jungalwala, F.B., R.J. Turel, J.E. Evans and R.H. McCluer (1975). Biochem. J. 145, 517.

Jungalwala, F.B., J.E. Evans and R.H. McCluer (1976). Biochem. J. 155, 55.

Jungalwala, F.B., V. Hayssen, J.M. Pasquini and R.H. McCluer (1979). J. Lipid Res. 20, 579.

Keenan, R.W. and P.J. Morre (1970). Biochemistry 9, 19.

Klein, R.A. (1972). J. Lipid Res. 13, 672.

Kuksis A. (1965). J. Am. Oil Chem. Soc. 42, 269.

Kuksis, A. (1971). Can. J. Biochem. 49, 1245.

Kuksis, A. (1972a). In R.T. Holman, W.O. Lundberg and T. Malkin (eds.), Progress in the Chemistry of Fats and Other Lipids. Pergamon Press, Oxford. Vol. 12, p.1.

Kuksis, A. (1972b). J. Chromatogr. Sci. 10, 53.

Kuksis, A. and L. Marai (1967). Lipids 2, 217.

Kuksis, A., W.C. Breckenridge, J.J. Myher and G. Kakis (1978). Can. J. Biochem. 56, 630.

Kuksis, A., J.J. Myher, W.C. Breckenridge and J.A. Little (1979). Lipid Profiles of Human Plasma High Density Lipoproteins. National Heart, Lung and Blood Institute Monograph, pp. 142-163.

Lindqvist, B., I. Sjogren and R. Nordin (1974). J. Lipid Res. 15, 65.

Luthra, M.G. and A. Sheltawy (1972a). Biochem. J. 126, 1231.

Luthra, M.G. and A. Sheltawy (1972b). Biochem. J. 128, 587.

Marai, L. and A. Kuksis (1973a). Can. J. Biochem. 51, 1248

Marai, L. and A. Kuksis (1973b). Can. J. Biochem. 51, 1365

Mavis, R.D., R.M. Bell and P.R. Vagelos (1972). J. Biol. Chem. 247, 2835.

Maine, J.W., B. Soltmann, J.F. Holland, N.D. Young, J.N. Gerber and C.C. Sweeley (1976). Anal. Chem. 48, 427.

Mitchell, R.H., R. Coleman and J.B. Finean (1973). Biochim. Biophys. Acta 318, 306.

Morrisett, J.D., H.J. Pownall, R.L. Jackson, R. Segura, A.M. Gotto, Jr. and O.D. Taunton (1976). In W-H. Kunau and R.T. Holman (eds.), Polyunsaturated Fatty Acids. American Oil Chemists' Society, p. 139. Champaign, Illinois.

Murata, T. (1977). Anal. Chem. 49, 2209.

Murata, T. and S. Takahashi (1973). Anal. Chem. 45, 1816.

Murata, T. and S. Takahashi (1977). Anal. Chem. 49, 728.

Myher, J.J. (1978). In A. Kuksis (ed.), Handbook of Lipid Research. Vol. 1 Fatty Acids and Glycerides. Plenum Press, New York, p. 123.

Myher, J.J. and A. Kuksis (1975). J. Chromatogr. Sci. 13, 138.

Myher, J.J. and A. Kuksis (1979). Can J. Biochem. 57,117.

Myher, J.J., A. Kuksis, L. Marai and S.K.F. Yeung (1978). Anal. Chem. 50, 557.

Ogino, H., T. Matsumura, K. Satouchi and K. Saito (1979). Biochim. Biophys. Acta. In Press.

Renkonen, O. (1965). J. Am. Oil Chemists' Soc. 42, 298.

Renkonen, O. (1966). Biochim. Biophys. Acta 125, 280.

Renkonen, O. (1967). Biochim. Biophys. Acta 137, 575.

Renkonen, O. (1967). Adv. Lipid Res. 5, 329.

Renkonen, O. (1971). In A. Niederwieser and G. Pataki, (eds.), Progress in Thin-Layer Chromatography and Related Methods. Vol. II. Ann Arbor Science Publishers, Ann Arbor, Michigan. p. 143.

Rothman, J.E. and J. Lenard (1977). Science 195, 743.

Rottem, S., Hasin, M. and Razin, S. (1973). Biochim. Biophys. Acta 323, 520.

Salem, N., Jr., L.G. Abood and W. Hoss (1976). Anal. Biochem. 76, 407.

Samuelsson, B. and K. Samuelsson (1969). J. Lipid Res. 10, 47.

Samuelsson, K. and B. Samuelsson (1970). Chem. Phys. Lipids 5, 44.

Satouchi, K. and K. Saito (1976). Biomed. Mass Spectrometry 3, 122.

Satouchi, K. and K. Saito (1979). Biomed. Mass Spectrometry. In Press.

Satouchi, K., K. Saito and M. Kates (1978). Biomed. Mass Spectrometry 5, 87.

Scanu, A.M. (1969). In E. Tria and A.M. Scanu (eds.), Structural and Functional Aspects of Lipoproteins in Living Systems. Academic Press, New York. p. 425.

Schomberg, G., Dielmann, H. Husmann and F. Waeke (1976). J. Chromatogr. 122, 55.

Shamgar, F.A. and F.D. Collins (1975). Biochim. Biophys. Acta. 409, 104.

Shamgar, F.A. and F.D. Collins (1975). Biochim. Biophys. Acta. 409, 116.

Skipski, V.P. (1972). In G.J. Nelson (ed.), Blood Lipids and Lipoproteins. Wiley-Interscience, New York. p.471.

Stahl, W.L. (1973). Arch. Biochem. Biophys. 154, 47.

Sundler, R. and B. Akesson (1973). J. Chromatogr. 80, 233.

Sundler, R., S.L. Sarcione, A.W. Albert and P.R. Vagelos (1977). Proc. Natl. Acad. Sci. USA 74, 3350.

Tagaki, T. and Y. Itabashi (1977). Lipids 12, 1062.

Tattrie, N.H. (1959). J. Lipid Res. 1, 60.

Waite, M. and P. Sisson (1976). In R. Paoletti, G. Porcellati and G. Jacini (eds.), Lipids, Vol. I, Raven, New York. p. 127.

Wood, G. and P.Y. Lau (1974). Biomed. Mass Spectrometry 1, 154.

Wood, G.W., P.Y. Lau, G. Morrow, G.N.S. Rao, D.E. Schmidt, Jr. and J. Tuebner (1977). Chem. Phys. Lipids 18, 316.

Wood, R. (1973). In R. Wood (ed.) Tumor Lipids: Biochemistry and Metabolism. American Oil Chemists' Society Press, Champaign, Illinois. p. 139.

Wood, R. (1975). Lipids 10, 736.

Wood, R. and R.D. Harlow (1969). Arch. Biochem. Biophys. 135, 272.

Yeung, S.K.F., A. Kuksis, L. Marai and J.J. Myher (1977). Lipids 12, 529.

DETERMINATION OF MOLECULAR SPECIES OF GLYCEROPHOSPHOLIPIDS BY A GC-MS SELECTED ION MONITORING TECHNIQUE

Kunihiko SAITO, Hirotaro OGINO and Kiyoshi SATOUCHI
Department of Medical Chemistry, Kansai Medical School
Moriguchi, Osaka, JAPAN

ABSTRACT

Four types of glycerophospholipids, i.e., 1,2-dalkyl, 1-alkenyl-2-acyl, 1-alkyl-2-acyl and 1,2-diacyl glycerols as TMS or TBDMS derivatives were separated and identified by monitoring three kinds of fragment ions, 1) M-15 or M-57 for determining molecular weight; 2) RCO + 74, R + 130 and RCH=CH + 56 for acyl, alkyl and alkenyl residues, respectively, and 3) base peaks corresponding to TMS-glycerol (m/e 129, 130) or O-TBDMS (m/e 131).

By this method, chronological changes in molecular species of choline glycerophospholipids of fetal and adult rat liver and brain were studied. The molecular species were composed mainly of the 1,2-diacyl type. In liver, during the period of gestation, $PC_{32:0}$ (16:0/16:0) and $PC_{34:1}$ (mainly 16:0/18:1) were the main components but after birth more unsaturated and longer chain species such as PC_{36} (18:0/18:1, 18:0/18:2, 18:0/18:3, 18:1/18:3) and PC_{38} (18:0/20:4, 18:1/20:4, 16:0/22:6) were predominant. In brain, $PC_{32:0}$ (16:0/16:0) accounted for 40 - 50% of the species throughout the period studied. $PC_{34:1}$ (16:0/18:1) decreased after birth and increased again with myelination.

INTRODUCTION

Natural glycerophospholipids are located mainly in biomembranes and related to many important biological or biochemical phenomena. They are divided into four types i.e., 1,2-dialkyl (I), 1-alk-1'-enyl-2-acyl (II), 1-alkyl-2-acyl (III) and 1,2-diacyl (IV) glycerophosphatides.

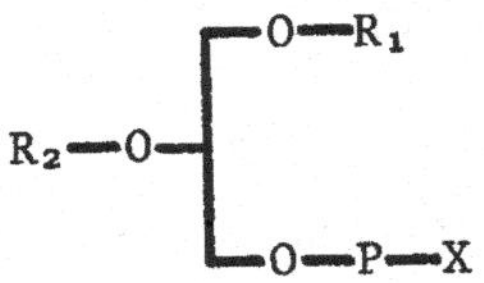

```
      ┌O—R1                         ┌O—CH=CH—R1
R2—O—┤                   O           │
      │                  ‖           │
      └O—P—X        R2—C—O—┤
                                     └O—P—X
      (I)                           (II)

                                       O
                                       ‖
          ┌O—R1                     ┌O—C—R1
   O      │                  O      │
   ‖      │                  ‖      │
R2—C—O—┤                 R2—C—O—┤
          └O—P—X                    └O—P—X

        (III)                       (IV)
```

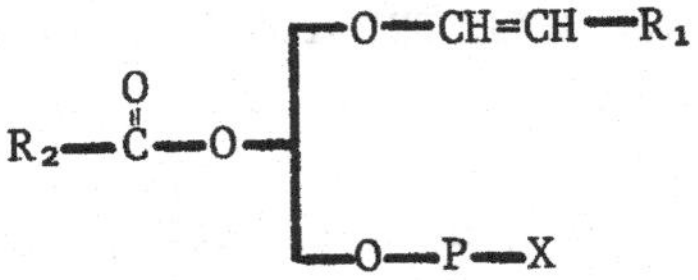

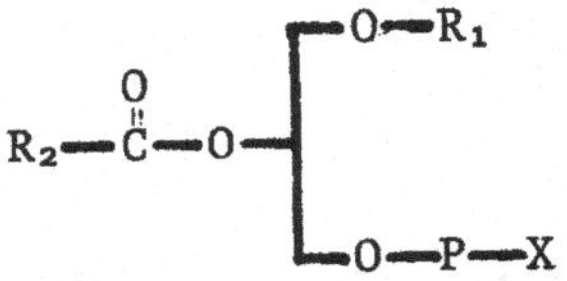

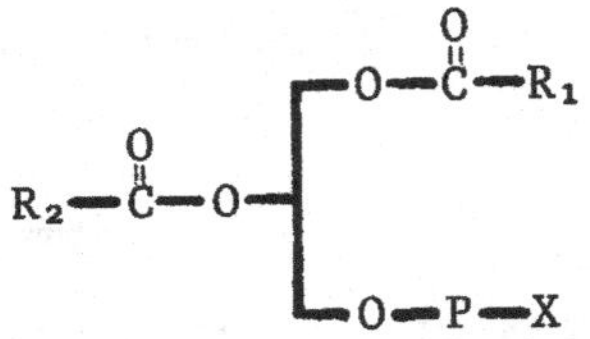

By varying the nature of the component fatty acids and/or fatty alcohols including vinyl alcohols, usually C_{16}-C_{22}, many molecular species are possible for each type. We are interested in the determination of the molecular species by gas chromatography-mass spectrometry (GC-MS), particularly selected ion monitoring (SIM) technique in relation to organ development.

GC-MS Analysis of Model Compounds

As model compounds of these four types, we prepared 1,2-dihexadecyl, 1-hexadec-1'-enyl-2-hexadecanoyl, 1-hexadecyl-2-hexadecanoyl and 1,2-dihexadecanoyl glycerols, which contained same chain length and degree of unsaturation ($C_{16:0}$) of alkyl, alk-1'-enyl and acyl residues. They were converted to trimethylsilyl (TMS) or t-butyldimethylsilyl (TBDMS) derivatives (Myher et al, 1978; Satouchi, et al, 1978; Satouchi and Saito, 1979). Shimadzu-LKB 9000 equipped with OKITAC computer, 4300S, was used. Electron energy was 22.5 eV. Principal fragment ions of TMS and TBDMS derivatives are summarized as shown in Fig. 1.

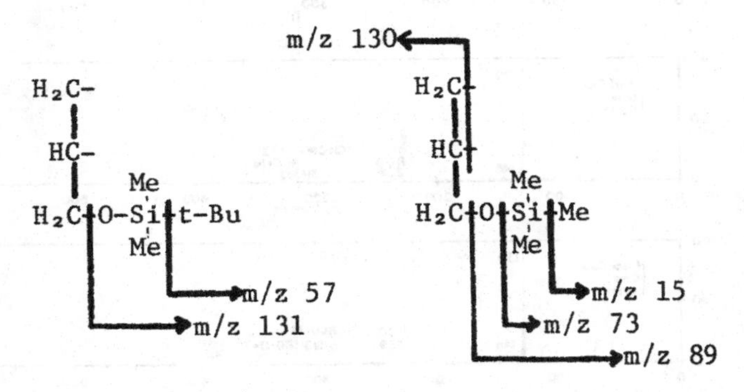

	TBDMS	TMS
Base peak	m/z 131, 171, 57	m/z 129, 130, 133, 145
Molecular	M-57	M-15
Acyl	RCO, RCO+74	RCO, RCO+74
Alkyl	R, R+130	R, RO+72
Alk-1'-enyl	RCH=CH+56, RCH=CH+130	RCH=CH-1, RCH=CHO+72

Fig. 1. Summary of Proposed Fragmentation Patterns of TBDMS and TMS Derivatives of Glycerol.

On TMS derivatives, ion at (M-15) was weak except 1,2-diacyl type. 1-Alk-1'-enyl-2-acyl type gave a very strong ion at m/z 385 (M-RCH=CHO). 1-Alkyl-2-acyl type showed no characteristic strong ions. On the other hand, TBDMS derivatives gave stronger ion,(M-57), than (M-15) of TMS derivatives. 1-Alk-1'-enyl-2-acyl and 1-alkyl-2-acyl types, compared to those of TMS derivatives, gave more valuable fragment ions concerning acyl, alk-1'-enyl and alkyl residues which corresponded to m/z 239 (RCO) and 313 (RCO+74), m/z 279 (RCH=CH+56) and 353 (RCH=CH+130), and m/z 355 (R+130), respectively (Fig. 2).

In general, TBDMS derivatives give more information than TMS derivatives particularly for alkyl and alk-1'-enyl types.

Among fragment ions, three kinds of principal ions were chosen as being diagnostic of TMS or TBDMS derivatives for SIM technique.

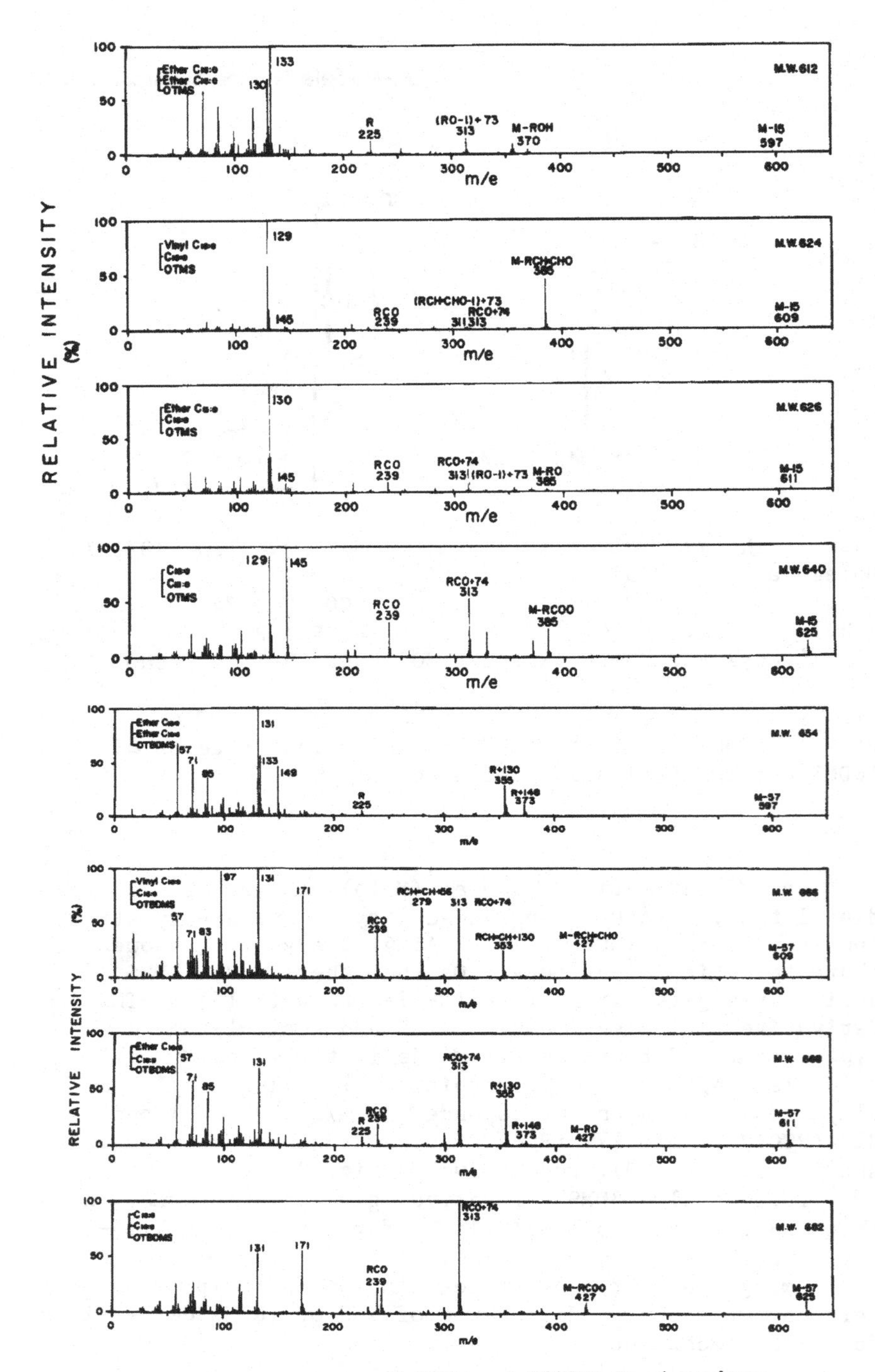

Fig. 2. Mass Spectra of TMS and TBDMS Derivatives.

For TMS derivatives, they were (1) (M-15) for molecular weight, (2) (RCO), (RCO+74) and (M-RCH=CHO, -RO or -RCOO) for acyl residue, (3) (RCH=CHO+72) for alk-1'-enyl residue, (4) (RO+72) for alkyl residue and (5) the base peaks, m/z 129, 130, 133 and 145, which derive from 1-alk-1'-enyl-2-acyl, 1-alkyl-2-acyl, 1,2-dialkyl and 1,2-diacyl glycerols, respectively.

For TBDMS derivatives, (1) (M-57) for molecular weight, (2) (RCO), (RCO+74) and (M-RCH=CHO, -RO or -RCOO) for acyl residue, (3) (RCH=CH+56) for alk-1'-enyl residue, (4) (R+130) for alkyl residue, and (5) m/z 171, 131 and 57 as common fragment ions were selected. A mixture of these TMS or TBDMS derivatives were separated by gas liquid chromatography (GLC) on 1 % OV-1 and eluted in the following order: 1,2-dialkyl, 1-alk-1'-enyl-2-acyl, 1-alkyl-2-acyl, and 1,2-diacyl types (Fig. 3).

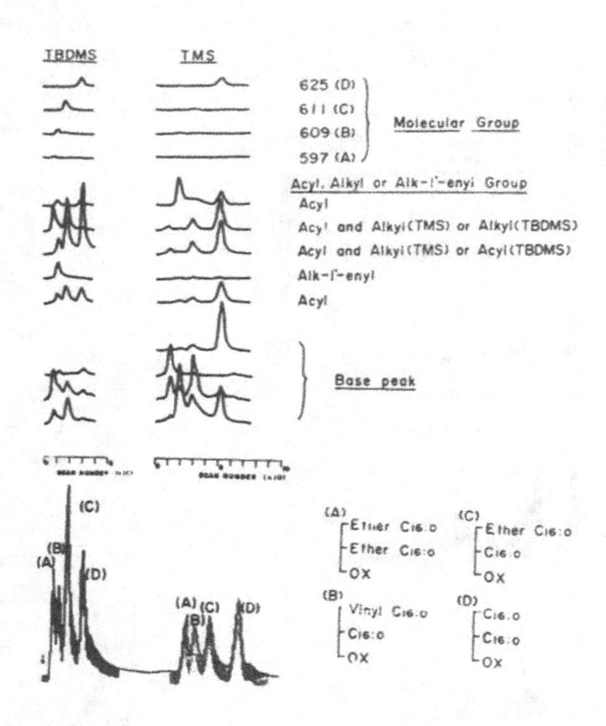

Fig. 3. SIM of TMS and TBDMS Derivatives of Four Types

Chronological Changes in Molecular Species of Choline Glycerophospholipids of Rat Liver and Cerebrum with Age

The protocol of the experiment was as follows: (1) extraction of total lipids by the method of Bligh and Dyer, 1959, (2) purification of choline glycerophospholipids by successive chromatography on silicic acid, preparative thin layer plate and alumina, (3) enzymatic hydrolysis of the choline glycerophospholipids with phospholipase C, (4) conversion of the resulting "diglycerids" to TMS or TBDMS derivatives, (5) GC-MS analysis by selected ion monitoring, and (6) integration of peak areas monitored by (M-15) or (M-57) by computer. Choline glycerophospholipids of rat liver and cerebrum were composed almost entirely of 1,2-diacyl type and TMS derivatives were used (Fig. 4).

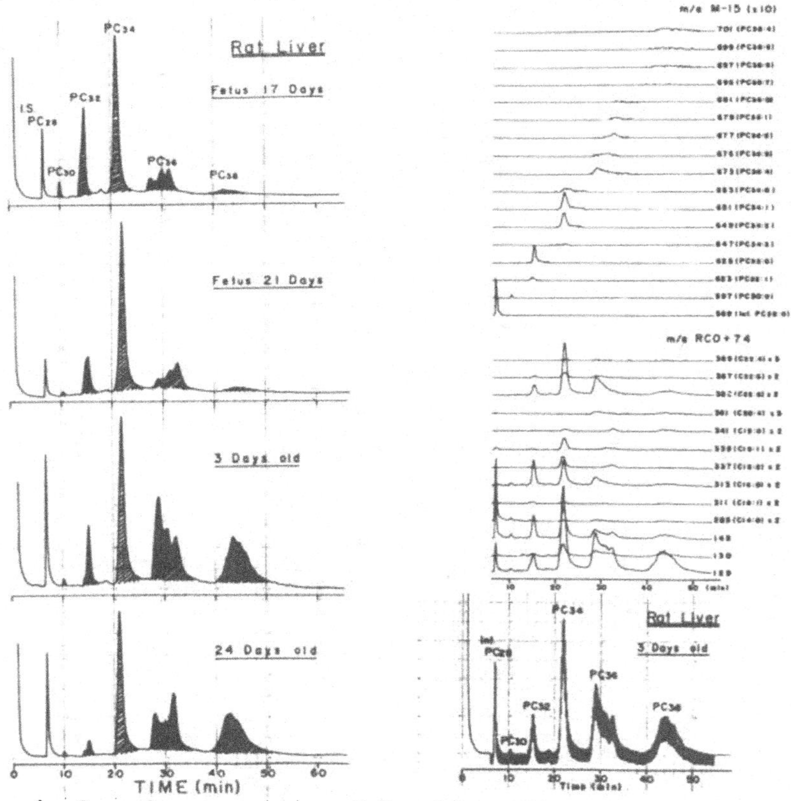

Fig. 4, Gas chromatograms of Rat Liver PC with Age (left) and SIM of 3-day-old Rat Liver PC (right).

In liver, during the period of gestation, $PC_{32:0}$ (mainly 16:0/16:0), $PC_{34:1}$ (mainly 16:0/18:1) and $PC_{34:2}$ (mainly 16:0/18:2) were the main compoments, but after birth more unsaturated and longer chain species such as $PC_{36:4}$, and $PC_{38:4}$, $PC_{38:5}$ and $PC_{38:6}$ were predominant (Fig. 5).

In cerebrum, main molecular species were $PC_{32:0}$ and $PC_{34:1}$. The former accounted for about 45 % throughout the period tested and the latter decreased soon after birth and increased again with myelination (Fig. 6).

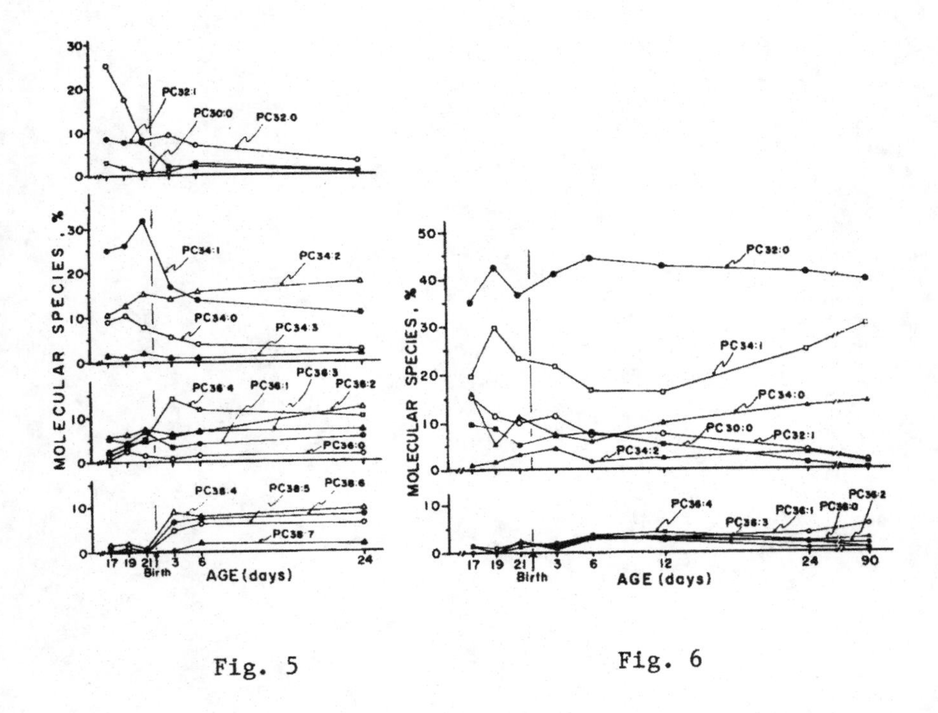

Fig. 5

Fig. 6

Fig. 5 and Fig. 6 Changes in Composition of Major Molecular Species with Development.
Liver (Fig. 5) and Cerebrum (Fig. 6).

The SIM technique particularly on TBDMS derivatives was found to be very useful to determine the complex molecular species such as animal glycerophospholipids on a few microgram levels of lipid-phosphorus (Kuksis, 1979; Satouchi and Saito, 1979). To determine the positional specificity of acyl residues in glycerophospholipids and absolute quantification of polyunsaturated species, the combination of enzymatic hydrolysis with phospholipase A_2 and direct inlet MS techniques are now being investigated.

REFERENCES

Bligh, E.G. and W.J. Dyer (1959). Can. J. Biochem. Physiol. 37, 911.

Kuksis, A. and J.J. Marai (1979). Satellite Symposium of XIth International Congress of Biochemistry on Control of Membrane Fluidity. This volume.

Myher, J.J., A. Kuksis, L. Marai and S.K.F. Yeung (1978). Anal. Chem. 50, 557.

Satouchi, K., K. Saito and M. Kates (1978). Biomed. Mass Spectrom., 5, 87.

Satouchi, K. and K. Saito (1979). Biomed. Mass Spectrom., 6, in press.

HIGH-PRECISION TLC-DENSITOMETRY OF MEMBRANE LIPIDS

Susumu ANDO, Kazuo KON and Yasukazu TANAKA

Department of Biochemistry, Tokyo Metropolitan Institute of Gerontology, Sakaecho, Itabashiku, Tokyo, Japan

A series of colorimetric determinations specific for each lipid are generally employed in a test tube for quantification of lipids isolated by conventional column chromatography or TLC. All classes of lipids could be analysed by such complicated methods, but appreciable amounts of samples would be required. If every kind of lipid is analysed based on a single principle, comparative studies on different classes of lipids may be performed in a more precise manner than employing many color reactions. In that sense, a flame ionization detector can be a versatile detector which gives the same responses per carbon atom for all organic molecules. Gas chromatography has been tried to delineate total lipid patterns by Kukusis et al.(1968). It may not always be applied to complex lipids because of difficulties in vaporization. HPLC employing a flame ionization detector was successfully applied to determination of neutral lipids and

Abbreviations: TLC,thin-layer chromatography; HPTLC,high-performance thin-layer chromatography; HPLC,high-performance liquid chromatography; TL,total lipids; NL,neutral lipids; PL,phospholipids; CE,cholesteryl ester; FC,free cholesterol; TG,triacylglycerol; FA,free fatty acid; PE,phosphatidylethanolamine; PC,phosphatidylcholine; LPC,lysophosphatidylcholine; SM,sphingomyelin; PS,phosphatidylserine.

The ganglioside nomenclature system of Svennerholm(1963) is used. The corresponding symbols of Korey and Gonatas (1963) are also included in Fig.6.

phospholipids by Stolyhwo and Privett(1973). The method, however, is not so popular yet probably due to some problems in the instrumentation. Another example using a flame ionization detector is thinchrography, which has been introduced and developed for lipid analysis(Tokunaga et al.,1973;Tanaka et al.,1977). Lipids are separated on a silicic acid-coated quartz stick in a similar manner as TLC and determined with the aid of the detector. On the thinchrography, however, it is difficult to identify samples with reference compounds and to leave an original chromatogram. On the other hand, TLC enables measurements of many samples in one run, and the chromatogram is kept for record. Mainly because of the simplicity, densitometry has been applied to TLC as an in situ analysis. TLC-densitometry is thought to be one of the ideal method to permit a simultaneous determination of all classes of lipids with the same accuracy in a minute quantity. Reported techniques are not likely the best yet in their accuracy and reproducibility. In order to achieve a high performance densitometry, uniform layer plate, fine-controlled visualization and internal standard method are required. We have developed an improved densitometry employing HPTLC plates, a strictly controlled heating system for visualization, and an accurate correction method. All possible variations due to irregularities of plates, separation, color development and others can be minimized by drawing a calibration curve for each plate. Output of a digital integrator connected to a densitometer is introduced to a computer in order to make a calibration formula and print out absolute values for each lipid of samples. Major neutral lipids and phospholipids were both determined on a single plate and gangliosides were analysed on another plate. This paper describes the methodology of TLC-densitometry and its application to analysis of rat erythrocyte membrane lipids, cultured cell membrane lipids, and ganglioside pattern analysis.

TLC-DENSITOMETRY

HPTLC plate(Nanoplate,10 x 20cm,E.Merck) was washed by developing with chloroform-methanol-water(65:35:8) and activated. The plate was covered with a clean glass plate to keep off moisture. Samples tested and a series of standard sample mixtures were applied to the plate in 5 mm streak. The standard mixtures were composed of oleoyl cholesterol, cholesterol, triolein, oleic acid, phosphatidylethanolamine, phosphatidylcholine, sphingomyelin and oleoyl alcohol.

1. Thin-layer chromatography

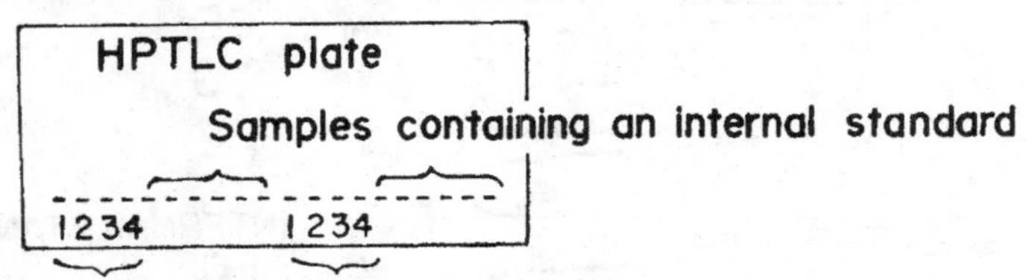

A series of standard mixtures

↓ 1st step development for separation of PL

↓ 2nd step development for separation of NL

↓ Visualization with heating on a hot plate

2. Densitometry

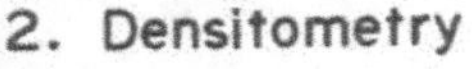

Scan with a transmittance mode

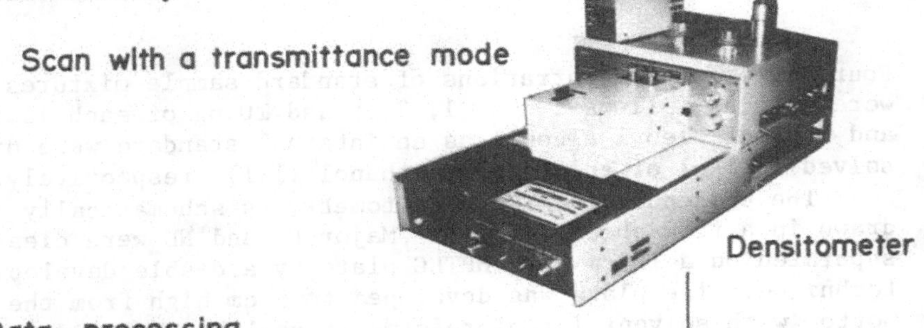

3. Data processing

Integrator

↓

Mini-computor

↓ Make calibration curves by approximation as a quadratic equation

↓ Print out absolute values for each component

Fig.1. Flow sheet of TLC-densitometry

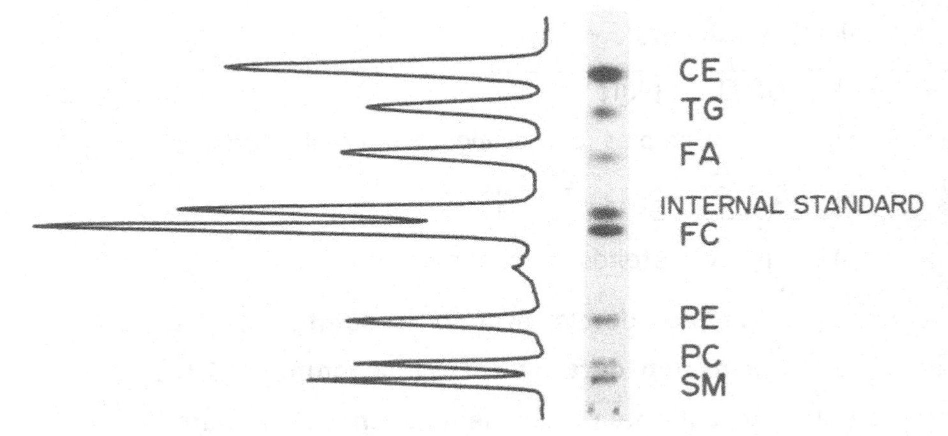

Fig.2. Thin-layer chromatogram of PL and NL, and its densitogram.

Four different concentrations of standard sample mixtures were made as follows : 1, 3, 5 and 10 ug of each lipid and 5 ug of oleoyl alcohol as an internal standard were dissolved in 5 ul of chloroform-methanol (1:1), respectively.

The procedure for TLC-densitometry is schematically drawn in a flow sheet (Fig.1). Major PL and NL were clearly separated on a 10 cm high HPTLC plate by a double developing technique. The plate was developed to 5 cm high from the bottom with solvent I (chloroform-methanol-acetic acid-formic acid-water, 35:15:6:2:1) and dried. The plate was then developed up to the top with solvent II (hexane-diisopropylether-acetic acid, 65:35:2) and dried completely. The internal standard, oleoyl alcohol, was more clearly separated from cholesterol with the solvent II than with a commonly used solvent system such as hexane-diethylether-acetic acid. Visualization of the plate is a critical step for densitometry. The plate was evenly sprayed with a diluted dichromate-sulfuric acid solution, and then heated on an aluminum hot plate maintained at 120 ± 1°. Heating on a hot plate was much better for reproducible and fine-controlled charring than in an oven as reported previously (Ando _et al_., 1978). Bands appeared as brown color and were measured by a Shimadzu Chromatoscanner CS-910 equipped with a digital integrator Chromatopac EIA (Shimadzu Seisakusho Ltd., Kyoto, Japan) under the following conditions : transmittance mode at 440 nm; slit width, 0.2 x 3 mm; scan speed, 20 mm/min.

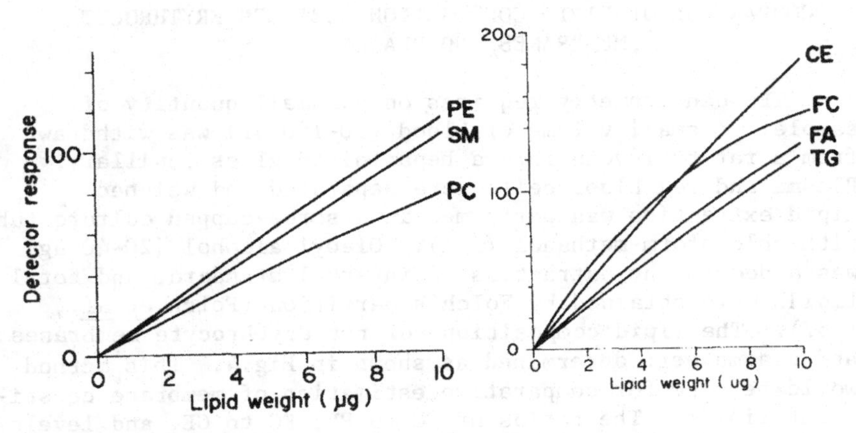

Fig.3. Calibration curves drawn by approximation as a second order equation.

A transmittance mode gave twice higher sensitivity and better linearity in the detector response than a reflectance mode, although Hezel claimed a preference for a reflection mode (1976). A typical thin-layer chromatogram and the corresponding densitogram are shown in Fig.2. The relative detector responses of each standard sample to that of oleoyl alcohol were introduced into a minicomputer 500 FDG system (Shimadzu Seisakusho Ltd.) to draw calibration curves for each plate. An ideal straight line was not obtained in every case as suggested by Goldman and Goodall(1969). Therefore, the calibration curves were made by approximation of the data to a quadratic equation(Fig.3). Lipids tested were determined in the range of 0.1-10 ug(approximately 0.1-10 nmoles). Reproducible data may be obtained in duplicate experiments, as shown in smaller standard error than ± 2.5% for lipids in the range of 0.5-10 ug. HPLC can compete in terms of accuracy with the present TLC-densitometry(Ullman and McCluer, 1978). However, consecutive separation of all kinds of lipids has not been achieved with current liquid chromatograph systems because some derivatives are highly sensitive to UV, but others are not. TLC-densitometry would be expected to become a practical , valid method of delineating total lipid patterns.

The procedure for ganglioside determination is discussed in the section on ganglioside pattern analysis.

COMPARISON OF LIPID COMPOSITION BETWEEN ERYTHROCYTE MEMBRANES AND PLASMA

TLC-densitometry requires only a small quantity of sample. A small volume of blood (50-100 ul) was withdrawn from a rat tail vein into a heparinized glass capillary. Plasma and red blood cells were separated and weighed. Lipid extraction was performed in a screw-capped culture tube with chloroform-methanol (2:1). Oleoyl alcohol (20-40 ug) was added to the extract as an internal standard, and total lipids were obtained by Folch's partition (Folch et al., 1957). The lipid composition of rat erythrocyte membranes and plasma were determined as shown in Fig.4. This method would be nice for comparative estimation of membrane constituent lipids. The ratios of FC to PL, FC to CE, and levels of minor components, FA and LPC, are immediately computed from the data. Amounts of TL are also estimated by summing up values of every component. The summation of the separate lipid components would give much closer values to the real

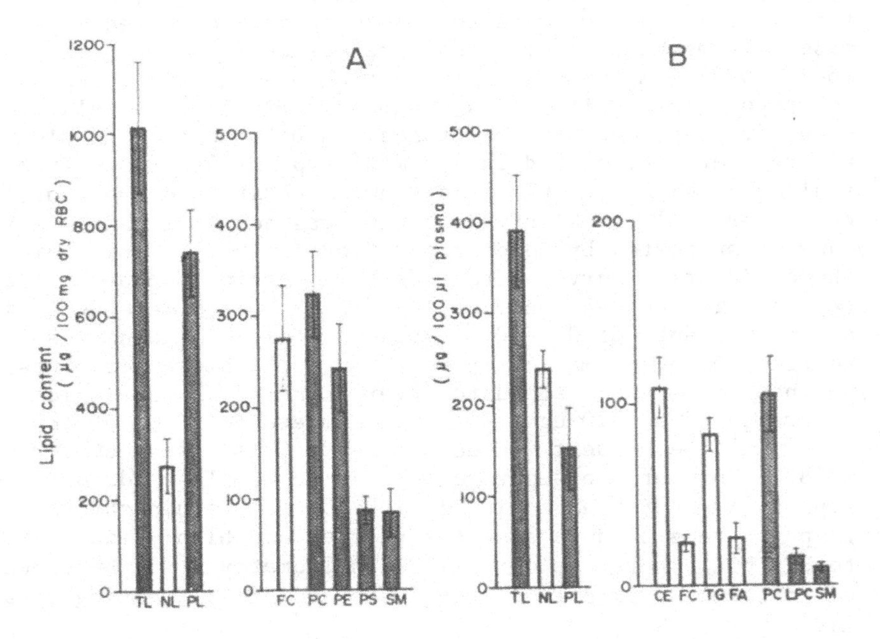

Fig.4. Lipid composition of rat erythrocyte membranes (A) and plasma (B).

quantity of TL than a weighing method with total lipid extracts which may contain non-lipid contaminants.

LIPID ANALYSIS OF CULTURED CELLS

TL of cultured fibroblast WI-38 were obtained by extraction with chloroform-methanol(1:1). The extract was applied to a DEAE-Sephadex column(0.5 ml)(Ledeen et al., 1973) to prepare neutral and acidic lipid fractions. The lipid patterns in the neutral fraction were determined by TLC-densitometry. Some changes in the amount of TL with passage and the ratio of FC to PL are shown in Fig.5. These data are not consistent with those of Kritchevsky and Howard (1970), who reported increasing TL levels with passage. They also present very high values for FA which are almost equivalent to FC levels. Our systematic analysis detects levels of FA as low as one tenth the levels of FC.

GANGLIOSIDE PATTERN ANALYSIS

Gangliosides are a family of glycosphingolipids containing sialic acid moieties. They are characterized by unusual solubility in water and are primarily located in plasma membranes, projecting the saccharide portions out of

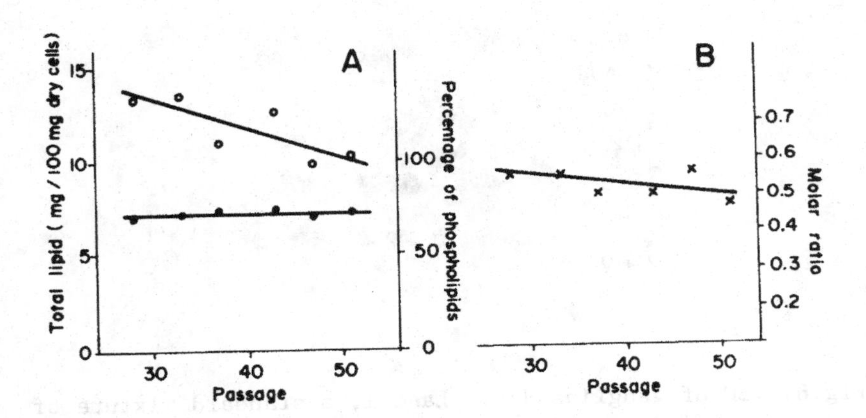

Fig.5. Changes in lipids of WI-38 cells with passage. (A),-o——o- ,TL;-•——•- ,percentage of PL in TL; (B), molar ratio of FC to PL.

cells. The lipids are widely distributed among many kinds of cells as well as neurons(Ng and Dain,1976). Each cell or organ has specific patterns of gangliosides which appear to be responsible for mechanisms of recognition and immunochemical reactions(Kohn et al.,1978). Comparative studies of ganglioside patterns among different species, organs and cells are of great interest in connection with their possible functions.

Quantification of individual ganglioside species has been carried out by the method of Suzuki(1964). The method

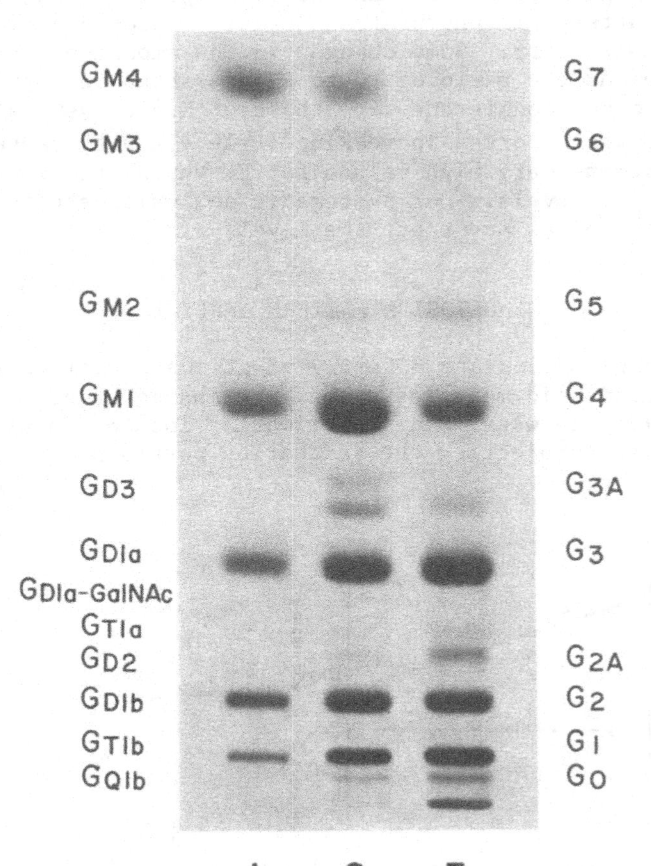

Fig.6. TLC of gangliosides. Lane 1, a standard mixture of purified gangliosides; lanes 2 and 3, normal human white matter and gray matter gangliosides, respectively. Data taken from Ando et al.,Anal.Biochem.89,437(1978), with permission of Academic Press Inc.

involves separation of gangliosides with TLC, followed by scraping the gel of each ganglioside band from the plate, and direct colorimetric determination of the lipid-bound sialic acid. Relatively large quantity of samples(at least 40 ug of lipid-bound sialic acid) are required for a single determination, and minor ganglioside species are frequently omitted due to the limited sensitivity of the method(MacMillan and Wherrett,1969). Mainly to simplify the tedious procedure, TLC-densitometry has occasionally been tried as an alternative to Suzuki's method(Sandhoff et al.,1968; Smid and Reinisova, 1973; Max and Quarles,1973). The densitometry has not been satisfactory yet for detection of minor components, probably as the results of the poor resolution and high background noise. Introduction of HPTLC plate has made it possible to

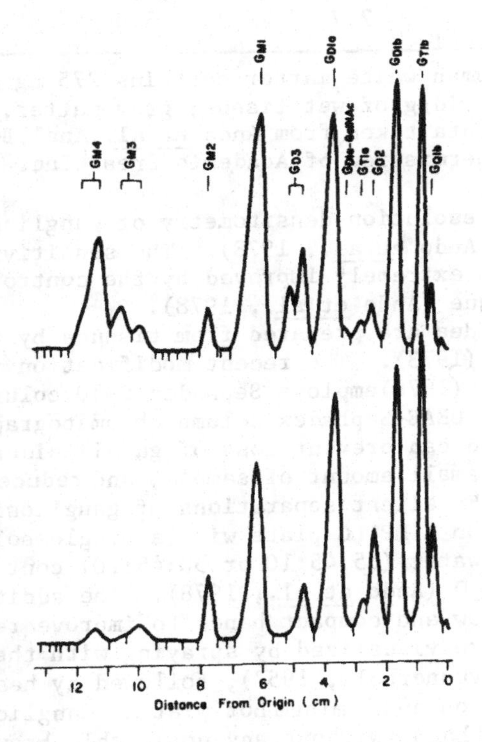

Fig.7. Densitograms of (A) normal human white matter and (B) gray matter gangliosides.
Data taken from Ando et al.,Anal.Biochem.89,437(1978), with permission of Academic Press Inc.

Table 1. Percentage Distribution of ganglioside sialic acid in human brain.[a)]

	White matter	Gray matter
G_{M4}	8.6	1.5
G_{M3}	4.8	2.7
G_{M2}	2.5	4.1
G_{M1}	21.6	14.9
G_{D3}	8.8	5.4
G_{D1a}	16.6	21.7
G_{D1a}-GalNAc	1.1	0.4
G_{T1a}	2.2	1.8
G_{D2}	3.1	8.0
G_{D1b}	16.9	18.2
G_{T1b}	11.1	16.3
G_{Q1b}	2.7	5.0

a) Normal human white matter contains 275 ug of lipid-bound sialic acid/g of wet tissue; gray matter, 875 ug/g of wet tissue. Data taken from Ando et al.,Anal.Biochem.89,437 (1978), with permission of Academic Press Inc.

achieve high resolution densitometry of gangliosides (Zanetta et al., 1977; Ando et al., 1978). The sensitivity and reproducibility are extremely improved by the controlled visualization technique (Ando et al., 1978).

Gangliosides are prepared from tissues by the method of Ledeen et al. (1973). The recent modification for the method by Ueno et al. (1978)employs Sephadex G-50 column instead of dialysis after DEAE-Sephadex column chromatography. The modified method can prevent loss of gangliosides especially in handling a small amount of sample, and reduces the time for preparation. Excellent separations of ganglioside species are achieved on a HPTLC plate with a single solvent, chloroform-methanol-water (55:45:10 or 50:45:10) containing 0.02 % (w/v) $CaCl_2 \cdot 2H_2O$ (Ando et al., 1978). The addition of calcium ion gives narrow and compact bands to improve resolution. Gangliosides are visualized by spraying with the resorcinol-HCl reagent (Svennerholm, 1957), followed by heating for 30 min at 95 ± 2° on an aluminum hot plate. Gangliosides appear as deep purple bands without any noticeable browning due to nonspecific charring. The densitometric scanning is performed at 580 nm with a transmittance mode, which gives much higher sensitivity than a reflectance mode. All of the ganglioside species reveal the same detector response based on known amounts of lipid-bound sialic acid. The range of 0.03-3 ug (0.1-10 nmole) of sialic acid can be determined by the present method.

Typical thin-layer chromatograms of gangliosides are shown in Fig.6. Densitograms corresponding to the chromatograms are recorded as in Fig.7 and the percentage distribution data of gangliosides are listed in Table 1. The TLC-densitometry has been applied to fine analyses of spinal cord gangliosides (Ueno et al., 1978) and inbred mouse brain gangliosides (Seyfried et al., 1979).

DEAE-silica gel recently developed by Kundu and Roy (1978) might have a potentiality to be applied to TLC of gangliosides. The ganglioside mapping technique introduced by Iwamori and Nagai (1978) would be expected to become a new combined method with densitometry.

Acknowledgement. We would like to thank Dr. T. Osawa of this Institute for providing cultured WI-38 cells with wide range of passages. We would also like to express our thanks to Dr. Y. Nagai (Director of the Department) for his constant advice throughout the course of this investigation.

ABSTRACT

An improved densitometry employing high performance TLC plates has been developed for the analysis of major neutral lipids and phospholipids. The densitometry involves strictly controlled charring of the lipid samples on TLC plates and an accurate correction of the density of the bands. An internal standard method is employed to compute absolute amounts of various lipids. This TLC-densitometry is highly sensitive and accurate. Lipid compositional analyses of small quantities of erythrocyte membranes and cultured cells were successfully carried out. The analysis of brain gangliosides is also described.

REFERENCES

Ando,S.,Chang,N-C. and Yu,R.K.(1978).Anal.Biochem.89,437.

Folch,J.,Lees,M.B. and Sloane Stanley,G.M.(1957). J.Biol. Chem.226,497.

Goldman,J. and Goodall,R.R.(1969). J.Chromatog.40,345.

Hezel,U.(1976). In R.E.Kaiser(ed.), Einführung in die Hochleistungs-Dünnschicht-Chromatographie(HPDC), Institutes für Chromatographie Bad Durkheim.

Iwamori,M. and Nagai,Y.(1978). Biochim.Biophys.Acta,528,257

Kohn,L.D.,Lee,G.,Grollman,E.F.,Ledley,F.D.,Mullin,B.R., Friedman,R.N.,Meldolesi,M.F. and Aloj,S.M.(1978). In R.E.Harmon(ed.), Cell Surface Carbohydrate Chemistry. Academic Press, N.Y. 103pp.

Korey,S.R. and Gonatas,J.(1963). Life Sci.2,296.

Kritchevsky,D. and Howard,B.V.(1970). In E.Holečková and V.J.Cristofalo(ed.), Aging in Cell and Tissue Culture, Plenum Press, N.Y. 57pp.

Kukusis,A.,Breckenridge,W.C.,Marai,L. and Stachnyk,O.(1968), J.Amer.Oil Chemists' Soc.,45,537.

Kundu,S.K. and Roy,S.K.(1978). J.Lipid Res.,19,390.

Ledeen,R.W.,Yu,R.K. and Eng,L.F.(1973). J.Neurochem.21,829.

Ledeen,R.W. and Yu,R.K.(1978). In N.Marks and R.Rodnight(ed.), Research Methods in Neurochemistry, vol.4. Plenum Press, N.Y. 371pp.

MacMillan,V.H. and Wherrett,J.R.(1969). J.Neurochem.16,1621.

Max,S.R. and Quarles,R.H.(1973). In J.C.Touchstone(ed.), Quantitative Thin-Layer Chromatography. John Wiley & Sons, Inc. 235pp.

Ng,S-S. and Dain,J.A.(1976). In A.Rosenberg and C-L.Schengrund (ed.), Biological Roles of Sialic Acid, Plenum Press, 59pp.

Sandhoff,K.,Harzer,K. and Jatzkewitz,H.(1968). Hoppe-Seyler's Z.Physiol.Chem.349,283.

Seyfried,T.N.,Glaser,G.H. and Yu,R.K.(1979). Biochem.Genet. 17,43.

Smid,F. and Reinisova,J.(1973). J.Chromatog.86,200.

Stolyhwo,A. and Privett,O.S.(1973). J.Chromatogr.Sci.11,20.

Suzuki,K.(1964). Life Sci.3,1227.

Svennerholm,L.(1957). Biochim.Biophys.Acta,24,604.

Svennerholm,L.(1963). J.Neurochem.10,613.

Tanaka,M.,Itoh,T. and Kaneko,H.(1977). Yukagaku(Tokyo),26,454.

Tokunaga,M.,Ando,S. and Ueta,N.(1973). Proc.Jap.Conf.Biochem. Lipids,15,195.

Ueno,K.,Ando,S. and Yu,R.K.(1978). J.Lipid Res.,19,863.

Ullman,M.D. and McCluer,R.H.(1978). J.Lipid Res.,19,910.

Zanetta, J.-P., Vitiello, F. and Robert, J. (1977). *J. Chromatogr.* 137, 481.

THE PULMONARY SURFACTANT: CONTROL OF FLUIDITY AT THE AIR-LIQUID INTERFACE

Fred Possmayer, I. LeRoy Metcalfe and Goran Enhorning, Department of Obstetrics & Gynaecology, University of Western Ontario, London, Ontario, Canada, and Department of Obstetrics and Gynaecology, University of Toronto, Toronto, Ontario, Canada.

ABSTRACT

The mammalian alveolus is stabilized by the presence of a specialized material, the pulmonary surfactant, which reduces the surface tension at the air-liquid interface. The disaturated lecithin, 1,2-dipalmitoyl-sn-phosphatidylcholine, is a major constituent of the surfactant. Unsaturated lecithins, other phospholipids, neutral lipid and protein are also present. The properties of compressed films of natural surfactant strongly resemble those of a monolayer of dipalmitoylphosphatidylcholine. However, below the transition temperature of 41°C, liposomes of dipalmitoylphosphatidylcholine are in the condensed gel state from which monolayers form very slowly. The mechanism by which dipalmitoylphosphatidylcholine is transferred from bilayers to monolayers has been investigated with a pulsating bubble technique. Natural surfactant reduces the surface tension of the pulsating bubble to 25-30 dynes/cm at maximal radius (0.55 mm) and to 0 dynes/cm at minimal radius (0.4 mm). Evidence is presented which strongly indicates that the proteins associated with natural surfactant are not required for the rate or the extent of the reduction of surface tension. Divalent cations appear to be important for surfactant behaviour. Using combinations of pure lipids, artificial surfactant preparations have been obtained which possess the basic physical and biological properties of natural surfactant.

INTRODUCTION

It is generally known that air bubbles are inherently unstable. If one blows a soap bubble with a pipe and removes one's mouth, the bubble will collapse. This instability is due to the fact that there is a pressure difference across a bubble surface which is described by the equation (1) formulated by Laplace :

$$\Delta P = \frac{2\sigma}{r}$$

where ΔP = pressure (dynes/cm^2)

σ = surface tension (dynes/cm)

r = radius (cm)

Since the pressure difference is inversely related to the radius, the collapsing pressure will increase as the bubble becomes smaller. The radius of a human alveolus is approximately 0.2 mm. If the surface tension of the air-liquid interface of our alveoli were equal to that of saline (72 dynes/cm), the pressure gradient across our alveoli would be sufficient to promote alveolar collapse. Fortunately, this does not occur because the alveoli are lined with a specialized material, the pulmonary surfactant, which reduces the surface tension from the 72 dynes/cm of saline to approximately 27 dynes/cm at maximum alveolar size. When the alveolar radius decreases during exhalation, the surface tension is reduced to an exceedingly low value, possibly 0 dynes/cm (Pattle, 1965; Clements, 1973; Goerke, 1974; King, 1974; Notter and Morrow, 1975).

RESULTS

Pulmonary surfactant can be obtained by washing the lungs of newly sacrificed animals with saline through the trachea. The resulting milky suspension is centrifuged,

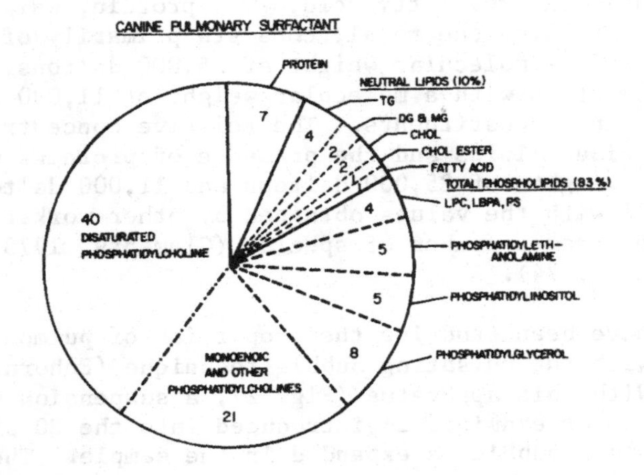

Fig. 1 Relative proportions (wt/wt) of the various components of natural surfactant obtained from dog lung. Abbreviations: CHOL, cholesterol; DG, diacylglycerol; LBPA, lysobisphosphatidic acid; LPC, lysophosphatidylcholine; MG, monoacylglycerol; PS, phosphatidylserine; TG, triacylglycerol.

first at 1,000 _g_ for 5 min to remove tracheal debris and whole cells and then at 12,000 _g_ for 30 min to yield a snow white pellet. This pellet, resuspended in saline or in 5 mM $MgCl_2$ and/or 10 mM $CaCl_2$, is referred to as natural surfactant. The composition of canine surfactant can be seen in Fig. 1. It consists of 83% phospholipid, 10% neutral lipid and 7% protein (wt/wt). The phospholipid constitutes 83% of the total: 61% as phosphatidylcholine, 8% as phosphatidylglycerol, 5% as phosphatidylinositol, 5% as phosphatidylethanolamine with smaller amounts as lysophosphatidylcholine, lysobisphosphatidic acid and phosphatidylserine. The phosphatidylcholine fraction contains a high proportion (78%) of palmitate and other saturated fatty acids. It can be calculated that approximately two-thirds of the phosphatidylcholine is disaturated. A high proportion of saturated fatty acids was not observed with the other phospholipids.

The neutral lipids account for 10% of the dry weight of natural surfactant: 4% as triacylglycerol, 2% as cholesterol, 2% as mono- and diacylglycerol, 1% as cholesterol ester and 1% as free fatty acid. The protein, which accounts for 7% of the total, consists primarily of a major fraction with a molecular weight of 35,000 daltons, a smaller fraction with a molecular weight of 11,000 daltons and some minor constituents. The relative concentrations of the various lipids and the presence of proteins with molecular weights of 35,000 daltons and 11,000 daltons agree well with the values obtained by other workers with surfactant from a number of species (Clements, 1973; Goerke, 1974; King, 1974).

We have been studying the properties of pulmonary surfactant with the pulsating bubble technique (Enhorning, 1977). With this apparatus (Fig. 2), a suspension of the material to be examined is introduced into the 20 μl test chamber and a bubble is expanded in the sample. The bubble, which communicates with the atmosphere, is continuously pulsated at a rate of 20 rpm with a minimal radius of 0.4 mm and a maximal radius of 0.55 mm. The pressure is

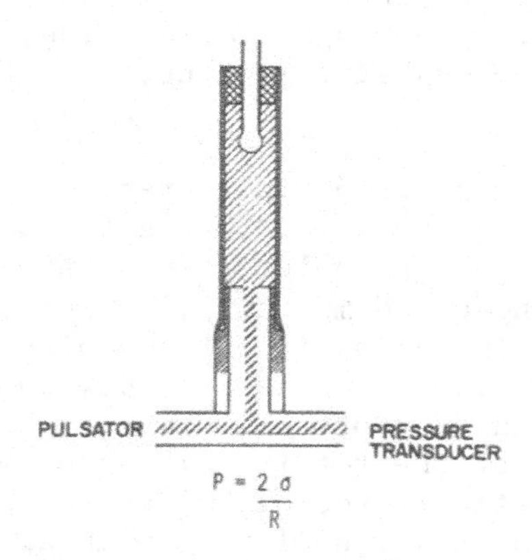

Fig. 2 Diagrammatical representation of the pulsating bubble apparatus. The bubble is pulsated with a maximal radius of 0.55 mm and a minimal radius of 0.4 mm. The surface tensions at maximal and minimal bubble volume are calculated from the recorded pressure by means of the Laplace equation.

recorded via a pressure transducer. Since the pressure across the bubble is monitored and the radius at any point can be determined, the surface tension can be calculated from the law of Laplace (equation 1).

With saline in the test chamber, a sinusoidal pressure tracing is obtained (Fig. 3a), from which it can be determined that the surface tension is 72 dynes/cm at both maximal and minimal bubble size. When the suspension contains natural surfactant at a concentration of 2% (wt/vol, based on phospholipid content), the pressure across the bubble is considerably reduced compared to saline (note the different scale used for 3b and c). The pressure tracing with surfactant is out of phase compared to the curve obtained with saline: the lowest pressure gradient is observed with the minimal bubble size. At the maximal radius, the surface tension is calculated to be 27.2 dynes/cm (25-30 dynes/cm with various preparations). At the minimal radius the surface tension falls to 0 dynes/cm. A similar pattern is

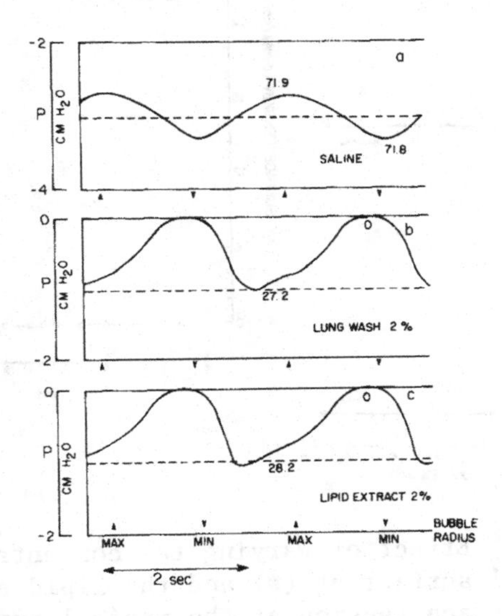

Fig. 3 Pressure tracings with the pulsating bubble apparatus for (a) saline, (b) a suspension of natural surfactant (2%, wt/vol), and (c) a suspension of the lipid extract of lung surfactant (2%, wt/vol). The surface tensions at the maximal and minimal bubble size are depicted.

obtained when a suspension of a protein-free chloroform-methanol extract of natural surfactant is examined (Fig. 3c). The surface tension at maximal bubble radius is 28.0 dynes/cm while at minimal bubble size the surface tension is 0 dynes/cm. These results suggest that the protein present in natural surfactant is not critical for the reduction of the surface tension.

When the concentration of natural surfactant or the chloroform-methanol lipid extract is varied, essentially identical curves are generated for the surface tensions at maximal and minimal radii (Fig. 4). Furthermore, when the

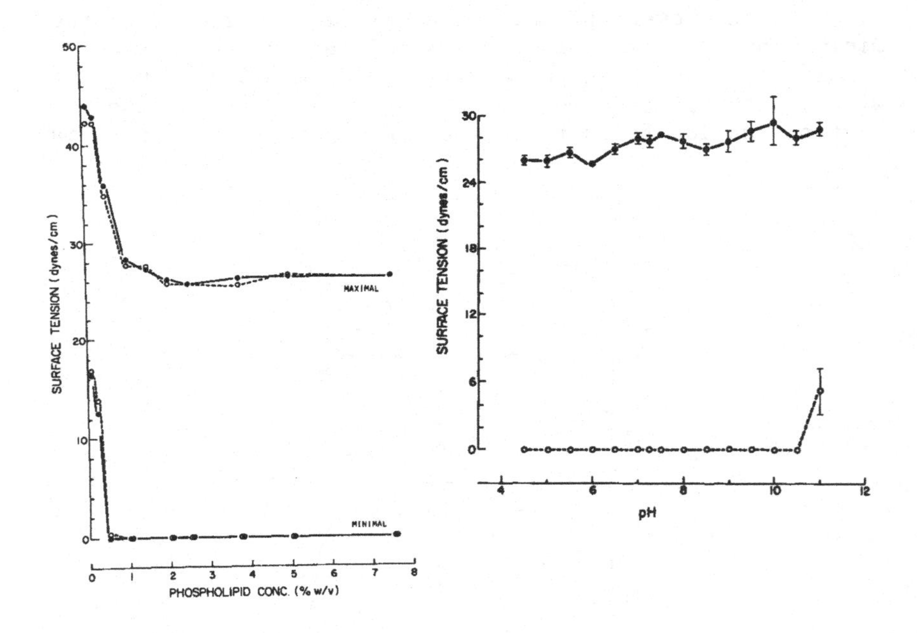

Fig. 4 (left) Effect of varying the concentration of natural surfactant (●) and the lipid extract (O) on the surface tension at the maximal and minimal radii of the pulsating bubble. The concentrations are based on the phospholipid content.

Fig. 5 (right) Effect of the pH of the subphase on the surface tension at maximal (●) and minimal (O) radii of the pulsating bubble.

pH of the subphase is varied between 4.0 and 10.5, there is no effect on the surface tension at minimal bubble volume (Fig. 5). A small increase is noted in the surface tension at maximal radius. The ability to lower the surface tension of the pulsating bubble is not affected by treating natural surfactant with trypsin or by boiling the lipid extract in methanol (Table 1). It is clear from these results that if the proteins associated with natural surfactant play a role in the reduction of the surface tension, they must be stable to heat, trypsin degradation, extraction into organic solvent and must function over a surprisingly wide pH range. Furthermore, since protein cannot be detected in the lipid extracts by the conventional Lowry procedure, the proteins must be effective at extremely low concentrations.

It is well known that the surface activity of the pulmonary surfactant is dependent on the presence of counter-ions. The effect of changing the cationic composition was investigated. It was observed that when the phospholipids were converted to the sodium salt form by passing the lipid extract through a Chelex-100 column, the surface tension reducing properties were altered (Table 2). Washing the lipid extracts with mild acid (0.1 N HCl in saline) using a biphasic distribution (Bligh and Dyer, 1959) also reduced the surface activity (Table 2).

The lipid extracts of natural surfactant can be fractionated by silicic acid, DEAE-cellulose or thin layer

Table 1: Effect of trypsin degradation of natural surfactant and boiling the lipid extract in methanol on the reduction of the surface tension of a pulsating bubble.

	Conc. % wt/vol	σ max dynes/cm	σ min dynes/cm
Natural surfactant	2%	26.5	0
Trypsin degraded*	2%	25.9	0
Lipid extract	2%	27.8	0
Boiled in methanol (10 min)	2%	28.9	0

*incubated overnight with trypsin (10 mg/ml) in potassium phosphate buffer (pH 7.4) at 37°C.

Table 2: Effect of treatment of the lipid extract of natural surfactant with Chelex-100 or mild acid on the surfactant properties.

	Conc. % wt/vol	σ max dynes/cm	σ min dynes/cm
Lipid extract	2%	26.5	0
After treatment with Chelex-100	2%	28.9	9.5
After washing with 0.1 N HCl	2%	37.5	17.6

chromatography. Recombinations of these fractions does not result in active preparations. However, it is possible to prepare suspensions containing a combination of pure synthetic and naturally derived lipids which mimic the basic properties of natural surfactant. One example of an artificial surfactant preparation, containing diphosphatidylcholine, egg phosphatidylcholine, soybean phosphatidylinositol and palmitic acid (60:30:10:1) is depicted in Fig. 6c. Not only does this preparation reduce the surface tension to approximately 27 dynes/cm at maximum radius and to 0 at minimum radius, but the rate of reduction of the surface tension after the initial formation of the bubble is similar to that observed with natural surfactant (Fig. 6c). With natural surfactant and the artificial preparation the reduction of the surface tension to 0 normally occurs slightly more rapidly than with the lipid extract (Fig. 6b). Nevertheless, with all three preparations, the surface tension is reduced to 0 within three to four cycles of the pulsating bubble.

These results strongly indicate that the proteins associated with natural surfactant are not required for surfactant adsorption or the reduction of the surface tension. The findings further suggest that it is possible to prepare immunologically inert artificial lipid mixtures with appropriate surfactant properties for clinical use. Preliminary results obtained with a biological model previously established by Enhorning and his colleagues (Enhorning et al., 1973, 1978) have demonstrated that artificial surfactant suspensions can be prepared which are just as effective as natural surfactant in promoting lung expansion and alveolar stabilization during pressure-volume loops on rabbit fetuses delivered prematurely at a gestational age of 27 days (term 31 days).

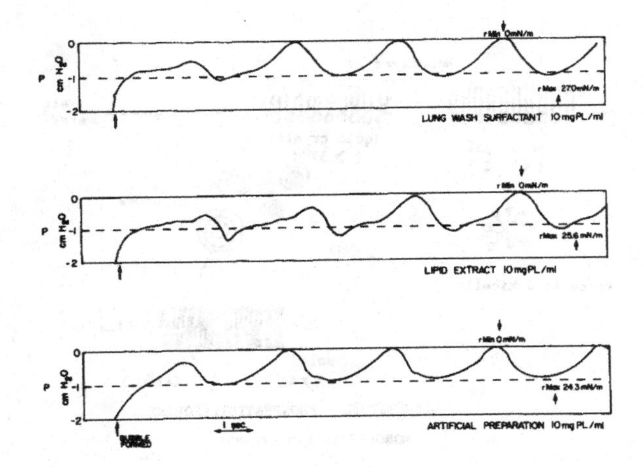

Fig. 6 Pressure tracings generated after the initial formation of pulsating bubbles with (a) natural surfactant, (b) lipid extract of natural surfactant, and (c) an artificial surfactant preparation containing dipalmitoylphosphatidylcholine, egg phosphatidylcholine, soybean phosphatidylinositol and palmitic acid (60:30:10:1, wt/wt). The surface tensions developed at the maximal and minimal radii of the pulsating bubble during the fourth pulsation are presented.

DISCUSSION

There is considerable evidence indicating that the monolayer which reduces the surface tension in the alveolus is principally composed of dipalmitoylphosphatidylcholine (Clements, 1973, 1977; King, 1974; Notter and Morrow, 1975). It is well known that below the transition temperature of 41°C liposomes of dipalmitoylphosphatidylcholine are in the condensed gel state from which monolayers form only very slowly (Fig. 7). The presence of phosphatidylcholines containing unsaturated fatty acids and the other lipids presumably promotes the transfer of dipalmitoylphosphatidylcholine from the stable bilayers to the monolayer. Interestingly, evidence has been obtained which indicates that once a monolayer is formed, dipalmitoylphosphatidylcholine has a gel-liquid crystal transition temperature of 37°C (Tinker, D.O.; personal communication). The mechanism for the transfer of dipalmitoylphosphatidylcholine molecules

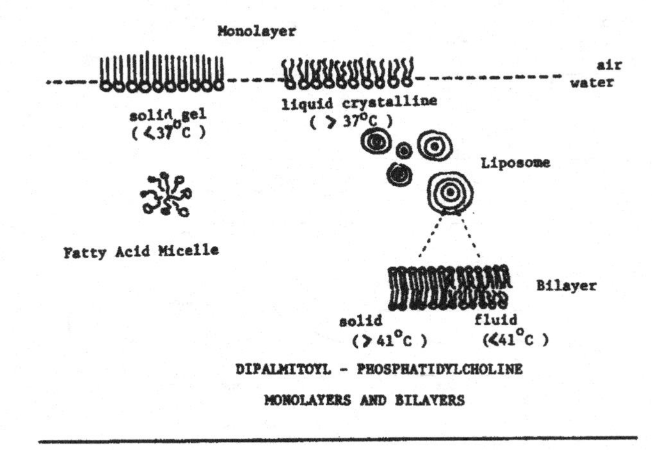

Fig. 7 Diagrammatical representation of dipalmitoyl-phosphatidylcholine bilayers and monolayers in the gel and liquid-crystal states.

from the stable bilayer to the stable monolayer remains unknown (Fig. 7). It is interesting to speculate that a nonbilayer phase, such as Hexagonal II (Cullis and Hope, 1978) may take part in this transfer. Studies are currently being conducted to investigate this possibility.

ACKNOWLEDGEMENTS

These studies were supported by grants from the Medical Research Council of Canada and the Canadian Lung Association.

REFERENCES

Bligh, E. G., and W. J. Dyer (1959). Can. J. Biochem. Physiol. 37, 911.

Clements, J. A. (1973). In C. A. Villee, D. B. Villee and J. Zuckerman (eds.), Respiratory Distress Syndrome. Academic Press, New York.

Clements, J. A. (1977). Am. Rev. Resp. Dis. (June Supplement) 115, 67.

Cullis, P. R., and M. J. Hope (1978). Nature 271, 672.
Enhorning, G. (1977). J. Appl. Physiol. 433, 198.
Enhorning, G., G. Grossman, and B. Robertson (1973). Am. J. Resp. Dis. 107, 921.
Enhorning, G., D. Hill, G. Sherwood, E. Cutz, B. Robertson, and C. Bryan (1978). Am. J. Obstet. Gynecol. 132, 529.
Goerke, J. (1974). Biochim. Biophys. Acta 334, 241.
King, R. D. (1974). Fed. Proc. 33, 2238.
Notter, R. H., and P. E. Morrow (1975). Annals Biomed. Eng. 3, 119.
Pattle, R. E. (1965). Physiol. Rev. 45, 48.

FLUIDITY OF MEMBRANE LIPIDS

William E. M. Lands

Dept. of Biological Chemistry, The University of Michigan, Ann Arbor, MI 48109, USA

As soon as our lab decided to examine the role of lipid biochemistry in controlling membrane fluidity, my colleagues and I had to face a fascinating challenge. Each of the terms, "lipid", "membrane", and "fluidity", is laden with allusions, elusions, and illusions. Each term is often regarded as a general (almost poetic) conceptual term without a specific, obligatory, precise definition. Thus the terms may indicate different things when used in different contexts or by different scientists. As we tried to design appropriate experiments on the behavior of membrane lipids, we set about defining more carefully those allusive terms to ensure that our measurements of the elusive properties did not lead us to illusive positions. The following comments briefly describe some of the underlying concepts of fluidity that we've shared with the younger workers in our lab as we tried to develop a deeper, more functional understanding. The comments provide a rather primitive overview that I hope will be helpful in assimilating further information.

Fluidity is the quality of ease of movement and is arithmetically the reciprocal of viscosity (Hildebrand, 1972). Viscosity is, in turn, the quality of resistance to movement in a fluid.

$$(1) \quad \text{Fluidity} = \phi = 1/\text{viscosity} = 1/\eta$$

The factors that create a resistance to molecular movement represent active events that are often more easily discussed and interpreted than their reciprocal properties. For example, the viscosity is a quantitative index of the absorption of kinetic energy, whereas fluidity represents a lack of that absorption.

(2) Viscosity = absorption of kinetic energy

Varied types of viscosity (and thereby, fluidity) originate in the basic fact that kinetic energy may be in three general forms: translational, rotational, and vibrational. Each of the forms of kinetic energy may be oriented either parallel or perpendicular to the plane of the membrane. This provides six separate forms of motion that may be impeded by interactions in the fluid lipid bilayer. We can thus expect to encounter at least six separate measurements of fluidity that may have greatly different quantitative values. Attempts to describe "the" fluidity of the bilayer lipids as a single entity seem illusive.

(3) Six types of kinetic energy are:

translational $\perp$	rotational $\perp$	vibrational $\perp$
translational $\parallel$	rotational $\parallel$	vibrational $\parallel$

There is, nevertheless, an important property of the fluid matrix that is common to the absorption of all six forms of energy. The density, mass per unit volume (m/V) of material, is the primary general feature which allows the microenvironment to absorb kinetic energy via molecular collisions. A more dense microenvironment will have more energy-absorbing collisions per minute for n molecules of m mass.

$$(4) \quad \frac{\Delta \text{ K.E. per unit}}{\text{Cross-sectional area}} = \frac{1/3\ n \cdot m \cdot (dx/dt)^2}{\text{Volume}}$$

The absorption of energy per collision is dependent upon the general density of the medium which affects the frequency of interaction, and upon the spatial interactions (or orientation) that can affect the efficiency of energy transfer during interactions (orientation factor = C).

$$(5) \quad \eta = \Delta\text{K.E.} = (m/V) \times C$$

The molal density decreases rapidly at the transition temperature between the solid and liquid states, and then progressively decreases with increased temperature above the transition. This temperature-dependent expansion can be expressed by the standard dilatometric relationship:

$$(6) \quad V = \beta \Delta T + V_0$$

In this case, V_0 is the molal volume of the liquid just above the melting point at which the fluidity is zero. We can then see that as the expansion is greater with temperature, the density is less, the viscosity is less, and the fluidity is greater. Thus melting point transition temperatures are most valuable in providing a reference point from which we may estimate the degree of expansion (and thus fluidity) of a lipid at a given temperature.

$$(7) \quad (V - V_0) = \beta\ (\Delta T)$$

We have found repeatedly that we could more easily interpret the role of fatty acids in affecting membrane fluidity when we examined the factors that influence the density of the lipid matrix (e.g., Holub & Lands, 1975). Thus the greater cross sectional area of unsaturated acyl chains (e.g., Demel et al. 1972) provides greater molal volumes and lower densities when these acids are esterified in membrane lipids. Also a low solid-liquid transition temperature for the phospholipid means that it will be more highly expanded at 30° (and thereby less dense) than would be a phospholipid with a higher transition temperature. The different relative contributions of saturated and unsaturated fatty acids to the mass and volume of the membrane may help us explain their contribution to membrane lipid fluidity. Hildebrand (1971,1972) modified Batschinski's (1913) earlier concept, and applied the concept of excess molal volume to establish a simple quantitative relationship between fluidity and molal volume:

$$(8) \quad \phi = 1/\eta = B\ (V - V_0)/V_0$$

This simple relationship accommodates the density features in the terms for excess molal volume and the spatial orientational features in the term, B. The six different fluidities noted above for movement within a membrane may be expected to all begin from zero at the same temperature at which V equals V_0 (Hildebrand, 1971). However, the coefficients accommodating spatial interactions can be expected to differ appreciably for diffusional (translational) movement in comparison to rotational (e.g., fluorescence depolarization) movement. Caution must always be exercised in trying to interpret the results on rotational fluidity in terms of absolute values for translational fluidity. The magnitude of the coefficients (B) for the different types of fluidity remain elusive. Nevertheless, the use of fluorescent molecules as "reporter probes" may be a convenient way to secure relative comparisons of the degree of expansion of similar fluids that affect the steady-state anisotropy by hindering the probe's rotation (Lakowicz, 1979). Thus, we can constructively regard all the general allusions to fluidity to refer in part to some aspect of fluid expansion or molal excess volume which describe the density of the lipid microenvironment. Recognition of density as a common factor in all fluidity considerations helps us interpret how a wide variety of instrumental approaches can give different measurements that depend upon the same basic fluid phenomenon. The challenge remains for each of us to pursue further the degree to which each of the six general types of fluidity indicated above may be reflected in the reported values from studies with electron spin resonance, fluoresence depolarizaton, differential scanning calorimetry, diffusivity, etc. Also much more needs to be known about the ways in which different fats, hydrocarbons and sterols interact with individual molecular species of the different types of membrane phospholipid to affect local densities of membrane lipid. As we learn the manner in which the different acyl chains contribute to the density of the membrane lipid matrix, we will be able to better discuss the role of fatty acid composition in controlling membrane fluidity.

ACKNOWLEDGEMENT

Research for this report was supported in part by a grant (AM-05310) from the United States Public Health Service.

REFERENCES

Batschinski, A. J. (1913). Z. Physik. Chem. 84, 643.

Demel, R. A., Geurts Van Kessel, W. S. M., and Van Deenen, L. L. M. (1972). Biochim. Biophys. Acta 266, 26-40.

Hildebrand, J. H. (1971). Science 174, 490-493.

Hildebrand, J. H. (1972). Proc. Nat. Acad. Sci. 69, 3428-3431.

Holub, B. J., and Lands, W. E. M. (1975). Can. J. Biochem. 53, 1262-1267.

Lakowicz, J. R., Prendergast, F. G., and Hogan, D. (1979). Biochemistry 18, 508-519.

PART II

CORRELATION OF MEMBRANE FLUIDITY WITH PHYSIOLOGICAL ACTIVITY

ADAPTIVE REGULATION OF MEMBRANE LIPID BIOSYNTHESIS IN BACILLI BY ENVIRONMENTAL TEMPERATURE

Armand J. Fulco and Dennis K. Fujii

Department of Biological Chemistry, UCLA
Medical School and Laboratory of Nuclear
Medicine, University of California,
900 Veteran Avenue, Los Angeles, CA 90024 USA

Some years ago, when we decided to study lipid metabolism in bacilli, we did so with several objectives in mind. One goal was to determine the mechanisms by which bacilli adaptively adjust the composition of their membrane lipids in response to changes in environmental temperature. We found that temperature-mediated variations in bacilli membrane lipids could be conveniently divided into slow, long-term alterations that required growth and net lipid synthesis, and short-term alterations that were rapidly implemented in response to temperature changes and did not necessarily involve new lipid synthesis. Slow alterations included shifts in fatty acid chain length and branching, increases in saturated to unsaturated fatty acid ratios, and changes in phospholipid head-group distribution. Fast alterations included changes in the asymetrical distribution of phospholipid types on the two sides of the membrane bilayer and decreases in the ratios of saturated to unsaturated fatty acids in these lipids. Taking *B. licheniformis* as a typical example among the bacilli, the data in Table I illustrate several effects of growth temperature on the fatty acid composition of membrane lipids. As growth temperature is decreased from 35° to 20°, three effects are noted (Quint and Fulco, 1973). There is a decrease in the average chain length of the fatty acids, there is a reversal in the ratio of iso to anteiso fatty acids (with anteiso branching being favored at the lower temperatures) and there is a marked increase in unsaturated fatty acids. In addition to the changes indicated in Table I, we have also found an increase in the ratio of phosphatidylethanolamine to phosphatidylglycerol as growth temperature decreases (Chang and Fulco, 1973). Although all of these modifications appear to be important for the maintenance of optimal membrane liquidity in bacilli, I will confine the remainder of this presentation to our work on the temperature-mediated mechanisms that regulate the ratio of saturated to unsaturated fatty acids in the membrane lipids.

TABLE I
FATTY ACID DISTRIBUTION IN B. LICHENIFORMIS 9259
GROWN AT THREE TEMPERATURES

Growth Temperature	Fatty Acid	Percent of total by Weight			Saturated to Unsaturated Ratio
		Saturated	Monounsaturated	Diunsaturated	
°C			%		
35	<C_{15}	12.0	0.5	0.0	24
	Iso-C_{15}	16.7	0.0	0.0	
	Anteiso-C_{15}	12.9	0.0	0.0	
	Iso-C_{16}	7.2	0.5	0.0	14.4
	n-C_{16}	9.8	2.7	0.0	3.6
	Branched C_{17}	32.2	1.6	0.0	20.1
	Other	3.8	0.0	0.0	
	Total	94.6	5.3	0.0	19.6
30	<C_{15}	12.9	trace	trace	
	Iso-C_{15}	15.8	0.0	0.0	
	Anteiso-C_{15}	15.7	0.0	0.0	
	Iso-C_{16}	5.1	2.3	0.1	2.1
	n-C_{16}	6.0	9.1	0.6	0.62
	Branched C_{17}	17.8	11.2	0.7	1.50
	Other	2.7	trace	trace	
	Total	76.0	22.6	1.4	3.16
20	<C_{15}	22.6	trace	trace	
	Iso-C_{15}	9.4	0.0	0.0	
	Anteiso-C_{15}	19.4	0.0	0.0	
	Iso-C_{16}	1.7	4.5	0.3	0.35
	n-C_{16}	1.4	11.8	2.8	0.10
	Branched C_{17}	5.2	17.5	1.5	0.27
	Other	1.8	trace	trace	
	Total	61.5	33.8	4.6	1.60

In theory, there are several ways for altering the saturated:unsaturated fatty acid ratio in response to temperature. These would include relative changes in the rates of β-oxidation, incorporation and biosynthesis. To our knowledge, there is no convincing evidence for temperature-mediated selective β-oxidation. Examples of the second process, selective incorporation, are known but probably do not represent a primary regulatory mechanism for controlling the ratio of saturated to unsaturated fatty acids in the membrane lipids of most bacteria (Fulco, 1974). The third process, temperature-mediated alteration in the relative rates of saturated and unsaturated fatty acid biosynthesis is the most efficient and the most general means of varying the ratio of saturated to unsaturated fatty acids. In eukaryotic organisms and in certain bacteria, including bacilli (Fulco et al., 1964), monounsaturated fatty acid biosynthesis involves the O_2-dependent removal of 2 hydrogens from a saturated fatty acid derivative to give the cis-unsaturated analog. Fig. 1 shows the common features that characterize O_2-dependent desaturation in both bacteria and higher animals (Fulco, 1977). This is easily the most efficient mechanism for increasing membrane fluidity since, in one step, a saturated acyl group is removed and a monounsaturated acyl group substituted. Furthermore, this can be a rapid process since neither cell growth nor new lipid synthesis is required for its implementation. The reverse reaction, hydrogenation of the double bond to increase the saturated:unsaturated fatty acid ratio, is not a significant reaction in aerobes and thus a change in this direction is a slow process involving de novo fatty acid synthesis.

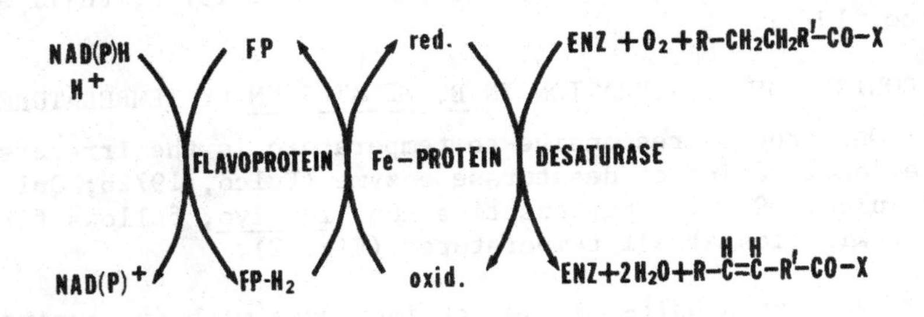

FIG. 1. General scheme for O_2-dependent desaturation of fatty acids.

To elucidate the biochemical mechanisms underlying the inverse relationship between temperature and desaturation, we surveyed a large number of *Bacillus* strains (Fulco, 1967; Fulco, 1969a; Fulco, 1969b; Fulco, 1970) and selected one, *B. megaterium* ATCC 14581, to study in detail. This strain exhibits several characteristics in its desaturation system which makes it well-suited for an investigation of this type. Desaturation activity is strongly temperature dependent; there is no desaturation at all in cultures growing at 35° but good desaturation in cultures growing at 20°. Unlike *B. licheniformis* which contains both a temperature-insensitive Δ^{10}-desaturase and a temperature-dependent Δ^5-system (Fulco, 1969b; Fulco, 1970a), *B. megaterium* 14589 contains only one fatty acid desaturating enzyme. It is O_2-dependent, membrane bound and desaturates only in the 5-position. It acts on the acyl groups (C_{15}-C_{19}) of intact membrane phospholipids but apparently not on the CoA or ACP thioesters. This type of substrate specificity is an advantage for *in vivo* work, since tracer amounts of labeled palmitate can be incorporated within seconds into membrane lipids and then retained indefinitely in this form during subsequent cell growth (Fulco, A. J., 1972a; Chang and Fulco, 1973; Chen and Fulco, 1978; Chen et al., 1978). We thus eliminated worries about abberations in desaturation caused by variations in substrate uptake, activation or utilization in competing reactions. Finally, we found that temperature-mediated regulation of desaturation appeared to involve effects only on the desaturase itself so that, under our experimental conditions, a change in desaturation rate reflects a change in desaturase enzyme concentration rather than variations in substrate, cofactors or the physical state of the membrane (Fulco, 1972a; Fulco, 1972b; Fujii and Fulco, 1977).

CONTROL OF DESATURATION IN *B. MEGATERIUM* BY TEMPERATURE

One process responsive to temperature is the irreversible inactivation of desaturase enzyme (Fulco, 1972b; Quint and Fulco, 1973). This inactivation, *in vivo*, follows first-order kinetics at all temperatures (Fig. 2).

The enzyme half-life, which increases with decreasing temperature, is determined solely by the incubation temperature and is independent of the previous growth history of the culture. In *B. megaterium* at least, temperature does not significantly affect the *in vivo* rate of the desaturation

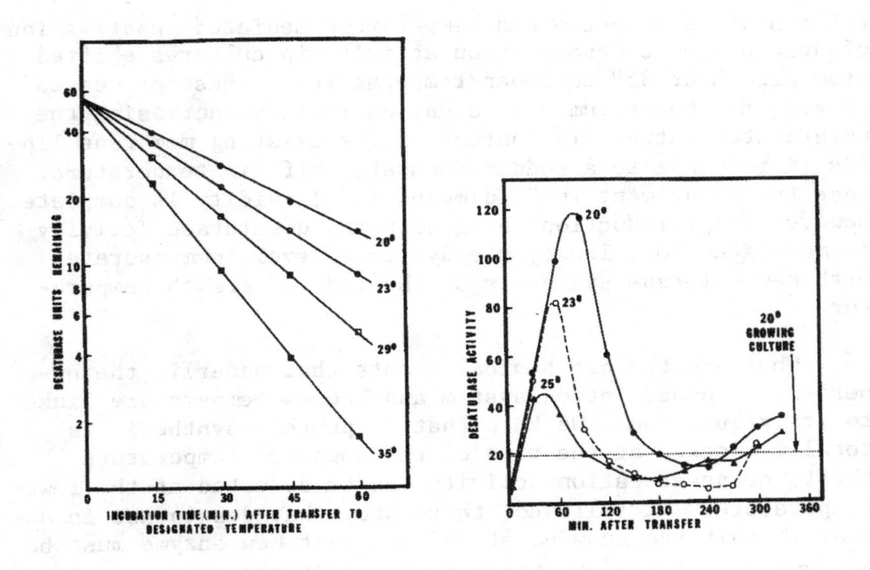

FIG. 2 (left). Half-life determinations of Δ^5-desaturase at 4 temperatures. Enzyme was induced by transfer of a 35°-culture to 20° for 1 hr. Protein synthesis was then abolished by the addition of chloramphenicol and the culture divided into portions. Each portion was then incubated at a different temperature (above). Desaturase activity was measured periodically in each portion to determine the amount of enzyme remaining.

FIG. 3 (right). Desaturase levels during hyperinduction and repression following culture transfer from 35° to three lower temperatures. The approximate steady-state level of desaturase in a culture growing from inoculum at 20° is shown for comparison.

reaction itself; at any given desaturase level, the initial rate of palmitate desaturation is relatively temperature-independent.

A second and more complex control system mediated by temperature is that of desaturase hyperinduction and repression (or modulation). Fig. 3 illustrates the combined effects of this system and temperature-mediated inactivation of desaturase on desaturation activity in cultures shifted from growth at 35° to lower temperatures. These processes provide the bacterium with a way of rapidly increasing the unsaturated fatty acid content of its existing membrane lipids in response to a sudden downward shift in temperature. Once the adjustment in "old membrane" liquidity is complete, however, hyperinduction shuts down and desaturase activity is modulated to a lower, steady-state level commensurate with new membrane synthesis at the reduced growth temperature.

What are the biochemical events that underlie the hyperinduction-modulation system and how is temperature linked to its regulation? We knew that if protein synthesis is totally blocked at the time of the downward temperature shift, no desaturation activity can be detected at the lower temperature. It followed, therefore, that desaturase is absent in cultures growing at 35° and that new enzyme must be synthesized to produce desaturation activity after shift-down. To proceed beyond this observation, however, it was necessary to measure parameters other than desaturase activity. These included cell growth, protein synthesis and desaturase half-life in all experiments, and DNA and RNA synthesis in some. Since, under the special assay conditions we employ (Fulco, 1972b; Quint and Fulco, 1973) desaturase activity is directly proportional to the amount of desaturase present per cell, we can readily calculate the rate of desaturase synthesis after correcting measured desaturase levels for _in vivo_ temperature-mediated first-order inactivation of enzyme. A simple derivation can also be obtained by dividing the absolute rate of desaturase synthesis by the concurrent rate of protein synthesis. This new value, which we call the "normalized rate" of desaturase synthesis is especially useful when comparing desaturation in cultures at different stages of growth or under different incubation conditions. Table II compares the normalized rates of desaturase synthesis in the three transfer cultures described in Fig. 3. We now see clearly that the initial rate of desaturase synthesis after shiftdown from 35° to a lower temperature is actually independent of the magnitude of the shift-down. However, hyperinduction continues in the 20° culture for about 75 min while it stops at 45 min in the 23° culture and turns off after 10-15 min in the culture transferred to 25°. Such data indicate that the total amount of desaturase synthesized

TABLE II

RATES OF NORMALIZED DESATURASE SYNTHESIS IN CULTURES TRANSFERRED FROM 35°, to 20°, 23° and 25°

Time After Transfer	Normalized Rate of Desaturase Synthesis		
	20°	23°	25°
Min		DU/Min/P	
15	6.14	6.31	5.60
45	7.26	6.76	2.59
75	6.53	0.00	0.90
105	0.36	0.47	1.30
135	0.00	0.48	0.86
165	0.48	0.23	1.00
195	0.48	0.51	1.72
225	0.73	0.23	2.50
255	2.13	0.78	1.77
285	3.45	3.60	3.31
315	4.43	4.07	4.30

after transfer is determined not by the rate of synthesis during hyperinduction but rather by the length of the hyperinduction period.

A second point of interest is illustrated in Figures 4 and 5. As Fig. 4 shows, the level of desaturase activity in a culture growing at 20° from inoculum to stationary phase is roughly constant while a similar culture transferred from 35° to 20° shows the hyperinduction phenomenon. However, if we look at the normalized rate of desaturase synthesis (Fig. 5) a different picture emerges for the culture growing at 20°. Normalized synthesis is constant only during the log phase of growth when the rate of protein synthesis (not shown) is constant. After 2 hr, growth becomes linear, the rate of protein synthesis decreases steadily and the normalized rate of desaturase synthesis increases. As the culture goes into early stationary phase (8 hr, 500 Klett units) the normalized rate actually approaches that attained during hyperinduction in the transfer culture. The absolute rate is, of course, much lower, commensurate with the very low rate of protein synthesis in stationary phase. It is important

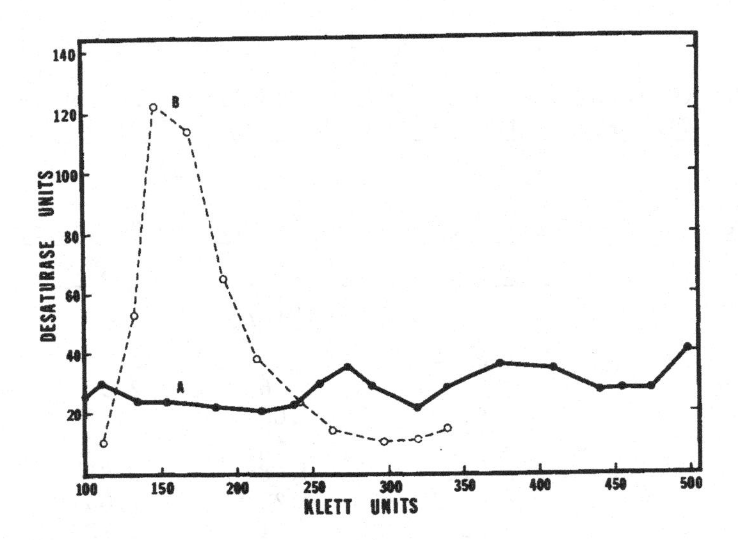

FIG. 4. Levels of desaturase activity in B. megaterium cultures at 20°. Curve A: culture growing from inoculum at 20°. Curve B: culture transferred from 35° to 20° when growth reached 112 Klett units (late log phase) [Fujii & Fulco, 1977].

to stress here that, for a culture transferred from 35° to 28° or less, at any stage of growth and under a variety of conditions, the maximum normalized rate of desaturase synthesis during hyperinduction is a constant (about 8 DU/min/P). This point is illustrated in Table III which shows the effect of partial inhibition of protein synthesis on hyperinduction maxima in cultures transferred from 35° to 20°. Complete inhibition of protein synthesis, of course, totally blocks hyperinduction. On the other hand, in the face of partial inhibition at any level, the maximum normalized rate of desaturase synthesis during hyperinduction remains constant. To put it another way, the maximum absolute rate of desaturase synthesis attained after temperature shift-down is simply proportional to the overall rate of protein synthesis. This maximum is not attained immediately after transfer to the lower temperature, however, but is always proceeded by a short lag period. As Table IV shows, sampling at 5 min intervals reveals that desaturase synthesis lags behind overall protein synthesis for the first 10

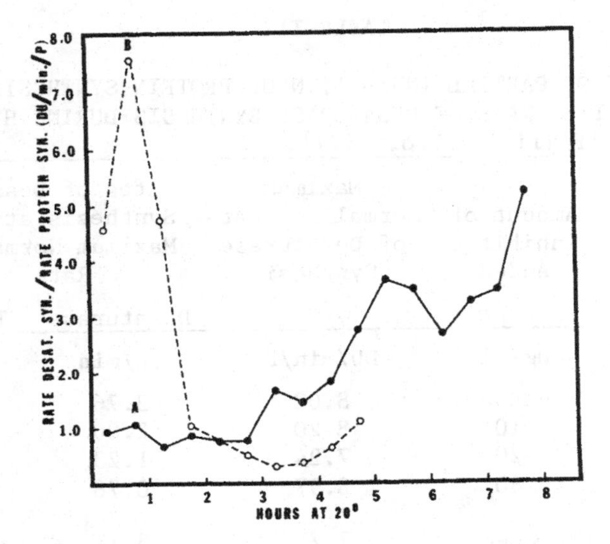

FIG. 5. Normalized rate of desaturase synthesis in B. megaterium cultures at 20°. This parameter was calculated for the two cultures described in Fig. 4 [Fujii & Fulco, 1977].

min. By 15 min, the normalized rate has reached a maximum and it remains at this level for about an hr. The rate then drops off in typical fashion (data not shown) after that time.

To this point, I've been describing phenomena without attempting to provide biochemical explanations. In fact, data such as I've presented did lead us to formulate a working hypothesis consisting of two basic tenets.

First of all, we proposed that the inability of B. megaterium, growing at 35°, to synthesize desaturase could be ascribed to the absence at this temperature of the messenger RNA coding for its synthesis. Hyperinduction, then, would require new RNA synthesis as well as new protein synthesis. Secondly, we suggested that the attenuation of hyperinduction as well as the relatively low rate of desaturase synthesis in cultures growing from inoculum at 20° could be explained by the action of a repressor or modulator at this temperature. It would follow that this modulator was absent in cultures growing at 35° since, otherwise, one should

TABLE III

THE EFFECT OF PARTIAL INHIBITION OF PROTEIN SYNTHESIS ON THE NORMALIZED RATE OF DESATURASE SYNTHESIS DURING HYPER-INDUCTION [Fujii & Fulco, 1977].

Experiment and Inhibitor	Amount of Inhibitor Added	Maximum Normalized Rate of Desaturase Synthesis	Rates of Desaturase Synthesis at the Maximum Normalized Rate	
			Desaturase	Protein
	μg/ml	DU/min/P	Du/min	P
1. Chlor-	none	8.02	3.76	0.469
amphen-	10	8.20	2.00	0.244
icol	20	7.24	1.23	0.170
	40	8.97	0.78	0.087
2. Nali-	none	7.71	3.73	0.484
Dixic				
Acid	(10 at -10 min)	6.67	1.32	0.198
	(10 at +10 min)	8.13	2.05	0.252

observe simple desaturase induction (rather than hyperinduction) when a culture is transferred to 20°. Hyperinduction, then, could be visualized as the unmodulated synthesis of desaturase initiated by transfer of a culture from 35° to 20° with a consequent stabilization or derepression of desaturase messenger. At the same time, however, synthesis of modulator would also begin and, once it attained an effective level, it would act to shut down hyperinduction and reduce desaturase synthesis to a rate commensurate with cultures growing from inoculum at 20°.

To test the hypothesis we've just outlined, a number of specific experiments were designed. Let us first consider the concept that desaturase messenger RNA is absent at 35° and hence must be induced after transfer of a culture to 20°. Certainly, the 10 min lag in desaturase synthesis relative to protein synthesis after culture transfer to the lower temperature (Table IV) is consistent with this idea. If active messenger were present at the time of transfer, no lag should appear. As a further test, we examined the effect of rifampicin, a specific RNA-synthesis inhibitor, on hyperinduction.

TABLE IV

KINETICS OF HYPERINDUCTION AT 20°*

Time Interval After Transfer to 20°	Desaturase Synthesized During Interval	Average Rate of Protein Synthesis During Interval	Normalized Rate of Desaturase Synthesis During Interval
Min	DU	P	DU/Min/P
0-5	0.6	0.235	0.51
5-10	3.1	0.274	2.26
10-15	10.6	0.279	7.60
15-20	11.6	0.282	8.23
20-25	9.0	0.284	6.34
25-30	8.8	0.287	6.13
30-45	30.2	0.297	6.78
45-60	33.7	0.318	7.06

* Data from Fujii & Fulco, 1977.

TABLE V

THE EFFECT OF RIFAMPICIN ON THE RATES OF DESATURASE AND PROTEIN SYNTHESIS IN A CULTURE TRANSFERRED FROM 35° to 20°

Time Interval	Rate of Desaturase Synthesis			Rate of Protein Synthesis		
	A	B	C	A	B	C
Min	DU/Min			P		
0-15	0.00	-	2.48	0.131	-	0.480
15-30	0.00	-	2.57	0.026	-	0.304
30-45	0.00	-	3.25	0.008	-	0.339
45-60	0.00	0.66	2.89	0.005	0.125	0.365
60-75	-	0.48	1.64	-	0.047	0.368
75-90	-	0.08	1.41	-	0.029	0.354
90-120	-	0.12	0.82	-	0.017	0.351
120-150	-	0.00	0.57	-	0.006	0.359

Note: A = Rifampicin at 0-time B = Rifampicin at 40 min
C = Control

Data from Fujii & Fulco, 1977.

As Table V shows, rifampicin totally blocked induction of desaturase when it was added at the time of culture transfer from 35° to 20°. Although RNA synthesis was totally eliminated after 30 sec, protein synthesis was not immediately shut down and had messenger been present, significant desaturase synthesis should have been observed. To emphasize this point, when rifampicin was added 40 min after transfer to 20° (Table V), significant desaturase synthesis could be detected for 30 min after addition. Although the absolute rate of desaturase synthesis decreased rapidly relative to the control culture, the normalized rate was about the same in both cultures after rifampicin addition.

The modulator postulate can also be tested in several ways. Of course, the very existence of the hyperinduction phenomenon implies that desaturase synthesis in a culture growing from inoculum at 20° is modulated in some way to maintain levels well below those theoretically obtainable. We at first thought that newly-synthesized unsaturated fatty acids themselves, or else a change in membrane liquidity due to their synthesis might be related to the rapid decline in desaturase synthesis following hyperinduction. However, preliminary experiments showed that incorporation of exogenous unsaturated fatty acids into membrane lipids of B. megaterium at 35° had absolutely no effect on the kinetics of hyperinduction-modulation after culture transfer to 20°. Furthermore, as we will show later, there is no direct involvement of either fatty acid composition or membrane liquidity in the control of desaturase synthesis.

The first test of the idea that desaturase synthesis is regulated at temperatures below 28° by a temperature-sensitive modulator involved the transfer of a culture, growing at 20° from inoculum, to 35° for short periods and then transfer back to 20°. We reasoned that if the modulator were a protein unstable at the higher temperature, then this manipulation should perturb or eliminate modulation. As Fig. 6 shows, this is exactly what happens when a culture growing at 20°, is given a 30 min pulse at 35°.

The resulting hyperinduction curve (D, Fig. 6) is essentially identical to those observed when cultures growing at 35° are shifted to 20°. At least 30 min at 35° is required for the complete (but transient) elimination of modulation. A 20-min pulse resulted in half-maximal hyperinduction while a 10-min pulse caused only small, cyclic

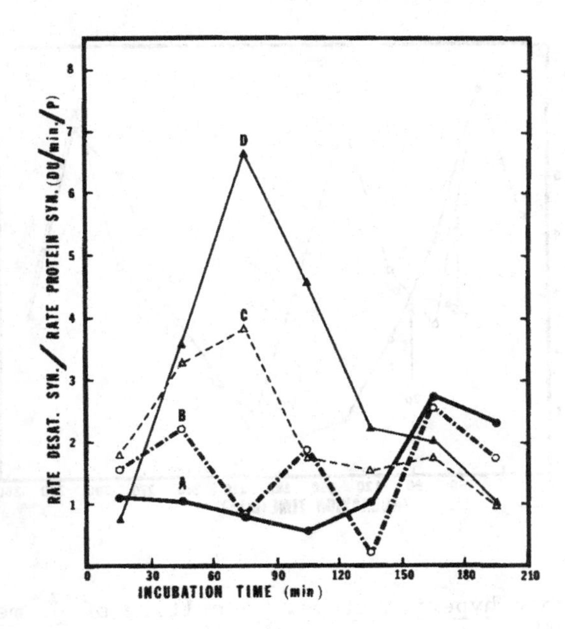

FIG. 6. Hyperinduction initiated in a culture growing at 20° by short-term shifts to 35°. A culture growing from inoculum at 20° was divided into 4 portions. The first portion (A) served as a control and remained at 20°. The other portions B, C and D were pulsed at 35° for 10, 20 and 30 min respectively and then transferred back to 20° (zero-time) and incubated at 20° for an additional 210 min [Fujii & Fulco, 1977].

perturbations. In cultures that had already undergone hyperinduction and attenuation after transfer from 35° to 20°, a second hyperinduction-attenuation cycle could be produced by a 30 min pulse at 35° (Fig. 7).

This "double hyperinduction effect is also consistent with our postulate and strongly suggests that modulation in a culture growing from inoculum at 20° and the attenuation of hyperinduction in a 35° to 20° transfer culture are caused by the same component. Finally, if we pulse at 30° for 30 min (Fig. 8), that is, at a temperature just above that where desaturase synthesis occurs, we see an effect similar to that observed with a 10 min pulse at 35° (Fig. 6). Again, such a result supports the existence of a temperature-sensitive modulator.

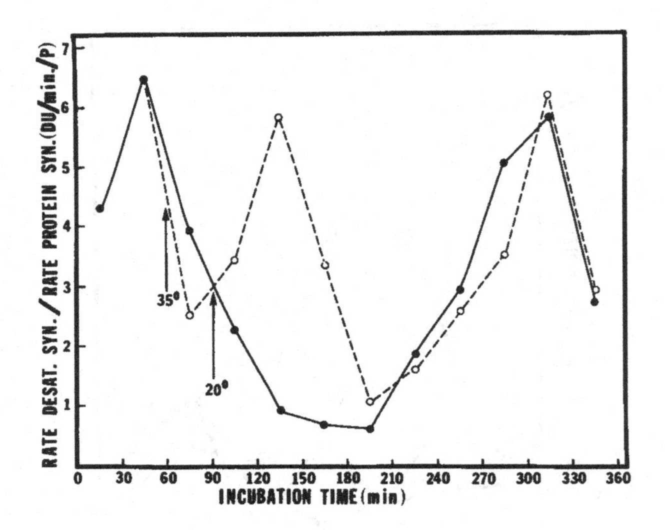

FIG. 7. Double hyperinduction. A culture of B. megaterium growing at 35° was transferred to 20° (zero-time) and allowed to hyperinduce. At 60 min, one-half of the culture (o) was pulsed at 35° for 30 min and then returned to 20°. The second half (•) was kept at 20° as a control [Fujii & Fulco, 1977].

These findings also argue against the possibility that alterations in fatty acid composition or membrane liquidity somehow mediate either hyperinduction or modulation of desaturase. Although the fatty acid composition of the B. megaterium membrane varies widely with growth temperature (Quint and Fulco, 1973), a culture grown at 20° and pulsed at 35°, a culture grown at 35° and then transferred to 20° and a culture subjected to multiple temperature shifts all give essentially the same hyperinduction-attenuation curves.

Although the data presented so far support the basic tenets of our hypothesis, we have not offered an explanation for the rather esoteric kinetics of these processes. We have consistently observed that the maximal rate of desaturase synthesis achieved during hyperinduction is determined by the rate of protein synthesis during the same period, that is, the maximum normalized rate of desaturase synthesis is always a constant. The evidence indicates that this represents desaturase synthesis in the absence of the

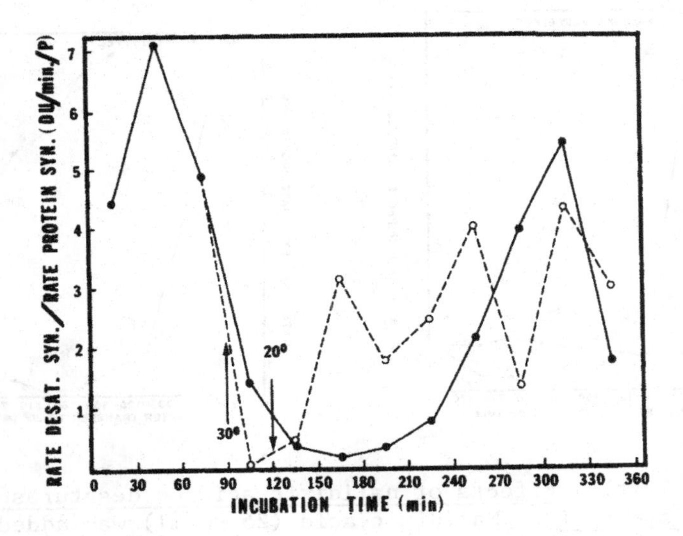

FIG. 8. The effect of transfer from 20° to 30° for 30 min. The experiment was similar to the one described in Fig. 7 except the pulse was at 30° rather than at 35°. The curves show the control culture (•) and the pulsed culture (o) [Fujii & Fulco, 1977].

temperature-sensitive modulator. If, however, the plateau or peak period of hyperinduction simply represents unmodulated desaturase synthesis, it is not clear why modulator action is delayed for periods up to several hr after shift-down, depending on conditions, and why, after maximum attenuation, the normalized rate of desaturase synthesis increases again. A related question is why a culture, growing from inoculum at 20°, shows a gradual loss of modulation of desaturase synthesis during linear growth.

We cannot yet answer these questions with certainty. We had originally suggested that the kinetics of modulation could be explained if the DNA that codes for modulator is not expressed at 35° and that even after cultures are transferred to lower temperatures, DNA replication is required before desaturase synthesis can be attenuated (Mead and Fulco, 1976). The demonstration that nalidixic acid, which inhibits DNA synthesis, also eliminates or depresses modulation, as well as the observed inverse relationship between growth rate (and hence rate of DNA synthesis) of a transfer

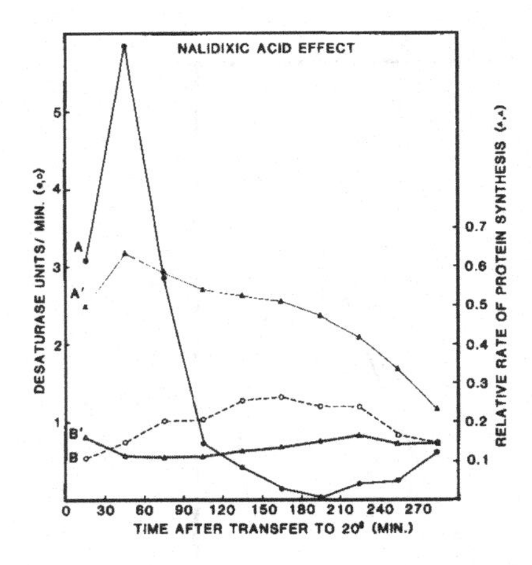

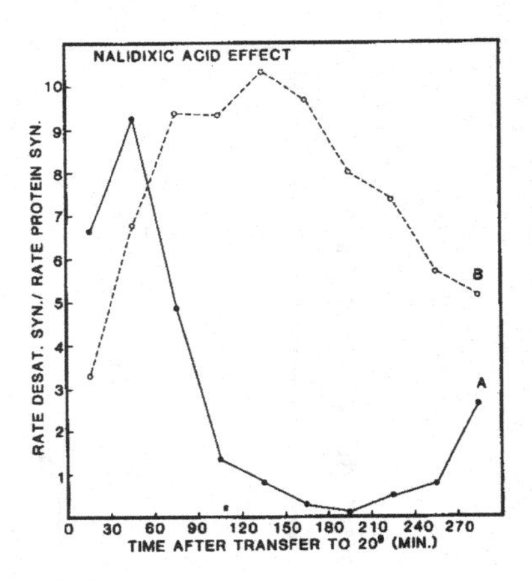

FIG. 9 (left). Effects of nalidixic acid on desaturase and protein synthesis. Nalidixic acid (25 µg/ml) was added at zero-time to a culture of B. megaterium transferred from 35° to 20°. Both the absolute rate of desaturase synthesis (B) and protein synthesis (B') were measured. For comparison, these parameters are also shown for the same transfer culture in the absence of nalidixic acid (A, desaturase; A', protein).

FIG. 10 (right). Effect of nalidixic acid on the normalized rate of desaturase synthesis. This parameter is shown for the control culture (A) and the nalidixic acid-treated culture (B) described in Fig. 9.

culture and the length of the hyperinduction period (Fujii and Fulco, 1977) could be interpreted as arguments in favor of such a hypothesis. Nevertheless, our thesis had to be discarded when we showed that 6-(p-hydroxyphenylazo)uracil, which quite specifically blocks DNA replication in bacilli, has no effect on the kinetics of either hyperinduction or modulation (Fujii and Fulco, 1977). We had noticed, however, that nalidixic acid, at the concentration needed to completely block DNA synthesis and prevent modulation also strongly inhibited protein synthesis (Figs. 9 and 10). Furthermore, the inverse relationship between growth rate and the duration of the hyperinduction period could also be interpreted in terms of protein synthesis. That is, the slower the rate of protein synthesis, the greater the delay in in the onset of modulation. To test this idea, we conducted a series of experiments which involved partially inhibiting protein synthesis to various degrees in transfer cultures by adding increasing amounts of chloramphenicol at the time of transfer from 35° to 20°. At 20 µg/ml of chloramphenicol, protein synthesis was inhibited about 60% and the attenuation of hyperinduction was both delayed and less intense. As Figs. 11 and 12 show, at a chloramphenicol concentration of 40 µg/ml, the maximum normalized rate of desaturase synthesis is achieved as usual but attenuation of desaturase synthesis is essentially eliminated . Indeed, from 135 to 225 min (Fig. 11) the absolute (as well as the normalized) rate of desaturase synthesis in the chloramphenicol-treated culture is greater than in the control, despite an 80% reduction in the rate of protein synthesis in the former. Thus, this experiment not only demonstrates the dependence of modulation upon the rate of protein synthesis,

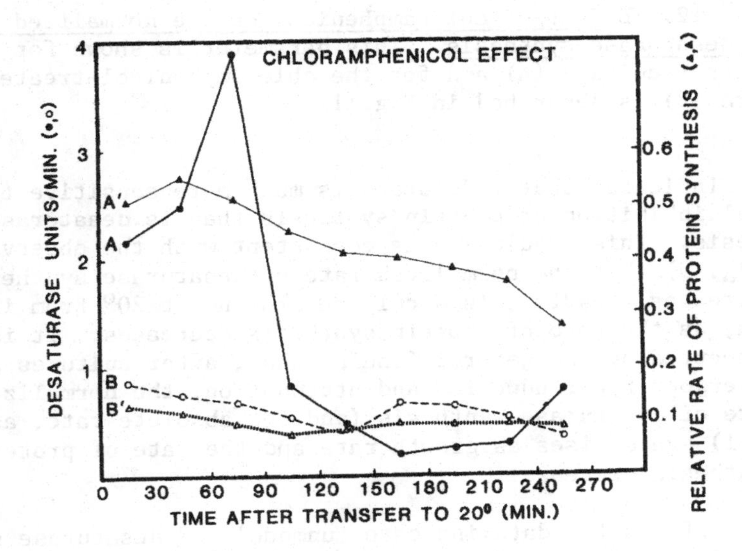

FIG. 11. Effect of chloramphenicol on desaturase and protein synthesis. Chloramphenicol (40 µg/ml) was added at zero-time to a culture of *B. megaterium* transferred from 35° to 20°. Both the rate of desaturase synthesis (B) and protein synthesis (B') are shown. These parameters are also shown for the control culture (no chloramphenicol added). (A), desaturase; (A'), protein.

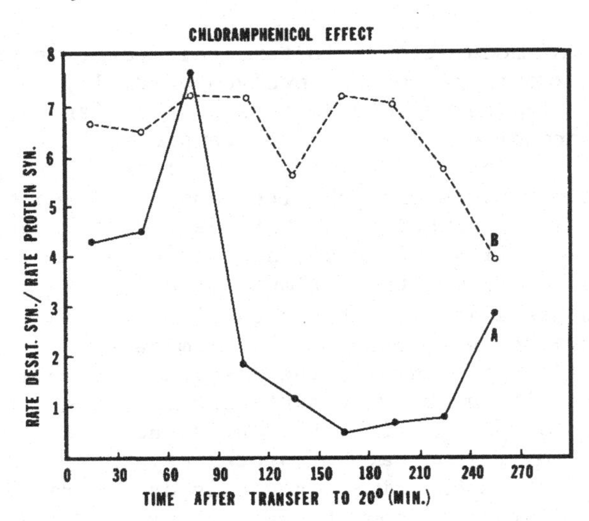

FIG. 12. Effect of chloramphenicol on the normalized rate of desaturase synthesis. This parameter is shown for the control culture (A) and for the chloramphenicol-treated culture (B) as described in Fig. 11.

but indicates that modulation is much more sensitive to partial inhibition of protein synthesis than is desaturase synthesis. This conclusion is consistent with the observation (Fig. 5) that the normalized rate of desaturase synthesis increases steadily, in a culture growing at 20° from inoculum, as the rate of protein synthesis decreases. It is also supported by the general finding that, after cultures have undergone hyperinduction and attenuation, the normalized rate of desaturase synthesis (and the absolute rate, as well) again rises as growth rate and the rate of protein synthesis decrease.

If, as the data indicate, unmodulated desaturase synthesis is simply proportional to the rate of overall protein synthesis, then modulation appears to be an exponential function of overall protein synthesis. We are not saying that modulator protein synthesis itself is an exponential function of overall protein synthesis. Rather, we might speculate that active modulator is an oligomer in equilibrium with an inactive monomeric precursor (Brown et al.). Oligomer concentration would thus be an exponential function of monomer concentration which, in turn, could be a simple function of overall protein synthesis. Such a scheme could account for the delay in effective modulation of desaturase synthesis after culture transfer from 35° tp 20° and also for the rapid acceleration of the attenuation of desaturase synthesis once modulation begins. Nevertheless, other explanations of modulation kinetics are possible and all such schemes involve considerable speculation.

The final experiment we wish to consider here is summarized in Fig. 13 which shows the effect of rifampicin on the attenuation of hyperinduction. It can be seen that the rate of decrease of desaturase synthesis in the presence of rifampicin (taken as a measure of the messenger half-life) is about the same regardless of whether this inhibitor is added during maximum hyperinduction (Table V), presumably in the absence of modulator, or during maximum attenuation (Fig. 13) when modulator is present. The half-life of the desaturase messenger RNA (9-11 min at 20°) calculated from these data is also consistent with the average half-life of messenger RNAs coding for general protein synthesis (i.e. 10-15 min) in the presence of rifampicin. The fact that the maximum rate of attenuation of desaturase synthesis in the control culture (curve A, Fig. 13) corresponds closely with the rate of decay of desaturase synthesis in the presence of rifampicin (curve B) suggests that the modulator may act at the transcription level by shutting down the synthesis of desaturase messenger RNA.

ABSTRACT

We have investigated the mechanisms in _Bacillus megaterium_ ATCC 14581 that regulate unsaturated fatty acid biosynthesis in response to changes in environmental temperature. One process involves the irreversible, strictly first-order, inactivation of desaturase. _In vivo_, the half-life of the enzyme is determined solely by the culture incubation temperature and increases with decreasing temperature. This mechanism is important in regulating the rate of fatty acid desaturation in _B. megaterium_ (and in other bacilli possessing a Δ^5-desaturating system) during the exponential phase of cell growth.

A more complex temperature-mediated control process involves the regulation of desaturase synthesis. Cultures growing from inoculum at 35° contain neither unsaturated fatty acids nor the desaturase necessary for their production.

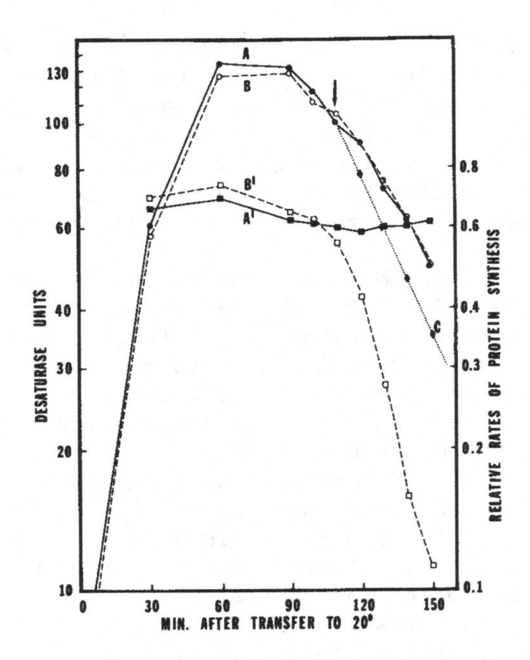

FIG. 13. Effect of rifampicin on desaturase activity. A culture of B. megaterium was transferred from 35° to 20° and then divided into 2 equal portions. The first portion (A) served as a control while the second portion (B) was treated with rifampicin (25 μg/ml) at exactly 109 min (arrow) after transfer from 35° to 20°. RNA synthesis was totally inhibited within 1 min of rifampicin addition. Desaturase activity (curves A and B) and protein synthesis (curves A' and B') were measured at 30 min intervals for the first 90 min and at 10 min intervals thereafter. A third portion (C) was obtained by withdrawing a sample from portion A at 109 min. Portion C was mixed immediately with chloramphenicol (100 μg/ml) and desaturase activity measured at 10 min intervals to determine desaturase half-life. There was no detectable protein synthesis in Portion C after chloramphenicol addition. Note that all data are plotted on a log scale [Fujii & Fulco, 1977].

When the culture temperature is lowered rapidly from 35° to 20°, however, synthesis of desaturase begins after a brief lag and attains a maximum rate shortly thereafter. This hyperinduction process, so-called because the rate of desaturase synthesis after culture transfer from 35° to 20° far exceeds the rate found in cultures growing from inoculum at 20°, requires new protein and RNA synthesis. The maximum rate of desaturase synthesis during hyperinduction, normalized for the rate of protein synthesis, is a constant, independent of experimental conditions. The hyperinduction phase is followed by a period of rapid attenuation until the rate of desaturase synthesis is at or below that of comparable cultures growing at 20° from inoculum. The shut-down of hyperinduction at 20° in transfer cultures as well as the lower rate of desaturase synthesis in cultures growing from inoculum at 20° appears to result from the action of a temperature-sensitive modulator protein which is absent at 35°. In the absence of modulator (i.e., during hyperinduction) the rate of desaturase synthesis is simply proportional to the overall rate of protein synthesis. Modulation, however, is an exponential function of protein synthesis. The modulator, which may be an oligomeric protein in equilibrium with an inactive monomer, seems to act at the level of transcription by selectively inhibiting the synthesis of the messenger RNA coding for desaturase. Finally, the first-order inactivation of desaturase as well as the hyperinduction and modulation of it's synthesis is directly mediated by temperature. Neither the state of membrane fluidity nor wide variations in O_2-tension affect these parameters.

ACKNOWLEDGMENTS

We wish to thank the editors of the Journal of Biological Chemistry for allowing us to reproduce here a number of figures and tables originally published in their journal (Fujii and Fulco, 1977). We wish to thank Ellin James and Joyce Adler for their excellent work in preparing the figures and tables for the manuscript and for secretarial assistance. Much of the research reported here was supported by Research Grant AI-09829 from the National Institute of Allergy and Infectious Diseases, NIH, USPHS.

REFERENCES

Brown, N. C., C. L. Wisseman III, and T. Matsushita (1972). Nature New Biol. 237, 72.

Chang, N. C. and A. J. Fulco (1973). Biochim. Biophys. Acta 296, 287.

Chen, S. L. and A. J. Fulco (1978). Biochem. Biophys. Res. Commun. 80, 126.

Chen, S. L., J. F. Lombardi, and A. J. Fulco (1978). Biochem. Biophys. Res. Commun. 80, 133.

Fujii, D. K. and A. J. Fulco (1977). J. Biol. Chem. 252, 3660.

Fulco, A. J., R. Levy, and K. Bloch (1964). J. Biol. Chem. 239, 998.

Fulco, A. J. (1967). Biochim. Biophys. Acta 144, 701.

Fulco, A. J. (1969a). J. Biol. Chem. 244, 889.

Fulco, A. J. (1969b). Biochim. Biophys. Acta 187, 169.

Fulco, A. J. (1970). J. Biol. Chem. 245, 2985.

Fulco, A. J. (1972a). J. Biol. Chem. 247, 3503.

Fulco, A. J. (1972b). J. Biol. Chem. 247, 3511.

Fulco, A. J. (1974). Annu. Rev. Biochem. 43, 215.

Fulco, A. J. (1977). In W. H. Kunau and R. T. Holman (eds.), Polyunsaturated Fatty Acids. American Oil Chemists' Society, Champaign, Illinois.

Mead, J. F. and A. J. Fulco (1976). The Unsaturated and Polyunsaturated Fatty Acids in Health and Disease, Thomas, Springfield, Illinois. 191 pp.

Quint, J. F., and A. J. Fulco (1973). J. Biol. Chem. 248, 6885.

ACTIVITY OF MEMBRANE-BOUND ENZYMES OF THE RESPIRATORY CHAIN DURING ADAPTATION OF FISH TO TEMPERATURE CHANGES.

A.D.F. Addink
Department of Animal Physiology
University of Leiden

Kaiserstraat 63, Leiden, The Netherlands

ABSTRACT

In tissues of the goldfish, Carassius auratus L., acclimated to 5°, 20° and 30°C, the changes in activity of succinate oxidase in red and white muscle and liver are determined. Red muscle activity decreases, liver activity increases, whereas in white muscle a low but constant activity was found. As this enzyme-complex is associated with an annulus of phospholipids in the inner membrane of mitochondria the unsaturation index of the membrane phospholipids in red muscle was found to decrease with increasing temperature. The changes in the **component** classes of these phospholipids and of the fatty acid composition are given. Whether the membrane fluidity is affected by these changes of the saturation at different temperatures is uncertain, but the effects on the annulus around succinate oxidase are reflected in the measured activity.

INTRODUCTION

Fish, being coldblooded or poikilotherm animals correspond in body temperature to their environment. Goldfish are active over a range from as high as 35°C to about 0°C. The temperature adaptation involves among others the regulation of enzyme activity over this temperature range, for instance the inner membrane bound respiratory chain proteins of mitochondria of the red muscle of the goldfish. The physiological responses in the fish to the change in temperature of the environment comprise the membrane fluidity. The lipid protein interaction on a molecular level is important as demonstrated by Hazel (1972) for mitochondrial succinate dehydrogenase. The phospholipids surround membrane bound protein(s) as an annulus (Lee, 1977; see Fig. 1). This is the case in succinate dehydrogenase, cytochrome b, c_1 and cytochrome oxidase.

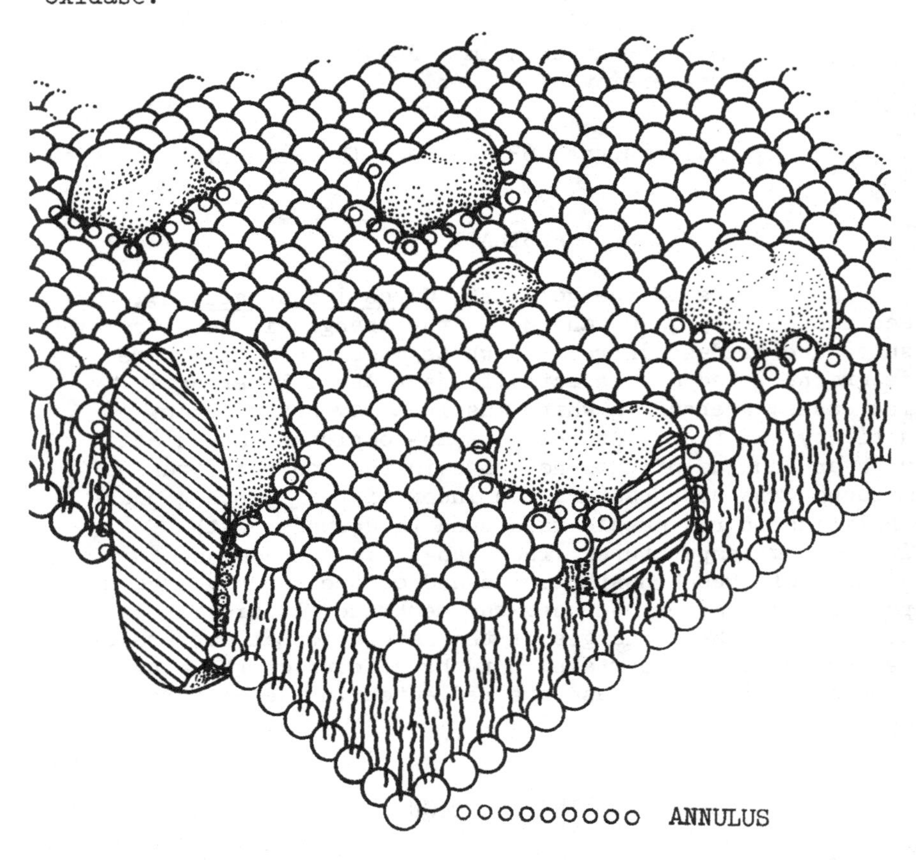

Fig. 1. The fluid-mosaic, lipid-bilayer model of a membrane modified after Singer and Nicolson, Science, 175, 720 (1972). The annulus of phospholipids surrounding the protein molecules is indicated by the small circles.

I like to report at this Symposium on Membrane Fluidity some physiological data on the research program of our department of the University of Leiden, more specific the results of temperature adaptation in the goldfish. The members of the research group include Dr. van den Thillart, Miss Kesbeke, Drs. van Waarde, Mourik and De Bruin; part of these results have been presented at the 2nd Symposium on fish physiology in Göteborg.

RESULTS and DISCUSSION

Our study of the activity of the succinate oxidase system at 15°, and 30° revealed that maximum activity of the complex enzyme system is affected by a temperature change after adaptation. See Table 1.

Table 1. Activity of Succinate Oxidase of Goldfish, Acclimated to 15° or 30° *

Temperature °C	Succinate oxidase activity μmoles.min^{-1}.g^{-1}		
	red muscle	white muscle	liver
15	18 ± 3	1.6 ± 0.2	5.3 ± 0.6
30	10 ± 0.5	1.5 ± 0.1	10.6 ± 0.8

* Data of van den Thillart (1977).

Table 2. Phospholipid composition of red muscle mitochondria from goldfish, acclimated to 5°, 20° and 30°C

Phosphatide species	Acclimation temperature				
	5°C		20°C		30°C
Phosphatidylcholine	43.7	→	48.1	→	56.9
Phosphatidylethanolamine	39.3	←	34.0	←	24.9
Diphosphatidylglycerol	10.1		10.8	←	8.3
Phosphatidylinositol	2.8		2.7		3.1
Phosphatidylserine	0.8		0.7	→	1.3
Phosphatidylglycerol	0.4	→	0.6		0.6
Sphingomyelin	0.3		0.4	→	1.7
Phosphatidic acid	0.2		0.0		0.0
Lyso phosphatide compounds	2.3		2.3		2.6
x (unknown)	0.2		0.4		0.6

The changes in the component classes of phospholipids are depicted in Table 2.
The separation of the above described species was achieved by thin layer chromatography. The individual phospholipids were sucked into pyrex glass tubes and, after digestion at 225°C, the amount of phosphorus was determined. The composition is expressed as phospholipid phosphorus in %. Significant differences ($p \leq 0.05$), tested according to Student's t-test, are indicated by arrows.

The fatty acid moieties of these membrane phospholipids become more unsaturated after adaptation to lower temperatures (Table 3).

Table 3. Fatty acid composition of total phospholipids from red muscle mitochondrial membranes.

Fatty acid shorthand name	Acclimation temperature				
	5°C		20°C		30°C
16 : 0	10.8		10.5		11.7
16 : 1	1.9	→	2.3		2.3
18 : 0	7.9		7.9		8.3
18 : 1	11.7	→	13.0	→	14.5
18 : 2	12.1	→	14.8		14.6
18 : 3	2.4		2.4		2.4
20 : 1	2.5		2.6	←	1.8
20 : 2	1.3		1.4		1.2
20 : 3	1.8		1.9		1.5
20 : 4	7.2		7.6		8.6
20 : 5	4.6		3.8		3.0
22 : 4	1.8		1.1		1.1
22 : 5	5.2		4.6		4.1
22 : 6	19.2	←	17.9	←	15.5

Mitochondria were isolated from red muscle of goldfish, acclimated to 5°, 20° and 30°C. Total lipids have been extracted, after which the phospholipids were purified by thin layer chromatography. The fatty acids were analyzed after hydrolysis of the phospholipids.
The composition is expressed as mol %; only values $\geq$ 1 are given.
Arrows indicate a significant difference ($p \leq 0.05$), tested according to Student's t-test.

Table 4. Fatty acid unsaturation index of red muscle mitochondrial phospholipids of goldfish, acclimated to 5°, 20° and 30°C

Unsaturation index	Acclimation temperature		
	5°C	20°C	30°C
Total	241.2	231.6	212.3
Mono-unsaturated	17.4	19.3	20.0
Poly-unsaturated	223.8	212.4	192.3

The unsaturation index is the summation of all double bonds per mol fatty acid, expressed in %.

The unsaturation index of the fatty acids of red muscle mitochondrial phospholipids of the goldfish adapted to 5°, 20° and 30°C is given in Table 4.
This index is the summation of all double bonds per molecule fatty acid expressed in percentage. A decrease in unsaturation can be observed at higher temperature.

Simultaneously Cossins and Prosser (1978, 1977) determined the unsaturation index of phosphatidylcholine of goldfish brain synaptosomes: with increasing temperature the index showed a decrease.

As our data are representative for the fatty acid of both outer and inner mitochondrial membranes one could imagine, that the changes in the annulus of phospholipid fatty acids of each inner membrane protein are minor or much more pronounced; resulting in some or no change in membrane fluidity in red muscle mitochondrial membranes of *Carassius auratus* L.

REFERENCES

Cossins, A.R. & C.L. Prosser (1978). Proc. Natl. Acad. Sci. 75, 2040-2043.

Cossins, A.R., M.J. Friedlander & C.L. Prosser (1977). J. Comp. Physiol. 120, 109-121.

Hazel, J.R. (1972). Comp. Biochem. Physiol. 43B, 863-882.

Kesbeke, F., G.J. de Bruin & G.E.E.J.M. van den Thillart (1979). Abstract from the 2nd Symposium on Fish Physiology, Göteborg.

Lee, A.G. (1977). TIBS, 231-233.

Van den Thillart, G.E.E.J.M. (1977). Thesis, Leiden, Influence of oxygen availability on the energy metabolism of goldfish, Carassius auratus L.

EFFECT OF MEMBRANE LIPID COMPOSITION ON MOBILITY OF LYMPHOCYTE SURFACE IMMUNOGLOBULINS

Marc J. Ostro, Beverley Bessinger, John Summers
and Sheldon Dray

Department of Microbiology and Immunology
University of Illinois at the Medical Center
Chicago, Illinois 60612

INTRODUCTION

Some types of cellular regulation are known to be associated with changes in membrane fluidity. For example, it has been shown that the neoplastic transformation of cells from a variety of origins is accompanied by a marked increase in cell surface fluidity (Inbar *et al.*, 1973; Shinitzky and Inbar, 1974; Ben-Bassat *et al.*, 1977; Inbar *et al.*, 1977a). Both cell cycle regulation (deLatt *et al.*, 1977) and contact inhibition of cells *in vitro* (Inbar *et al.*, 1977b) are also known to be linked with membrane fluidity changes. These changes in fluidity have been demonstrated to be a function of the membrane cholesterol/ phospholipid (c/p) ratio (Shinitzky and Inbar, 1974; Chen *et al.*, 1975; Borochov and Shinitzky, 1976; Cooper *et al.*, 1978), so that the relatively low ratios are associated with increased fluidity. The c/p ratio in cell membranes can be experimentally manipulated by mixing the cells with liposomes of various cholesterol content (Borochov and Shinitzky, 1976; Cooper *et al.*, 1978; Inbar and Shinitzky, 1974; Alderson and Green, 1975; Dunnick *et al.*, 1976; Chen and Keenan, 1977). This approach has been previously employed to study the relationship between membrane lipid composition and tumor development (Shinitzky and Inbar, 1974; Inbar and Shinitzky, 1974) and has also been used to study the role of lipids in the mitogenic stimulation of lymphocytes (Alderson and Green, 1975; Chen and Keenan, 1977; Ozato *et al.*, 1978), the ion flux across erythrocyte membranes (Wiley and Cooper, 1975) and the vertical motion of intrinsic erythrocyte membrane proteins (Borochov and Shinitzky, 1976).

Recently, work has been focused on the interrelationship between the fluidity of lymphocyte membranes, as affected by their lipid composition, and the stimulation and proliferation of the cells involved in the immune response. Inbar and Shinitzky (1975) have suggested that a controlled increase in the fluidity of lymphocyte surface membranes may occur *in vivo* during antigenic stimulation. Several laboratories have further investigated this problem by artificially manipulating membrane cholesterol levels with liposomes and assessing the effect of these changes on the mitogenic response. To date no consistent correlation between membrane lipid composition and the ability of lymphocytes to respond to mitogenic challenge has been established. Alderson and Green (1975) have demonstrated that liposome-induced increases in membrane cholesterol levels of bovine mesenteric lymphocytes lead to a marked suppression of concanavalin A (Con A) stimulated blast transformation. Chen and Keenan (1977) have shown the reverse to be true; i.e., reduced membrane cholesterol levels resulted in suppression of Con A-induced blastogenesis when using human peripheral blood lymphocytes. In addition, Ozato *et al*. (1978) have shown that murine thymocytes demonstrate an enhanced mitogenic response to Con A when treated with liposomes under conditions that would either raise or lower membrane cholesterol levels.

The interpretation of this controversial data may be facilitated by an understanding of the effect of membrane lipids on the lateral mobility of surface receptors, although such a correlation also appears nebulous. For example, Borochov and Shinitzky (1976), working with erythrocyte membranes, have shown that increased fluidity associated with liposome-induced cholesterol depletion, results in a substantially enhanced displacement of membrane proteins. Furthermore, we have found that reduction of cell membrane microviscosity with cholesterol-free liposomes, stimulates a capping of surface immunoglobulins (sIg) on rabbit spleen cells. Conversely, Ben-Bassat *et al*. (1977) have found that increased membrane fluidity leads to a decrease in the mobility of the Con A receptor and Hilgers *et al*. (1978) have demonstrated that increases in the fluidity of membranes of murine thymus-derived leukemia cells resulted in the reduced ability of histocompatability antigens to form caps. While no absolute explanation for the apparent contradiction described above is available, the differential response to enhanced membrane fluidity of distinct receptors may be a function of the orientation of the proteins in the membrane. Specifically, it has been shown that sIg does not appear to be a transbilayer protein (Walsh and Crumpton, 1977) and does not seem to be attached to the cytoskeleton of the cell in the absence of cross-linking agents (Flanagan and Koch, 1978). In contrast to this observation, it has been suggested that both the Con A receptor and the histocompatability antigens are anchored in either the microtubules or microfilaments underlying the cell membrane (Walsh and Crumpton, 1977; Albertini et al., 1977), possibly limiting their ability to move in response to changes in membrane lipid composition. Here we present data which indicates that alteration of membrane composition by cholesterol-free liposomes results in a capping of surface immunoglobulins, a process which is independent of cytoskeletal function and appears to be modulated by the reduction of the cholesterol/phospholipid ratio (c/p) in the lymphocyte membrane.

LIPOSOME-INDUCED CAPPING OF SURFACE IMMUNOGLOBULINS

Liposome Preparation

Large unilamillar liposomes were prepared by the ether infusion technique of Deamer and Bangham (1976) as adapted by Matthews et al. (1979), and were composed of egg yolk lecithin (L), cholesterol (C), and dicetylphosphate (D), in the following molar ratios: 8L:2D, 7L:1C:2D, 6L:2C:2D, 5L:3C:2D and 4L:4C:2D. In all experiments liposomes were gel filtered on Sepharose 2B and standardized by their absorption at 650nm.

Liposome-Mediated Loss of Spleen Cell sIg by Capping

Rabbit spleen cells were treated with 1 ml of 8L:2D liposomes for 1 hr at 37° and assessed for surface Ig by an immunoadherance assay (rosetting) which involves B-cell binding of antibody (Ab) coated sheep erythrocytes (Vyas et al., 1968; Molinaro and Dray, 1974). Erythrocytes were coated with either purified anti-light chain allotype Ab, goat anti-light chain Ab or goat anti-rabbit Ig isotype Ab. Following liposome treatment there was a decrease of between 50 and 96% in the number of cells bearing sIg. This loss was found to be dose dependent. Liposome treatment of cells did not appear to result in any appreciable cytotoxicity since after 1 hr of liposome-cell cocultivation, cells were found to be greater than 90% viable.

To follow the fate of spleen cell sIg after liposome addition, immunofluorescence (IF) using rhodamine-conjugated anti-light chain Ab was employed (Fig. 1). Spleen cells were treated for 5, 30 or 60 min with liposomes, washed with azide to arrest further movement of surface receptors, and assessed for sIg by either rosetting or IF. The original percentage of Ig-bearing cells was determined by rosetting spleen cells prior to liposome incubation and was determined to be similar to the percentage of rosette forming cells (RFC) obtained when spleen cells were incubated with buffer. After 5 min of liposome-cell incubation, 60% of the cells exhibited membrane fluorescence (MF) with no capped cells apparent. After 30 min of liposome-cell cultivation, only 17% of the cells exhibited membrane fluorescence. This loss (43%) in membrane fluorescence was associated with an increase in the percentage of capped cells (0 to 23%) and nonfluorescent cells (40-60%). After 60 min of incubation, the percentage of membrane fluorescent cells remained constant, while the percentage of capped cells decreased from 23 to 12% and the percentage of nonfluorescent cells increased from 60 to 70%. The percentage of the cells forming rosettes at 5, 30 and 60 min was similar to the percentage of Ig-bearing cells which exhibited general membrane fluorescence (5 min - 54% RFC, 60% MF; 30 min - 13% RFC, 19% MF; 60 min - 16% RFC, 18% MF). Cells treated with liposomes and labeled with fluorescent Ab are shown in Figure 2.

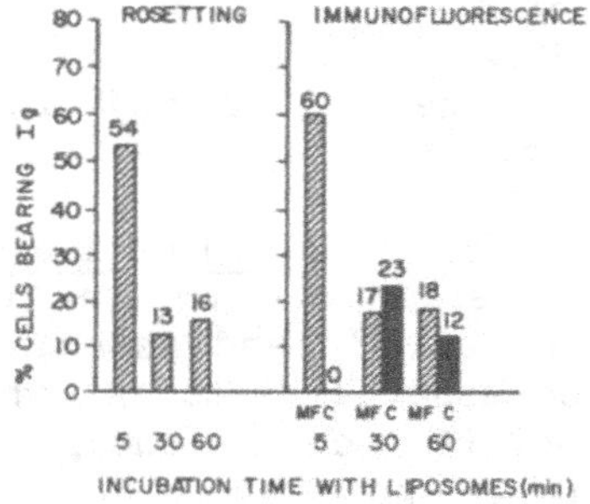

Figure 1. Time dependent study of liposome-induced loss of rabbit spleen cell surface Ig. Spleen cells (2×10^7) were mixed with 1 ml of 8L:2D liposomes for either 5, 30, or 60 min at 37°C, washed with 0.02% azide and assessed for surface Ig by both rosetting and immunofluorescence. MF-membrane fluorescence, C-capped cells.

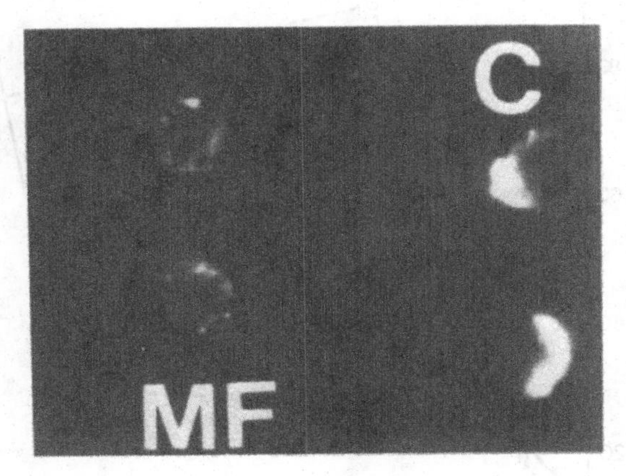

Figure 2. Fluorescent micrograph of rabbit spleen cells incubated with 8L:2D liposomes for 60 min at 37°C. Spleen cells were washed free of liposomes with 0.02% azide and treated with rhodamine-conjugated anti-light chain Ab for 30 min at 4°C. MF-membrane fluorescence, C-capped cells.

Recovery Kinetics of Capped sIg

Liposome-induced capping of sIg was compared to capping stimulated by crosslinking Ab in a effort to determine if the capping observed following the addition of liposomes to spleen cells was due to a crosslinking mechanism. Since it was difficult to evaluate this possibility directly, it was predicted that if the liposomes were functioning as cross-linking agents, the kinetics of recovery of sIg on spleen cells which had been treated with liposomes should closely parallel the recovery rate of cells which had been induced to cap by Ab. The results of this experiment can be seen in Figure 3. Immediately after spleen cells were incubated with either liposomes or Ab, the cells were washed and put into culture with 50μg/ml of dextran sulfate, a B-cell mitogen. Aliquots of cells were removed from the culture over a 24 h period, washed with azide and rosetted with anti-Ig coated erythrocytes. Loss of sIg was induced by both the liposome and Ab treatment. The Ab-treated cells recovered 90% of their sIg after 6.5 h (the time of protein turnover) while substantial recovery of the sIg on liposome-treated cells was not complete until 24 h, implying that the two phenomena are mechanistically distinct. This conclusion was further supported by assessing the effect of lidocaine on both Ab and liposome-induced capping. Lidocaine has been shown to totally inhibit capping by Ab (Poste et al., 1975)

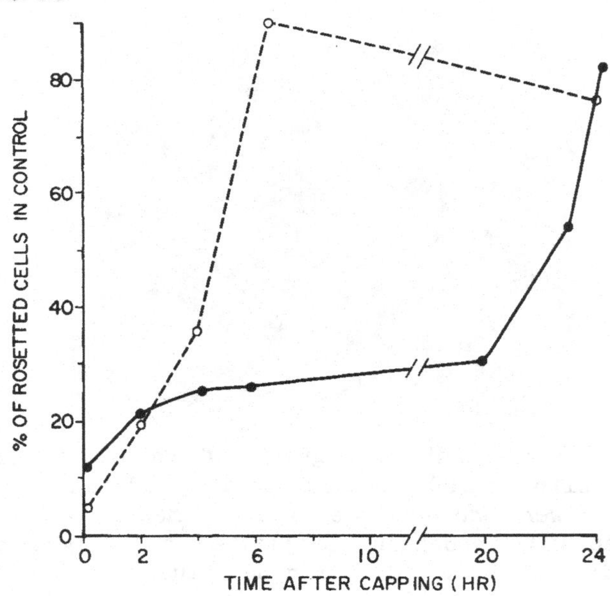

Figure 3. Kinetics of recovery of rabbit spleen cells after treatment with either 8L:2D liposomes (●—●) or anti-b4 Ab (○--○). Spleen cells (2 X 10^7) were treated with 1 ml of liposomes or 50μg of Ab for 1 h at 37°C, after which they were washed and assessed for surface Ig over time by rosetting with anti-Ig coated erythrocytes.

by destroying microfilaments and microtubules. When 1 x 10^{-3} M lidocaine was added to spleen cells 30 min prior to the addition of crosslinking Ab or liposomes, the Ab-induced capping was completely inhibited while the capping stimulated by liposomes was unaffected.

DEPENDENCE OF LIPOSOME CAPPING ON MEMBRANE CHOLESTEROL/PHOSPHOLIPID RATIOS

Since capping of sIg by liposomes appeared to be independent of cytoskeletal function, it was hypothesized that the observed capping was caused by a decrease in the lymphocyte c/p ratio resulting from lipid exchange between cholesterol-free liposomes and cells as well as fusion of the vesicle bilayer with the lymphocyte membrane, and that the recovery rate of liposome-capped sIg was dependent upon the restoration of a normal membrane c/p ratio. Based upon this theory it was predicted that the liposome-induced capping should be abrogated if the cholesterol content of the liposome was increased to a level where the ability of the vesicle to affect a net reduction in the c/p ratio of the lymphocyte membrane was impaired. To that end, spleen cells were treated with liposomes of increasing cholesterol content and the loss of sIg was monitored by rosetting (Fig. 4). The addition of 8L:2D liposomes resulted in a decrease in the number of RFCs from 53% in the Hepes-treated control to 17%. As the mole % of cholesterol in the liposomes was increased, there was a simultaneous increase in the % RFCs until the control level was reached. While this data is consistent with the above hypothesis, the possibility that the cholesterol dependent elimination of capping is due to a reduced tendency of cholesterol-containing liposomes to fuse with lymphocytes cannot be totally discounted, although the difference between the ability of 8L:2D and 4L:4C:2D liposomes to fuse with rabbit lymphocytes has been found to be minimal (Ostro, unpublished observation).

The effect of cell membrane cholesterol levels on the capping and recovery of sIg was further investigated by studying the consequence of altering the rate of endogenous cholesterol biosynthesis on the rate of recovery of liposome-capped sIg. It was found that cholesterol biosynthesis

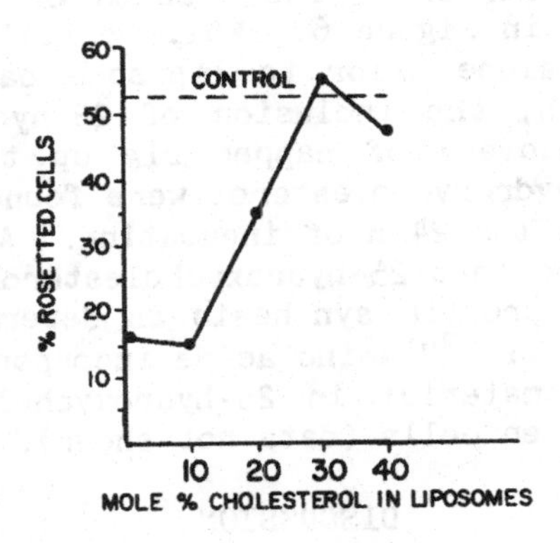

Figure 4. Determination of the loss of spleen cell surface Ig as a function of liposome cholesterol content. Spleen cells (2 X 10^7) were treated for 1 h at 37°C with either 8L:2D, 7L:1C:2D, 6L:2C:2D, 5L:3C:2D, or 4L:4C:2D liposomes. Following the incubation period, cells were washed and assessed for surface Ig by rosetting. Control consisted of cells treated with 1 ml of Hepes buffer.

could be stimulated by dextran sulfate and that this stimulation could be specifically abrogated by the addition of 1μg/ml of 25-hydroxycholesterol to the preincubation mixture (Fig. 5). [25-hydroxycholesterol has been shown to specifically inhibit cholesterol biosynthesis by preventing the production of 3-hydroxy-3-methylglutaryl Co A reductase, a regulatory enzyme required for the production of mevalonic acid (Kandutsch *et al.*, 1978).] Using this information, the effect of altered cholesterol biosynthesis on the recovery of liposome-capped sIg was assessed. Spleen cells were pretreated withdextran sulfate 10 h prior to liposome addition and the rate of recovery of capped surface Ig was compared to that observed when mitogen pretreatment was eliminated. It was found that cells which had been pretreated with dextran sulfate recovered their sIg in approximately 8 h compared to the normal 24 h recovery rate (Fig. 6).

The observed increase in recovery rate induced by dextran sulfate pretreatment could be due to an unassayed effect of the drug on the lymphocytes. Since 25-hydroxycholesterol specifically inhibits the dextran sulfate stimulation of cholesterol biosynthesis (Fig. 5), the enhanced recovery rate of sIg in the presence of dextran sulfate should be eliminated when 25-hydroxycholesterol is added to the pretreatment mixture. The results of this experiment can also be seen in Figure 6. While cells pretreated with dextran sulfate alone prior to liposome capping recovered their sIg in 8 h, the inclusion of 25-hydroxycholesterol prevented the recovery of capped sIg up to 20 h. Cells treated with 25-hydroxycholesterol were found to be greater than 90% viable after 24 h of incubation. As an added control, it was found that 25-hydroxycholesterol had no appreciable effect on protein synthesis as determined by comparing the amount of [^{3}H]amino acids incorporated into TCA precipitable material in 25-hydroxycholesterol-treated and untreated spleen cells (data not shown).

DISCUSSION

We have demonstrated that treatment of rabbit spleen cells with 8L:2D liposomes results in a substantial decrease in Ig-bearing cells when assayed by either rosetting or IF and that this decrease is due to a capping phenomenon. The liposome-induced capping was shown to be distinct from capping by antibody in that 1) cells capped with Ab recovered their sIg in approximately 6 h while liposome

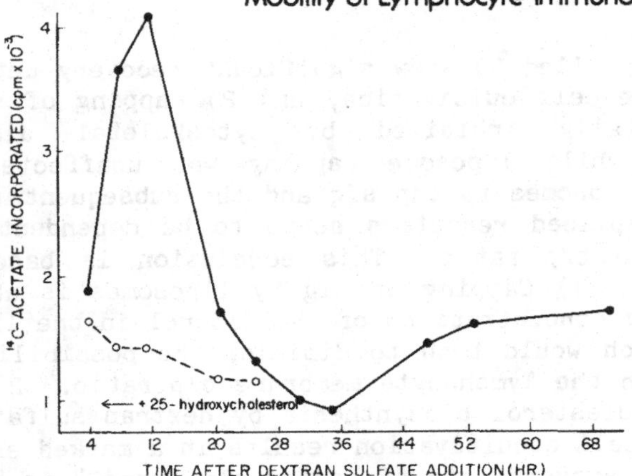

Figure 5. Evaluation of the rate of cholesterol biosynthesis following treatment of spleen cells with dextran sulfate (●—●) or Hepes buffer (○- -○). Cells were maintained in RPMI-1640 medium and at time intervals pulsed for 2 h with ^{14}C-acetate. Cholesterol was extracted as a digitonide and assessed for radioactivity. The arrow indicates the level of cholesterol biosynthesis at 12 h post dextran sulfate addition when 25 hydroxycholesterol was included in the incubation mixture.

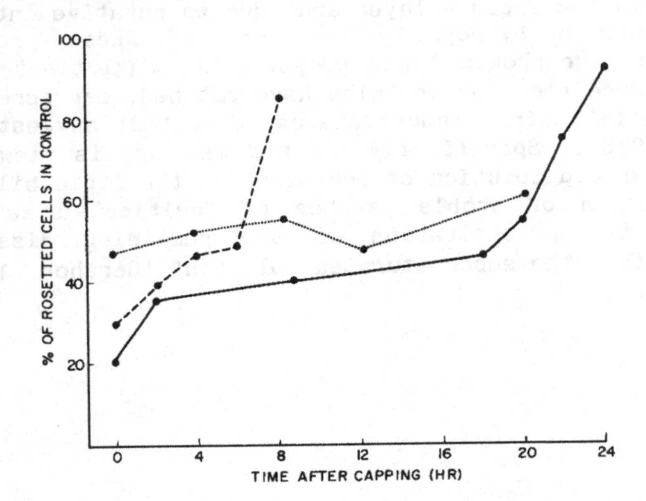

Figure 6. Kinetics of recovery of 8L:2D liposome-capped sIg on spleen cells pretreated for 10 h with dextran sulfate (●- -●), dextran sulfate and 25-hydroxycholesterol (●····●), and Hepes buffer (●—●).

capped cells failed to show significant recovery until 24 h post liposome-cell cultivation, and 2) capping of sIg with Ab was totally inhibited by cytoskeletal disruption (lidocaine) while liposome capping was unaffected. The ability of liposomes to cap sIg and the subsequent recovery of these displaced receptors seems to be dependent on the cell membrane c/p ratio. This conclusion is based on 3 observations. 1) Capping of sIg by liposomes is abrogated by increasing concentrations of cholesterol in the liposome membrane which would tend to diminish the possibility of a net change in the lymphocyte membrane c/p ratio. 2) Stimulation of cholesterol biosynthesis by dextran sulfate prior to liposome-cell cocultivation results in a marked enhancement in the recovery rate of capped sIg from 24 to 8 h. 3) Elimination of the dextran sulfate stimulation of cholesterol biosynthesis by 25-hydroxycholesterol delays the rate of sIg recovery without affecting protein synthesis.

While the mechanism for liposome-induced capping is not as yet clear, it seems likely that liposome enhancement of membrane fluidity would result in an increased lateral mobility of intrinsic membrane proteins in the plane of the membrane (Singer and Nicolson, 1972). We suggest that this enhanced mobility could facilitate collisions between proteins within the lipid bilayer and, due to putative interactions between hydrophilic segments of these proteins within the hydrophobic lipid compartment, agglutinate into patches. Once receptor proteins have patched, cap formation may be initiated in a manner analogous to that suggested by Gershon (1978). Specifically, if the membrane is viewed as a supersaturated solution of receptors in the lipid bilayer, then formation of stable patches of "critical size" may "grow by the precipitation of the remaining dissolved receptors from the supersaturated solution" (Gershon, 1978).

ABSTRACT

Treatment of rabbit spleen cells with lecithin liposomes caused a substantial decrease in the number of cells bearing surface immunoglobulin (Ig) as assessed by rosetting and immunofluorescent techniques. Immunofluorescent studies demonstrated that the decrease of surface Ig was the result of a capping phenomenon. A comparison of liposome versus antibody-induced capping was made by monitoring the respective rates of recovery of displaced surface Ig on spleen cells. Cells treated with specific antibody recovered their surface Ig 6.5 hours after antibody-cell incubation, whereas cells treated with liposomes required 24 hours for a comparable recovery. Thus, the two phenomena appeared to be mechanistically distinct. This conclusion was supported by the fact that disruption of cytoskeletal function with lidocaine completely inhibited antibody-induced capping while capping stimulated by liposomes was unaffected. Our data indicate that liposome-induced capping may be modulated by the cholesterol/phospholipid ratio in the lymphocyte membrane. Data in support of this evaluation are as follows: 1) Capping was inhibited by incremental addition of cholesterol to the liposomal membrane. 2) Stimulation of the rate of cholesterol biosynthesis in spleen cells with dextran sulfate prior to the addition of liposomes reduced the time needed for the recovery of surface Ig on liposome-treated cells from approximately 24 to 8 h. 3) Inhibition of cholesterol biosynthesis by 25-hydroxycholesterol resulted in the failure of surface Ig to recover. Our data suggest that there is a relationship between the expression of lymphocyte surface Ig and cholesterol biosynthesis. Furthermore, the failure of lidocaine to inhibit liposome-induced capping suggests that surface Ig, under normal conditions, exists free in the lipid bilayer rather than anchored to the lymphocyte cytoskeleton.

REFERENCES

Albertini, D.F., R.D. Berlin, and J.M. Oliver (1977). J. Cell Sci. 26, 57.

Alderson, J.C.E., and C. Green (1975). FEBS Lett. 52, 208.

Ben-Bassat, H., A. Polliak, S.M. Rosenbaum, E. Naparstek, D. Shouval, and M. Inbar (1977). Cancer Res. 37, 1307.

J. Borochov, H., and M. Shinitzky (1976). Proc. Natl. Acad. Sci. 73, 4526.

Chen, S. S-H., and R.M. Keenan (1977). Biochem. Biophys. Res. Com. 79, 852.

Chen, H.W., H-J. Heiniger, and A.A. Kandutsch (1975). Proc. Natl. Acad. Sci. 72, 1950.

Cooper, R.A., M.H. Leslie, S. Fischkoff, M. Shinitzky, and S.J. Shattil (1978). Biochemistry 17, 327.

Deamer, D., and A.D. Bangham (1976). Biochim. Biophys. Acta 443, 629.

deLaat, S.W., P.T. van der Saag, and M. Shinitzky (1977). Proc. Natl. Acad. Sci. 74, 4458.

Dunnick, J.K,. R.F. Kallman, and J.P. Kriss (1976). Biochem. Biophys. Res. Com. 73, 619.

Flanagan, J., and G.L.E. Koch (1978). Nature 273, 278.

Gershon, N.D. (1978). Proc. Natl. Acad. Sci. 75, 1357.

Hilgers, J., P.J. Van Der Sluis, W.J. Blitterswijk, and P. Emmelot (1978). Br. J. Cancer 37, 329.

Inbar, M., and M. Shinitzky (1974). Proc. Natl. Acad. Sci. 71, 2128.

Inbar, M., and M. Shinitzky (1975). Eur. J. Immunol. 5, 166.

Inbar, M., M. Shinitzky, and L. Sachs (1973). J. Mol. Biol. 81, 245.

Inbar, M., R. Goldman, L. Inbar, I. Bursuker, B. Goldman, E. Akstein, P. Segal, E. Ipp, and I. Ben-Bassat (1977a). Cancer Res. 37, 3037.

Inbar, M, I. Yuli, and A. Raz (1977b). Exp. Cell Res. 105, 325.

Kandutsch, A.A., H.W. Chen, and H-J. Heiniger (1978). Science 201, 498.

Matthews, B.F., S. Dray, J. Widholm, and M.J. Ostro (1979). Planta 145, 37.

Molinaro, G.A., and S. Dray (1974). Nature 248, 515.

Ozato, K., L. Huang, and R.E. Pagano (1978). Membrane Biochemistry 1, 27.

Poste, G., D. Papahadjopoulos, and G.L. Nicolson (1975). Proc. Natl. Acad. Sci. 72, 4430.

Shinitzky, M., and M. Inbar (1974). J. Mol. Biol. 85, 603.

Singer, S.J., and G.L. Nicolson (1972). Science 175, 720.

Vyas, G.N., H.H. Fudenberg, H.M. Pretty, and E.R. Gold (1968). Immunol. 100, 274.

Walsh, F.S., and M.J. Crumpton (1977). Nature 269, 307.

Wiley, J.S., and R.A. Cooper (1975). Biochim. Biophys. Acta 413, 425.

PLASMA MEMBRANE Mg^{2+}ATPase ACTIVITY IS INVERSELY RELATED TO LIPID FLUIDITY.

John R. Riordan

Departments of Biochemistry and Clinical
Biochemistry, University of Toronto and
Research Institute, The Hospital for Sick
Children, 555 University Avenue
Toronto, Canada M5G 1X8

ABSTRACT

The negative temperature dependence of the Mg^{2+}ATPase of liver plasma membranes led to the postulate that decreased activity may result from increased fluidity of membrane lipids. To test this hypothesis bulk lipid fluidity was manipulated in two ways. First, membrane phospholipids were removed by phospholipase A_2 treatment and replaced by specific exogeneous phospholipids with different liquid crystaline to gel transition temperatures (T_c). The cholesterol content was also decreased or increased in similar reconstitution experiments. Second, the fluidity of the whole membrane was increased by treatment with *cis*-vaccenic acid. Enzyme activity and fluorescence polarization of diphenyl hexatriene or β-parinaric acid were determined after these modifications. Activity was inhibited at temperatures above the T_c of the reconstituted phospholipids and elevated below. Increased cholesterol enhanced activity. Activity of *cis*-vaccenic acid-treated membrane was directly proportional to the degree of polarization of fluorescence of diphenyl hexatriene. This somewhat unusual relationship between enzyme activity and lipid fluidity may reflect a requirement for a certain degree of lipid order either for optimal catalytic function or adequate exposure of the active site at the lipid-water interface.

INTRODUCTION

Those lipid-sensitive membrane enzymes examined thus far have shown a requirement for fluid membrane lipid (Kimelberg, 1977). In an earlier study we observed that the Mg^{2+}ATPase of liver plasma membranes exhibited a negative temperature dependence above 31° which could be overcome by cross linking of membrane proteins (Riordan et al, 1977). Hence, it was suspected that the disorder of a highly fluid environment may be detrimental to the activity of this enzyme. This report describes the effects of alterations in lipid fluidity on this enzyme and provides evidence for optimal activity in rigid rather than fluid membrane lipid.

METHODS

Membranes were treated with phospholipase A_2 (bee venom) at 37° for 1 h in 50 mM Tris-HCl, pH 7.6 containing 1 mM $CaCl_2$. After pelleting at 105,000 g for 20 min these membranes were washed twice with this buffer, recentrifuging in the same way each time. For reconstitution with pure phospholipids this hydrolysed membrane was sonicated for 5 min in a bath sonicator with dried lipid at a ratio of 2 mg of phospholipid per mg membrane protein. The temperatures at which sonication was performed were 23° for dilinoleoyl phosphatidyl choline (dlpc) and dioleoyl phosphatidyl choline (dopc), 30° for dimyristoyl phosphatidyl choline (dmpc), 48° for dipalmitoyl phosphatidyl choline (dppc) and 60° for distearoyl phosphatidyl choline (dspc). Controls were sonicated at the same temperatures in the absence of added lipid. The reconstituted membranes were separated from excess free lipid by centrifugation through a 30% (w/v) sucrose cushion at 160,000 for 30 min in a Beckman Airfuge.

The isolation of rat liver plasma membranes and assay of Mg^{2+}ATPase were carried out as described previously (Riordan _et al_, 1977). Protein was determined according to Lowry _et al_ (1951). Lipid analyses including separation (Rouser _et al_, 1967) and quantitation (Bartlett, 1959) of phospholipids, cholesterol (van Hoeven and Emmelot, 1972) and fatty acid determinations have recently been described (Riordan _et al_, 1979). The fluorescence polarization of diphenyl hexatriene and β-parinaric acid were measured using a SLM System 4000 Scanning Polarization Spectrofluorometer with excitation and emission at 357 nm and > 399 nm for the former and 321 nm and > 386 nm for the latter. These probes were allowed to partition into the lipid phase of the membranes by the procedures of Shinitzky and Inbar (1976).

RESULTS

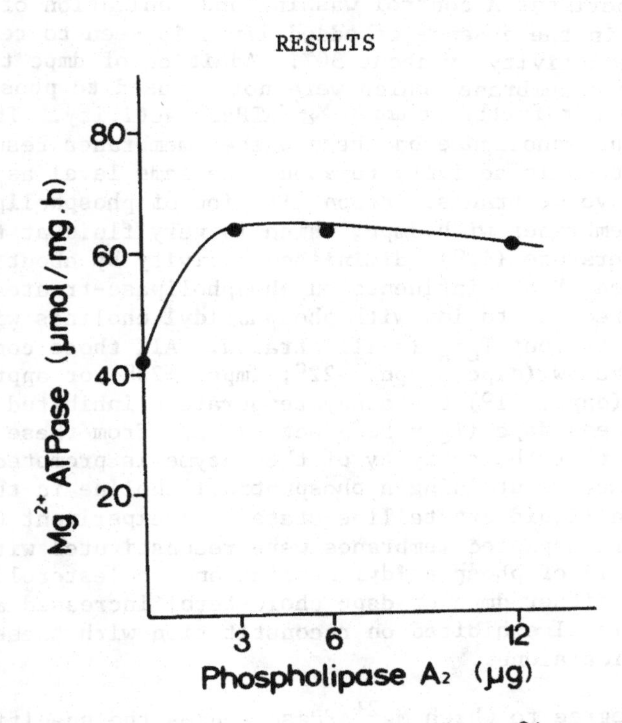

Fig. 1. Rat liver plasma membrane Mg^{2+}ATPase as a function of treatment with phospholipase A_2. 1.2 mg of rat liver plasma membrane was treated with the indicated amounts of enzyme for 1 h at 37° in 50 mM Tris-HCl, 1 mM $CaCl_2$, pH 7.6, washed and assayed for Mg^{2+}ATPase at 40°.

Fig. 1 shows the effect of treatment of membranes with increasing amounts of phospholipase A_2. Mg^{2+}ATPase activity is increased to a maximum of about 50% over the control on exposure to 2.5 µg of enzyme per mg membrane protein. Two larger amounts of phospholipase had virtually the same effect. This removal of phospholipids (primarily phosphatidyl choline; Riordan et al, 1977) increased the fluorescence polarization of DPH substantially (Table I).

This fluidity decrease presumably reflects the increased protein to lipid ratio.

The influence of reconstitution of the phospholipase treated membranes with pure phospholipids is shown in Fig. 2. In experiment A control washing and sonication of native membranes in the absence of added lipid is seen to reduce Mg^{2+}ATPase activity by about 30%. Addition of dmpc to these washed membranes which were not exposed to phospholipase did not further change Mg^{2+}ATPase activity. The action of phospholipase on these washed membranes resulted in an increase in activity to about the same level as that of the native membranes. Reconstitution of phospholipid depleted membranes with dmpc, which is very fluid at the assay temperature (40°), diminished activity by about 85%. In experiment B the influence on phospholipase-treated membranes of reconstitution with phosphatidyl cholines with markedly different T_c's is illustrated. All those compounds with T_c's below (dlpc; dopc, -22°; dmpc, +23°) or approximately at (dppc, 41°) the assay temperature inhibited the enzyme whereas dspc (T_c = 58°) activated. From these data it appears that the activity of the enzyme is promoted by an environment containing a phosphatidyl choline in the gel rather than liquid crystalline state. In experiment C phospholipid-depleted membranes were reconstituted with a mixture (1:1) of phosphatidyl choline and cholesterol. When added with either dmpc or dspc cholesterol increased activity above the level exhibited on reconstitution with these phospholipids alone.

The degree to which Mg^{2+}ATPase senses the specific phospholipid reconstituted is illustrated in Fig. 3 which shows a sharp break at about 23° in the slope of the line relating temperature and specific enzyme activity on reconstitution with dmpc. The rate at which activity rises above the T_c of dmpc is considerably less than that at temperatures below this point.

The changes in bulk lipid fluidity which actually accompanied the above reconstitutions were monitored by fluorescence polarization (Table I). Experiment A demonstrates that dopc and dmpc substantially reduce the higher degree of polarization caused by phospholipase whereas dppc does so only slightly; dspc increases the polarization beyond that of phospholipase-treated membranes. Addition of cholesterol on reconstitution with dmpc increased β-PnA polarization to

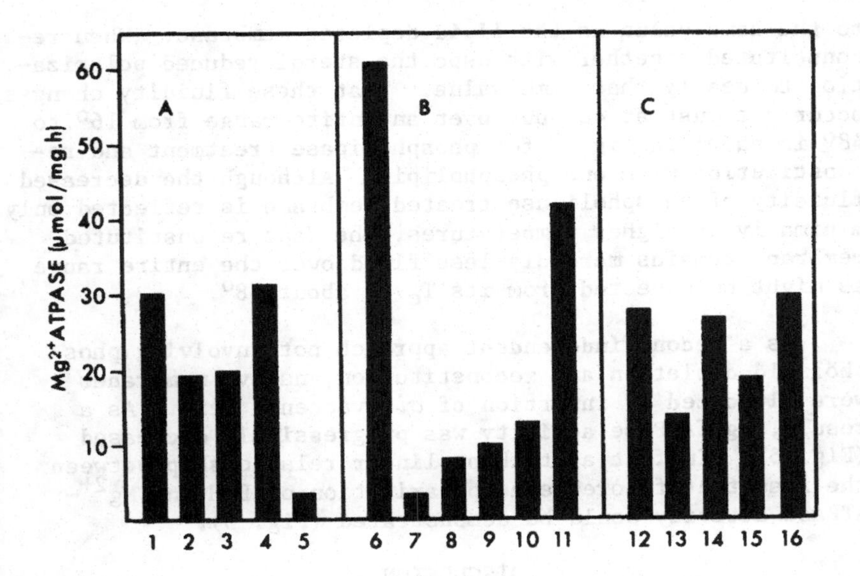

Fig. 2 Membranes were treated with phospholipase A_2 and reconstituted as described in 'Methods'. 1. native membrane, 2. washed membrane, 3. washed membrane + dmpc, 4. p'lipase treated membrane, 5. p'lipase + dmpc, 6. p'lipase treated, 7. p'lipase + dlpc, 8. + dopc, 9. + dmpc, 10. + dppc, 11. + dspc, 12. p'lipase treated, 13. + dmpc, 14. + dmpc + cholesterol, 15. + dspc, 16. + dspc + Cholesterol.

Table I

Fluorescence polarization changes on reconstitution.

Exp. A	DPH (Po = 0.45)	Exp. B	β-PnA (Po = 0.48)
Treatment	P(40°)	Treatment	P(40°)
control	0.32	control	0.31
p'lipase	0.36	p'lipase	0.34
dopc	0.32	dmpc	0.31
dmpc	0.32	+cholesterol	0.34
dppc	0.35	dspc	0.41
dspc	0.37	+cholesterol	0.35

to the same value as the lipid-depleted membrane. When reconstituted together with dspc the sterol reduced polarization to nearly that same value. That these fluidity changes occur not just at 40° but over an entire range from 16° to 48° is shown in Fig. 4 for phospholipase treatment and reconstitution with one phospholipid. Although the decreased fluidity of phospholipase-treated membrane is reflected only minimally at higher temperatures, the dspc reconstituted membrane remains markedly less fluid over the entire range as might be expected from its T_c of about 58°.

As a second independent approach not involving phospholipid depletion and reconstitution, native membranes were fluidized by insertion of *cis*-vaccenic acid. As a result, Mg^{2+}ATPase activity was progressively decreased (Fig. 5). In fact a striking linear relationship between the resultant fluorescence polarization of DPH and Mg^{2+}-ATPase activity could be demonstrated (Fig. 5).

DISCUSSION

Two independent methods of changing the bulk lipid fluidity of liver plasma membranes provided evidence that the activity of Mg^{2+}ATPase is modulated by the state of the lipid phase. Furthermore, this modulation differed in direction from that of most other membrane enzymes which have been studied from this point of view (Kimelberg, 1977). In other words, increased lipid fluidity decreased rather than increased activity of this enzyme. This relationship could be demonstrated in a strictly quantitative manner on fluidization with *cis*-vaccenic acid. It also held true in the experiments in which endogeneous phospholipids were depleted with phospholipase A_2 followed by reconstitution with pure phosphatidyl cholines with a wide range of T_c's. However, in the case of dppc reconstitution, when the enzyme was assayed at 40°, very near the T_c of this phospholipid, activity decreased as it did with more fluid lipids such as dmpc and dopc. On reconstitution with cholesterol-phospholipid mixtures, activity was greater than when only the phospholipid alone was used. This was true even with a

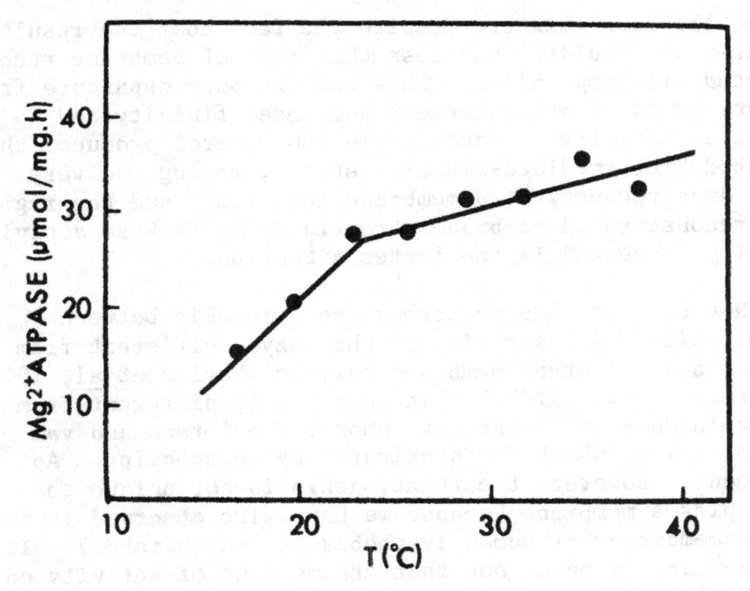

Fig. 3 Temperature dependence after dmpc reconstitution.

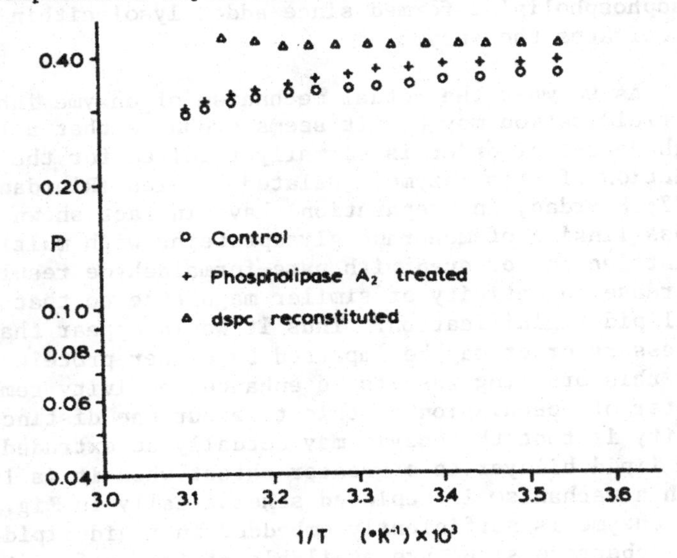

Fig. 4 Arrhenius plots of fluorescence polarization of DPH in native, p'lipase A_2-treated and dspc-reconstituted membranes.

dspc-cholesterol mixture despite the fact that the resultant membrane fluidity was less than that of membrane reconstituted with dspc alone. This was the only departure from the strict correlation between decreased fluidity and increased activity. Hence, while cholesterol produced the expected "intermediate-fluidity state", making the very fluid dmpc-reconstituted membrane less fluid and the rigid dspc-reconstituted membrane more fluid, Mg^{2+}ATPase activity actually increased in the latter situation.

Not only is this reciprocal relationship between lipid fluidity and activity of the enzyme different from that of several other membrane enzymes (Farias et al, 1975; Sandermann, Jr., 1978) but it is probably different from the Mg^{2+}ATPase of erythrocyte ghosts (Roelofson and van Deenen, 1973), which is inactivated by phospholipase A_2 treatment. However, the relationship is not unique to liver plasma membrane because we have also observed it in plasma membranes of human lymphoblasts (unpublished). It is important to point out that enhancement of activity on phospholipase treatment is apparently not an effect of the lysophospholipids formed since added lysolecithin strongly inactivates the enzyme.

As to what the actual mechanism of enzyme inhibition on fluidization may be, it seems probable that a fairly high degree of order is normally required for the optimal function of this enzyme. Related studies (Riordan et al, 1977; Riordan, in preparation) have in fact shown that cross linking of membrane glycoproteins with multivalent plant lectins or even with para-formaldehyde results in an increase in activity of similar magnitude to that caused by lipid rigidification. Thus it would appear that the necessary order may be imparted by either protein or lipid. How this ordering results in enhanced activity remains a matter of speculation at this time but one distinct possibility is that the enzyme may actually be extruded from the lipid bilayer to a greater extent when it is less fluid. Such a mechanism is depicted schematically in Fig. 6 where the enzyme is sufficiently embedded in fluid lipid that not all substrate sites are available at the surface to the water soluble substrate. When fluidity is decreased, however, the more highly ordered lipid phase may displace the enzyme to some extent so that all substrate sites are exposed at the surface. This possibility is currently

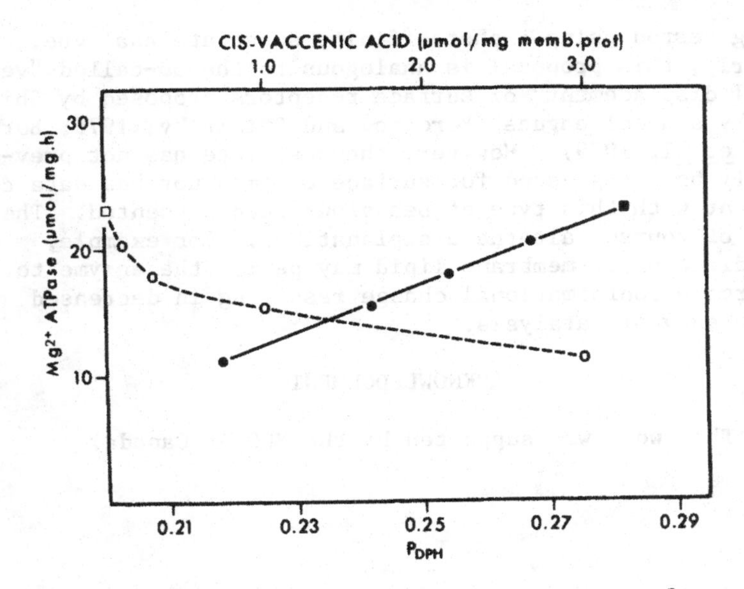

Fig. 5 Influence of *cis*-vaccenic acid on Mg^{2+}ATPase (o) and fluorescence polarization of DPH (●).

Fig. 6 Cryptic and exposed active sites of Mg^{2+}ATPase.

being tested using a photoaffinity substrate analogue. Clearly, this proposal is analogous to the so-called "vertical displacement" of surface receptors proposed by Shinitzky and colleagues (Borochov and Shinitzky, 1976; Borochov *et al*, 1979). However, the postulate has not previously been suggested for surface enzymes nor has data consistent with this type of behaviour been presented. There are, of course, alternate explanations. For example, fluidization of membrane lipid may permit the enzyme to undergo a conformational change resulting in decreased efficiency of catalysis.

ACKNOWLEDGEMENTS

This work was supported by the MRC of Canada.

REFERENCES

Bartlett, G.R. (1959). J. Biol. Chem. 234, 466.

Borochov, H. and Shinitzky, M. (1976). Proc. Natl. Acad. Sci. 73, 4526.

Borochov, H., Abbott, R.E., Schachter, D. and Shinitzky, M. (1979) Biochemistry 18, 251.

Farias, R.N., Bloj, B., Moreno, R.D., Sineriz, F. and Trucco, R.E. (1975). Biochim. Biophys. Acta 415, 231.

Kimelberg, H. (1977) In G. Poste and G.L. Nicolson (eds.) Cell Surface Reviews, Elsevier, Amsterdam.

Lowry, O.H., Rosebrough, N.J., Farr, A.L. and Randall, R.J. (1951). J. Biol. Chem. 193, 265.

Riordan, J.R., Slavik, M. and Kartner, N. (1977). J. Biol. Chem. 252, 5449.

Riordan, J.R., Alon, N., Buchwald, M. (1979). Biochim. Biophys. Acta 574, 39.

Roelofsen, B. and van Deenen, L.L.M. (1973). Eur. J. Biochem. 40, 245.

Rouser, G., Kritchevsky, G. and Yamamoto, A. (1967). In G.V. Marinetti (ed.), Lipid Chromatographic Analysis. Marcel Dekker, New York.

Sandermann, Jr., H. (1978). Biochim. Biophys. Acta 515, 209.

Shinitzky, M. and Inbar, M. (1976). Biochim. Biophys. Acta 433, 133.

Van Hoeven, R.P. and Emmelot, P. (1972). J. Memb. Biol. 9, 105.

ALKALINE PHOSPHATASE IN RED CELL MEMBRANE: INTERCONNECTION OF ACTIVITIES AND MEMBRANE LIPID FLUIDITY

Christina Ziemann and Guido Zimmer

Gustav-Embden-Zentrum der Biologischen Chemie

Universität Frankfurt, Theodor-Stern-Kai 7,
6000 Frankfurt/Main, German Federal Republic

ABSTRACT

A p-nitrophenylphosphate hydrolysing enzyme bound to the human red cell membrane was studied over a temperature range of 12-30°C under conditions of high pH (9.8) and high ionic strength (1 M diethanolamine/HCl buffer) which were recommended for determination of alkaline phosphatase activity in human serum. This enzymatic activity revealed some remarkable characteristics, in particular concerning its behaviour towards thiol reagents. Thus it is completely inactivated by dithiothreitol, L(+) cysteine and 2-mercaptoethanol,while it is only slightly inhibited by mersalyl, and not affected by the SH-oxidizing agent diamide. Furthermore, the enzyme appears not to be identical with the p-nitrophenylphosphatase activity, which was studied previously at lower pH and lower ionic strength by other authors.

For each temperature, K_m and v_{max} were determined from Lineweaver-Burk plots. Discontinuities at about 17-18° C were found for the temperature dependencies of the enzyme's activities as well as of the enzyme's Michaelis constants.

These discontinuities thus occurred at the same temperature at which the phase transition of the red cell membrane lipids was reported.

Moreover, electron spin resonance investigations were carried out, using a stearic acid spin label for polar membrane regions. In accordance with previous findings, application of an order parameter disclosed discontinuities at a similar temperature range.

INTRODUCTION

There are different methods available which serve to indicate a phase transition of the red cell membrane phospholipids at about 17-19° C, e.g. direct viscosity measurements (Zimmer and Schirmer, 1974; Simmons and Naftalin,1976); X-ray diffraction (Gottlieb and Eanes, 1974); Raman spectroscopy (Verma and Wallach, 1976); NMR spectroscopy (Cullis, 1976). Another proof for the existence of this lipid phase transition by help of ESR spectroscopy is given in the present paper. We have used the spin label 2-(3-carboxypropyl)-4,4-dimethyl-2-tridecyl-3-oxazolidinyloxyl (No.618, Syva) over a temperature range and calculated the order parameter S(Gaffney, 1976). Moreover, we were interested to find a membrane-bound enzyme system which is sensitive to mobility and structure of the membrane lipids since such interdependence has been reported previously in other membranes as reviewed by Sandermann (1978); as an example of this kind, alkaline phosphatase was investigated.

Spectroscopic measurements were also carried out in a milieu similar to activity estimations of the membrane alkaline phosphatase. This was done, because transition temperature should be dependent on the milieu, for example on pH,and ionic strength (Sandermann, 1978). Alkaline phosphatase activity was determined by hydrolysis of p-nitrophenylphosphate at pH 9.8.

Another example of a red cell membrane bound system which is dependent on lipid structure was previously given by Lacko et al. (1973) for glucose transport.

MATERIALS & METHODS

1) Preparation of Red Cell Membrane. Fresh human red cell concentrates were obtained from the Blutspendedienst Hessen, Frankfurt. Hemolysis was carried out overnight with 7.5 vol 15 mM sodium phosphate buffer pH 7.5. Subsequently, the membranes were washed with 15 mM and 10 mM phosphate buffer according to the method of Dodge *et al.* (1963). Thereafter, the membranes were frozen overnight and, after thawing, centrifuged. The membrane pellets were suspended with a small quantity of dist. H_2O and lyophilized.

2) Estimation of Alkaline Phosphatase Activity. A red cell membrane suspension was prepared at 5.33 mg/ml in 1M diethanol amine/HCl buffer pH 9.8. This suspension was sonicated by

means of a Branson-S-75 sonifier for 5 times 15 sec at about 3.5 amperes. Foaming was avoided at this step of the preparative procedure. The temperature was held below 10° C by means of crushed ice.

Suspensions treated in this manner gave highly reproducible results. Treatment with an ultraturrax and homogenization with a Potter-Elvehjem device did not give reproducible measurements since foaming could not be avoided.

Before starting the measurements, membrane suspensions were preincubated at the desired temperature for at least 15 min. This preincubation was necessary since activities in general increased during the first minutes and stayed constant from about 10 min preincubation time onward. The milieu for the estimations of enzyme activities was varied in order to optimize the conditions. Thus the pH was varied from 8.0 - 10.0, also ionic strength between 0.1 and 1.0 M, KCl concentration from 1-300 mM, Mg^{++} from 5-10 mM. Other ions, for example Ba^{++}, Sr^{++}, Ca^{++} were without effect. ATP (2.5 - 30 mM), EDTA (2.5 mM) were also without effect on the activities.

The substrate was p-nitrophenylphosphate (Merck), and solutions (1 M diethanolamine -HCl, 10 mM KCl, 5 mM Mg^{++} pH 9.8) were freshly prepared every day.

12 concentrations of substrate ranging from 0.5 - 40 mM were prepared. 2 ml of each substrate solution was added to 1 ml membrane suspension and increase in absorbance at 405 nm was measured for at least 10 min. Stopping of the reaction by means of 0.5 ml 1 n NaOH, centrifugation (ultraturrax experiments) or clearing with 200 µl 10% sodium dodecylsulfate gave similar results as measurements obtained in suspension when continuously measured. Reproducibility, however, was much better when measurements were carried out in suspension, and continuously. Increase in absorbance was taken for estimation of relative enzyme activity. The measurements were carried out in an Eppendorf photometer using a thermostated cuvette and a recorder No. 4410 Eppendorf. The temperature range was 12 - 30° C.

As inhibitors of the reaction 2-mercaptoethanol (2.3 mM), cysteine (1 mM), dithiothreitol (1 mM), mersalyl (1-10 mM) were used. Activators were not found for the reaction.

3) ESR Spectroscopic Measurements. 20 mg of red cell membrane were weighed and suspended with 20 ml of either sodium

phosphate 10 mM pH 7.4 or 1 M diethanolamine HCl buffer, pH 9.8. Thereafter, centrifugation with an Eppendorf type 3200 centrifuge was carried out for 2 min. Supernatants were discarded and 1/20 of the pellets was suspended in 100 µl suspension buffer. 1 µl of the spin label in 5 mM solution in ethanol was added and mixed vigorously. After 30 min. standing at room temperature spectra were taken over a temperature range of 7-25° C. A Bruker ESR spectrometer type B-MN, 155-45 SI 6 and a cooling device type P-WT 100/700 were used.

RESULTS & DISCUSSION

Activity of alkaline phosphatase is dependent on the preparative procedure used in isolation of red cell membrane. Membranes were either prepared following the method of Dodge et al. (1963), using 15 and 20 mM sodium phosphate pH 7.5, or the procedure described by Schrier (1967) was used. The latter method is characterized by washing with hypotonic NaCl solutions in order to hemolyse the cells. In our hands, the hypotonic washing procedure yielded only very low activities, while the preparation following Dodge et al. (1963) ended up with membranes which were appreciably active. Activities were the higher, the smaller the dilution of the original red blood cell concentrate. It was suspected, that the membrane-bound enzyme might be in equilibrium with solubilized protein in the suspension or washing solution. Possibly the membrane-bound enzyme originated from plasma or serum with the following equilibrium :

$$\text{phosphatase}_{\text{membrane bound}} \rightleftharpoons \text{phosphatase}_{\text{serum}}$$

This suspicion is reinforced, since optimal conditions for measurement of the alkaline phosphatase are similar to those which are recommended for estimation of the activity in human serum by Walter and Schütt (1970).

Furthermore, enzymatic activity in red cell membrane differs from p-nitrophenylphosphatase activity, estimated by other authors in a Tris-HCl buffer system at pH 7.8 (Rega et al., 1968). No activation was found by either potassium-, magnesium-, calcium - ions, nor by ATP. This behaviour also is similar to alkaline phosphatase from human serum.

Enzymatic activity was not changed by addition of either barium, strontium or phosphate ions. Addition of NaOH (1 ml 1 N) or foaming immediately destroyed activity. Substrate cleavage per minute, however, was low compared with activities in human serum. We would tend to conclude that content of the membrane in this enzymatic activity is low as well. Increases of absorbance, however, were linear over a time of about 15 minutes. Thus, in spite of low absolute activities, exact measurements could be taken. Even at temperatures of 30° C, spontaneous cleavage of substrate was only 2% of enzymatic increase. Substrate concentrations were varied from about 0.33 mM to 26.6 mM for each temperature in order to avoid artificial discontinuities in the Arrhenius diagrams because of temperature dependencies of the Michaelis constants (Silvius et al., 1978). Cleavage of substrate was in line with Michaelis-Menten kinetics for each temperature measured. Occasionally, at concentrations of substrate over 20 mM substrate inhibition was found. Lineweaver-Burk diagrams allowed v_{max} and k_m estimations. K_m values proved to be dependent on temperature (see Figure 2). Eadie-Hofstee plots also yielded good linearities (not shown).

In Fig. 1, Arrhenius diagram is shown for the enzymatically catalysed cleavage of p-nitrophenylphosphate. A discontinuity is found in the region of 17-19°.

It should be noted that the activity of alkaline phosphatase was not appreciably altered by additions of sodium dodecylsulfate up to 4 mg/mg protein. This may imply the following possibilities:

1) The boundary lipid, which is bound to hydrophobic protein domains is not influenced even by high concentrations of SDS, and according to its phase transition thus causes different mechanisms of the enzymatic reaction.

2) The enzyme protein itself undergoes a conformational change (Webb, 1963), which is dependent on temperature, and which is independent of membrane lipid.

A correct decision thereon could not yet be made.

Fig. 2 shows that the minimum for the Michaelis constant is found at about 18° C. This also would speak for a conformational change of the enzyme at this temperature range.

Fig. 3 and 4 show results of spin labeling experiments with the stearic acid label. On the ordinate, the order parameter S according to Gaffney (1976) is plotted against temperature. The order parameter, giving an indication of the mean angular deviation of the fatty acid chain from its average orientation in the membrane also exhibits a discon-

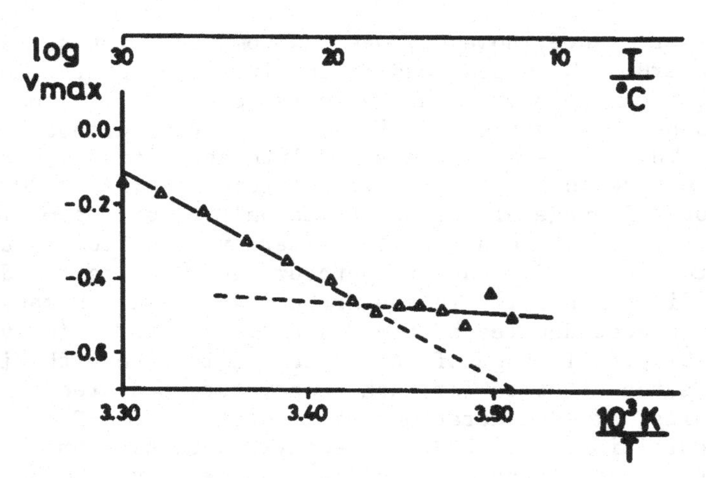

Figure 1
Arrhenius plot of hydrolytic cleavage of p-nitrophenylphosphate by membrane-bound alkaline phosphatase;
1 M diethanolamine/HCl buffer pH 9.8;
each experimental point was separately determined by means of a Lineweaver-Burk diagram from which v_{max} was estimated
$|S.D.| \leq 0.03$

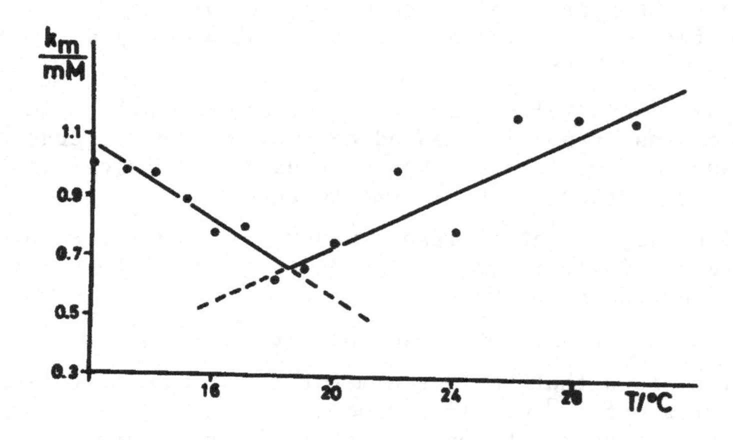

Figure 2
Temperature dependencies of Michaelis constants of membrane-bound alkaline phosphatase; experimental conditions similar to those in Fig. 1); regression lines are shown, the correlation coefficients are

$T < 19^\circ C$: $r = -0.96$, S.D. $\pm$ 0.06 mM
$T > 19^\circ C$: $r = 0.94$, S.D. $\pm$ 0.08 mM

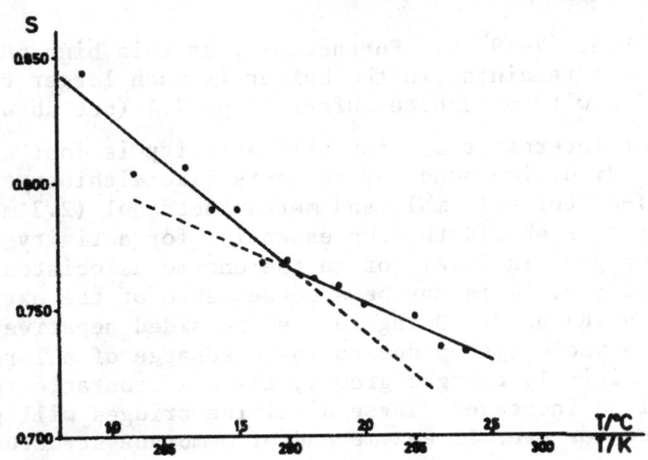

Figure 3
Temperature dependency of order parameter S of spin label 618 (Syva) in red cell membrane;
10 mM sodium phosphate buffer pH 7.4
T ≤ 17° C : S.D. ±0.01
T ≥ 17° C : S.D. ±0.003

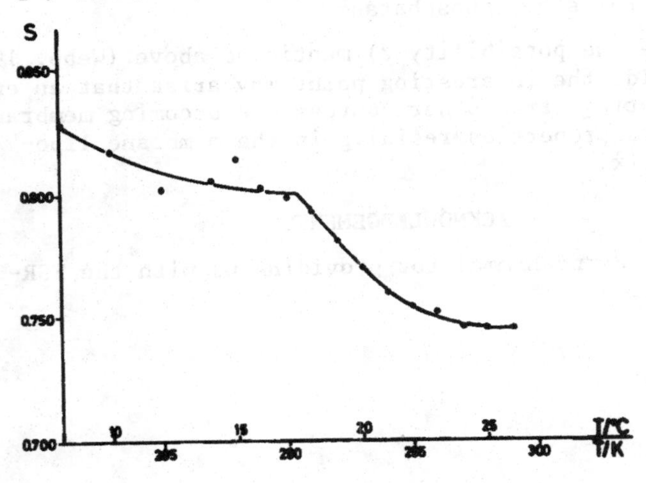

Figure 4
Temperature dependency of order parameter S of spin label 618 (Syva) in red cell membrane;
1 M diethanolamine/HCl buffer pH 9.8
The experimental points fit on differently bent lines. There is a greater decline of the values above 17° C.

tinuity at about 17-19° C. Furthermore, at this high pH the amount of label remaining in the buffer is much larger than in experiments with phosphate buffer of pH 7.4 (not shown).

It is of interest that enzymatic activity is lost after reduction of disulfide bonds by reagents like dithioerythritol (1 mM), L(+)-cysteine (1 mM), and mercaptoethanol (2.3 mM). Disulfide bridges should thus be essential for activity either in the enzyme protein itself or in the enzyme associated membrane proteins. This may be a consequence of the extreme alkaline pH value used. Owing to the one-sided negative charge of the whole system due to the discharge of a large number of positively charged groups, the electrostatic repulsion will be increased. These disulfide bridges will gain in importance for overall enzyme and/or membrane structure.

It is of interest that the points of measurement do not follow straight lines in Fig. 4. This behaviour cannot be explained positively but may be due to some mechanism dependent on the extreme pH.

A problem which is obviously not solved in the present paper is whether or not the phosphatase activity measured is derived from serum phosphatase.

In case the possibility 2) mentioned above (Webb, 1963) does not hold, the interesting point may arise that an enzyme originating from other sources, in becoming membrane-bound, adopts properties residing in the membrane lipoprotein matrix.

ACKNOWLEDGEMENTS

We thank Dr. Bernd Lammel for providing us with the ESR-Spectrometer.

REFERENCES

Cullis, P.R. (1976). FEBS Letters 68, 173.

Dodge, J.T., Mitchell, C., and Hanahan, D.J. (1963). Arch. Biochem. Biophys. 100, 119.

Gaffney, B.J. (1976). In L.J. Berliner (ed.), Spin Labeling Theory and Applications. Academic Press, New York, San Francisco, London.

Gottlieb, M.H. and Eanes, E.D. (1974). Biochim. Biophys. Acta 373, 519.

Lacko, L., Wittke, B., and Geck, P. (1973) J. Cell Physiol. 82, 213.

Rega, A.F., Garrahan, P.J., and Pouchan, M.J. (1968). Biochim. Biophys. Acta 150, 742.

Sandermann, H., Jr. (1978). Biochim. Biophys. Acta 515, 209.

Schrier, S.L. (1967). Biochim. Biophys. Acta 135, 591.

Silvius, J.R., Read, B.D., and McElhaney, R.N. (1978). Science 199, 902.

Simmons, N.L., and Naftalin, R.J. (1976). Biochim. Biophys. Acta 419, 493.

Verma, S.P., and Wallach, D.F.H. (1976). Biochim. Biophys. Acta 436, 307.

Walter, K., and Schütt, C. (1970). In H.U. Bergmeyer (ed.), Methoden der enzymatischen Analyse. Verlag Chemie, Weinheim, Bergstr., FRG

Webb, J.L. (1963). Enzyme and Metabolic Inhibitors, Vol. 1, General Principles of Inhibition, Academic Press, New York, London.

Zimmer, G., and Schirmer, H. (1974). Biochim. Biophys. Acta 345, 314.

MEMBRANE PERMEABILITY IN PORCINE MALIGNANT HYPERTHERMIA

K.S. Cheah and A.M. Cheah

Agricultural Research Council Meat Research

Institute, Langford, Bristol, BS18 7DY, U.K.

INTRODUCTION

Malignant hyperthermia syndrome can be induced in stress-susceptible pigs by anaesthesia with halothane (Harrison et al., 1968; Berman et al., 1970; Nelson et al., 1972; Hall et al., 1972; Gronert and Theye, 1976). The predominant clinical symptoms of this syndrome are gross muscular rigidity, rapid rise in body temperature, tachycardia, hyperventilation, severe metabolic acidosis and elevated levels of serum metabolites (Brucker et al., 1973; van den Hende et al., 1976). The genetic inheritance was suggested to be due either to a single autosomal dominant gene with incomplete penetrance (Hall et al., 1966), or to an autosomal recessive gene with variable penetrance (Christian, 1972) or to a single recessive gene with incomplete (Ollivier et al., 1975) or complete (Eikelenboom et al., 1978) penetrance. In spite of the well documented etiology of porcine malignant hyperthermia syndrome the lesion responsible for the series of biochemical events leading to this syndrome is unknown. The inherent defects are not apparently confined to skeletal muscle, as it is believed that neuronal and hormonal abnormalities (LaCour et al., 1971) may also exist. Porcine malignant hyperthermia appears to be a primary disorder of skeletal muscle (Britt, 1974; Gronert and Theye, 1976), and the sympathetic nervous system is implicated only as a secondary response in the syndrome (Gronert et al., 1978).

This paper shows that the lesion in porcine malignant hyperthermia is most likely to be due to differences in membrane integrity, a conclusion derived from studies on the osmotic fragility of erythrocytes and on the effect of temperature on mitochondrial calcium transport.

MATERIALS AND METHODS

Mitochondria were isolated from M.longissimus dorsi (LD) immediately post-mortem using B.subtilis proteinase (Cheah, 1970). The Ca^{2+}-stimulated respiration for succinate oxidation was measured polarographically with a Clark oxygen electrode /Yellow Spring Oxygen Monitor (Model 53)/ in a total volume of 2.50 ml (Cheah and Cheah, 1978). Ca^{2+} efflux was measured with murexide at 20°C using the Aminco-Chance dual-wavelength/split-beam spectrophotometer operating in the dual-wavelength mode at 540-510 nm. Erythrocyte fragility was determined by the amount of haemoglobin released per ml of red blood cells in the form of reduced pyridine haemochromogen at 556 nm, after the cells had been subjected to osmotic shock at 0.60% NaCl at 25°C for 5 minutes.

RESULTS AND DISCUSSION

Halothane sensitive (i.e. malignant hyperthermic) pigs can easily be differentiated from normal pigs by the anaerobic rate of Ca^{2+} efflux of LD muscle mitochondria (Fig. 1). This difference in Ca^{2+} efflux probably reflects the existence of an alteration in membrane permeability in affected pigs, a conclusion supported by the difference in osmotic fragility of erythrocytes and in the transition temperature of the Arrhenius plots of the Ca^{2+}-stimulated respiration of LD muscle mitochondria.

A striking difference in membrane stability was observed when erythrocytes of both types of pigs were subjected to osmotic shock in 0.60% NaCl at 25°C. Erythrocytes of halothane sensitive pigs were more fragile than those of normal as shown by the release of more haemoglobin (Fig.2). The amount of haemoglobin released per ml of red blood cells also follows closely the rate of Ca^{2+} efflux of LD muscle mitochondria, and also shows a good relationship to halothane sensitivity.

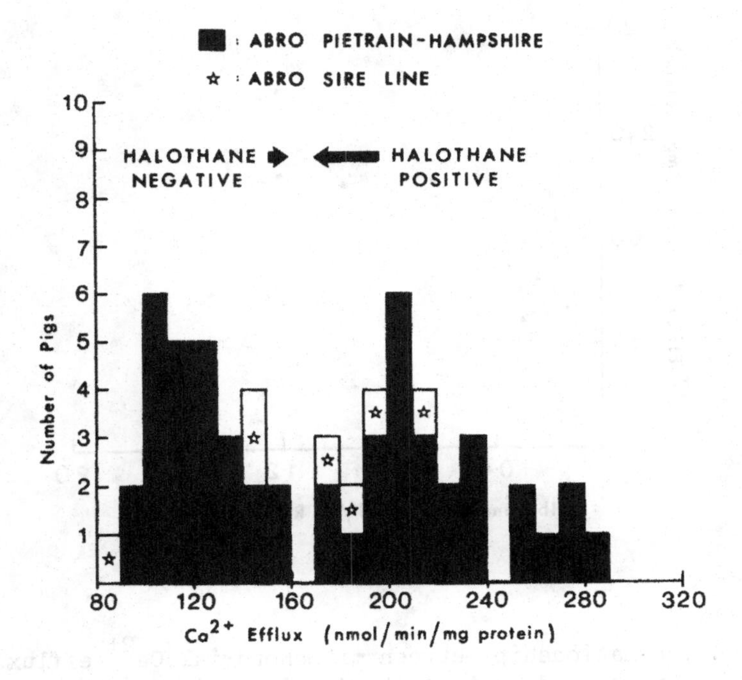

Fig. 1 : Relationship between rates of Ca^{2+} efflux and halothane sensitivity.

The anaerobic rate of Ca^{2+} efflux was estimated from the fast phase of the biphasic efflux (Cheah and Cheah, 1976) in the presence of 2.50 mM P_i. 120-150 nmol Ca^{2+}/mg protein used in all experiments at 20°C. Prediction of halothane sensitivity was made from the mitochondrial Ca^{2+} efflux rates prior to knowledge of the results of the halothane test carried out by Dr. Webb. An efflux rate of 165 nmol Ca^{2+}/min/mg protein clearly demarcates the halothane sensitive from the halothane insensitive pigs.

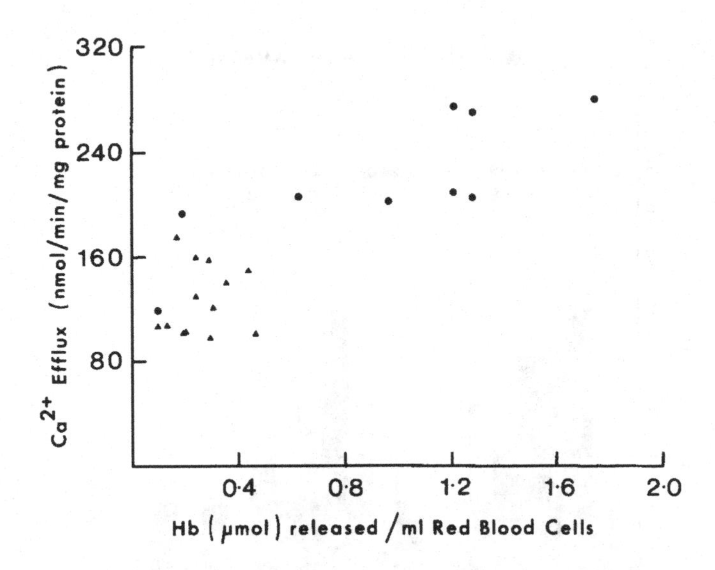

Fig. 2 : Relationship between mitochondrial Ca^{2+} efflux rates and erythrocyte fragility in halothane sensitive (●) and halothane insensitive (▲) pigs.

LD muscle mitochondria of halothane sensitive pigs are more sensitive to Ca^{2+} than those of normal at high temperature. At 40°C, a marked difference in the mitochondrial coupling integrity was observed, in that mitochondria of halothane sensitive pigs were more easily uncoupled by Ca^{2+} (Fig. 3). Uncoupling was observed after a total addition of 1297 ± 163 (n = 4) nmol Ca^{2+} per mg protein (Trace B). Under the same conditions, no uncoupling was observed with mitochondria from normal pigs even after a total addition of 2297 ± 288 (n = 3) nmol Ca^{2+} per mg protein (Trace A). At 25°C, however, no significant difference was observed in the coupling integrity of the mitochondria of halothane sensitive or normal pigs (Fig. 4). No significant difference was observed in the mitochondrial coupling integrity when ADP was used instead of Ca^{2+}.

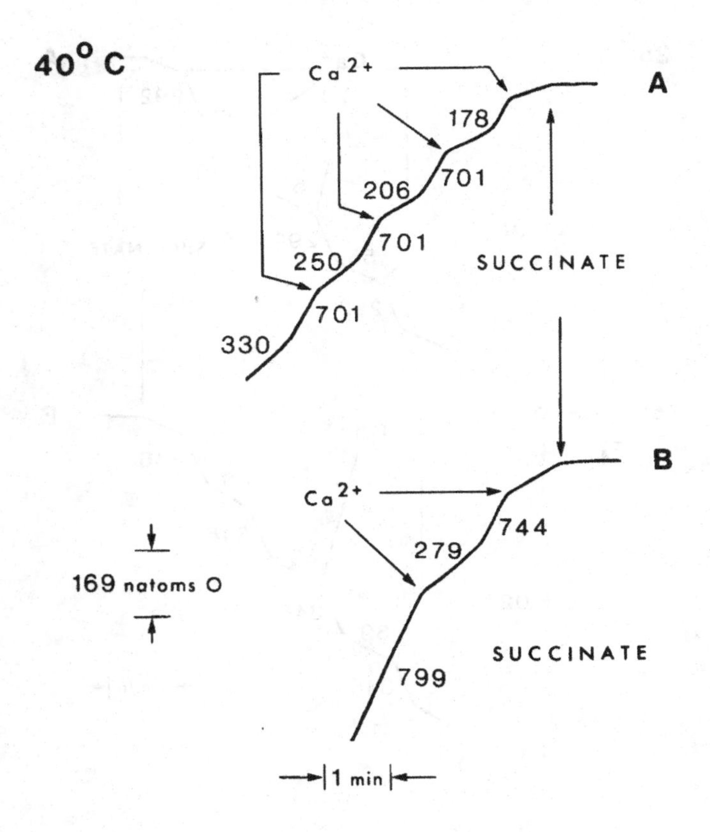

Fig. 3 : Effect of Ca^{2+} on succinate oxidation by LD muscle mitochondria of halothane insensitive and sensitive pigs at 40°C.

Trace A illustrates a typical experiment showing the state 3 - state 4 transition induced by Ca^{2+} during succinate oxidation by mitochondria of halothane insensitive pig. Four additions of 300 nmol were added without causing uncoupling of mitochondria. Total protein, 0.50 mg; total Ca^{2+} added, 2400 nmol/mg protein. Trace B represents a typical experiment showing the effect of Ca^{2+} on LD muscle mitochondria of halothane sensitive pig during succinate oxidation. The second addition of 300 nmol Ca^{2+} completely uncoupled the mitochondria. Total protein, 0.48 mg; total Ca^{2+} added, 1250 nmol/mg protein.

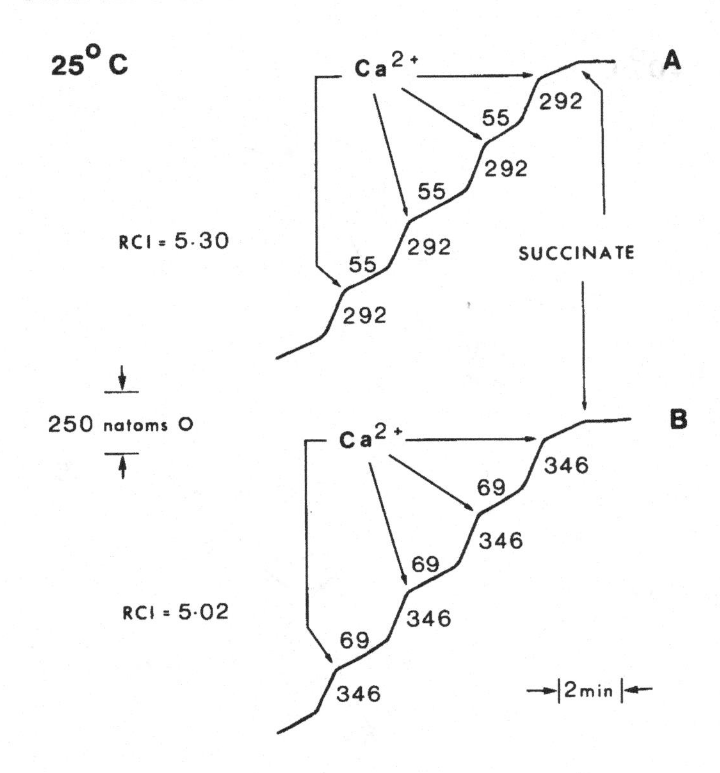

Fig. 4 : Effect of Ca^{2+} on succinate oxidation by LD muscle mitochondria of halothane insensitive and halothane sensitive pigs at 25°C.

Experimental details are similar to those described in the legend to Fig. 3 except that each addition represents 600 nmol Ca^{2+} instead of 300 nmol Ca^{2+}. Trace A and trace B represent LD muscle mitochondria from halothane insensitive and halothane sensitive pigs respectively.

Total protein: Trace A, 1.26 mg; Trace B, 0.96 mg.

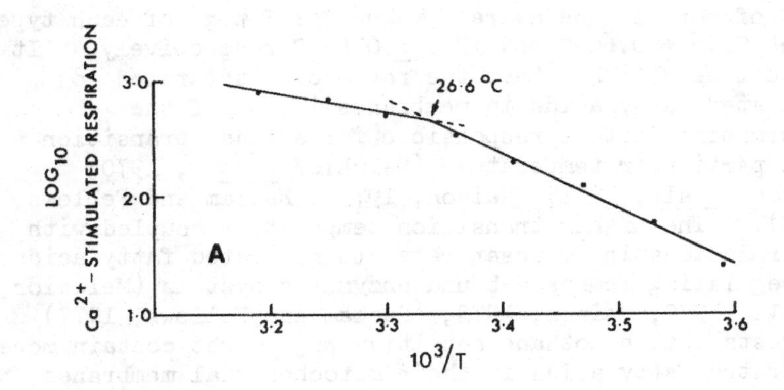

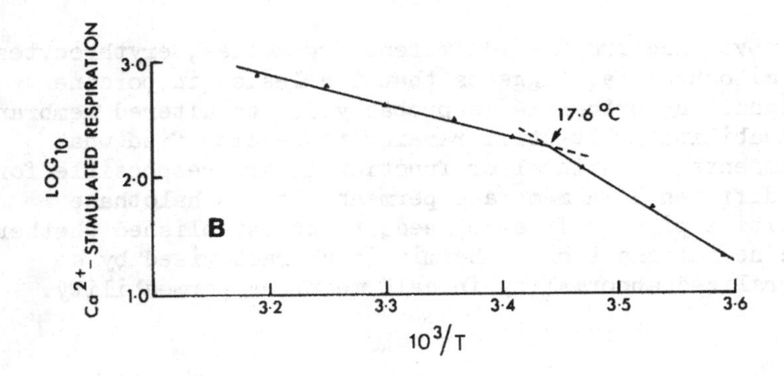

Fig. 5 : Arrhenius plots of Ca^{2+} – stimulated respiration of LD muscle mitochondria from halothane sensitive and halothane insensitive pigs.

The Ca^{2+}-stimulated respiration was estimated with a Clark oxygen electrode. Fig. 5 represents typical results obtained with mitochondria from halothane sensitive (A) and halothane insensitive (B) pigs.

The existence of a difference either in the structural or functional integrity of the mitochondrial membranes of halothane sensitive pigs could be demonstrated by the Arrhenius plots of Ca^{2+}-stimulated respiration (Fig. 5). LD muscle mitochondria of halothane sensitive pigs (A) showed a transition temperature 9°C higher than that of normal, the average value for 3 pigs of each type being 26.6 ± 0.60°C and 17.6 ± 0.65°C respectively. It is well established that the ratio of unsaturated to saturated fatty acids in membranes is one of the determining factors responsible for a phase transition at a particular temperature (Melchior _et al._, 1970; Raison _et al._, 1971; Raison, 1973; Haslam and Fellows, 1973). The higher transition temperature coupled with the relationship of unsaturated to saturated fatty acids in regulating membrane-bound enzymatic systems (Melchior _et al._, 1970; Raison, 1973; Haslam and Fellows, 1977) suggests that halothane sensitive pigs might contain more saturated fatty acids in their mitochondrial membranes than those of normal.

Evidence from two different organelles, erythrocytes and mitochondria, suggests that the lesion in porcine malignant hyperthermia is probably due to altered membrane permeability. It still remains to be clarified what components, structural or functional, are responsible for the difference in membrane permeability in halothane sensitive pigs. It also needs to be established whether porcine malignant hyperthermia is characterised by a generalized abnormality in cell membrane permeability.

ABSTRACT

Mitochondria of _M.longissimus dorsi_ (LD) from malignant hyperthermic pigs showed a much higher anaerobic rate of calcium ions efflux ($P<0.001$; $n = 50$), and were also more easily uncoupled by calcium ions at 40°C than those from normal pigs. The Arrhenius plots of calcium-stimulated respiration showed that mitochondria from LD muscle of malignant hyperthermic pigs had a transition temperature 9°C higher (26.6°C ± 0.60; $n = 3$) than that of normal (17.6°C ± 0.65; $n = 3$). Significant difference ($P<0.001$; $n = 22$) in osmotic fragility of erythrocytes between malignant hyperthermic and normal pigs was also observed.

ACKNOWLEDGEMENTS

The authors are grateful to Dr. A.J. Webb Agricultural Research Council Animal Breeding Research Organisation, Edinburgh, for supplying the halothane-sensitive and -insensitive pigs.

REFERENCES

Berman, M. C., G.G. Harrison, A. A. Bull, and D. D. Kench (1970). Nature 225, 653.

Britt, B. A. (1974). New Eng. J. Med. 290, 1140.

Brucker, R. F., C. H. Williams, J. Popinigis, T. L. Glavez, W. J. Vail, and C. A. Taylor (1973). In R.A. Gordan, B. A. Britt and W. Karlow (eds), Int. Symp. Malignant Hyperthermia. Charles C. Thomas Publishers, Springfield, 238 pp.

Cheah, K. S. (1970). FEBS Lett. 10, 109.

Cheah, K. S., and A. M. Cheah (1976). J. Sci. Fd. Agric. 27, 1137.

Cheah, K. S., and A. M. Cheah (1978). FEBS Lett. 95, 307.

Christian, L. L. (1972). Proc. Pork Quality Symp. University of Wisconsin Press, Madison, 91 pp.

Eikelenboom, G., D. Minkema, P. van Eldik, and W. Sybesma (1978). In J. A. Aldrete, and B. A. Britt (eds.), 2nd Int. Symp. Malignant Hyperthermia. Grune and Stralton Publishers. 141 pp.

Gronert, G. A., and R. A. Theye (1976). Anaesthesiol. 44, 36.

Gronert, G. A., J. H. Milde, and R. A. Theye (1978). In J. A. Aldrete and B. A. Britt (eds.), 2nd Int. Symp. Malignant Hyperthermia. Grune and Stralton Publishers. 159 pp.

Hall, L. W., C. M. Trim, and N. Woolf (1972) Brit. Med. J. 2, 145.

Hall, L. W., N. Woolf, J. W. P. Bradley, and D. W. Jolly (1966). Brit. Med. J. 2, 1305.

Harrison, G. G., J. F. Biebuyk, J. Terblanche, D. M. Debt, R. Hickman, and J. J. Saunders (1968). Brit. Med. J. 3, 594.

Haslam, J.M., and N. F. Fellows (1977). Biochem. J. 166,565.

LaCour, D., P. Juul-Jensen, and E. Reske-Neilson (1971). Acta Anaesth. Scand. 15, 299.

Melchior, D. L., H. J. Morowitz, J. M. Sturtavant, and T. Y. Tsong (1970). Biochim. Biophys. Acta. 219, 114.

Nelson, T. E., E. W. Jones, J. H. Venable, and D. D. Kerr. (1972). Anaesthesiol. 36, 52.

Ollivier, L., P. Sellier, and G. Monin (1975). Ann. Genet. Sel. Anim. 7, 159.

Raison, J. K. (1973). Bioenerg. 4, 285.

Raison, J. K., J. M. Lyons, R. J. Melhorn, and A. D. Keith (1971). J. Biol. Chem. 246, 4036.

van den Hende, C., D. Lister, E. Muylle, L. Ooms, and W. Oyaert (1976). Brit. J. Anaesth. 48, 821.

PART III

FATTY ACID CHANGES ACCOMPANYING PHYSIOLOGICAL EVENTS

ROLE OF PHOSPHOLIPID DESATURASES IN CONTROL OF MEMBRANE FLUIDITY

M. Kates and E. Pugh,

Department of Biochemistry,
University of Ottawa,
Ottawa, Canada.

ABSTRACT

Liver microsomes from rats fed a "normal" Purina Chow diet were compared with those from starved (48 h) rats and starved rats refed a fat-free diet, with respect to 20:3-CoA desaturase, 20:3-phosphatidylcholine desaturase, fatty acid composition and membrane fluidity (fluorescence polarization). Microsomal membranes of starved rats had a higher degree of unsaturation (due to increased 20:4 and 22:6 acids) while those of starved-refed rats had a lower degree of unsaturation (decreased 18:2 and 20:4 acids), compared to the "normal" controls. Activities of the two desaturases were increased 2-3 fold in the microsomes from starved-refed rats, compared to the control but were not detectable in the microsomes of starved rats. Fluorescence polarization measurements with diphenylhexatriene as probe showed no significant differences between any of the microsomal membranes in the temperature range 11-37°; the respective total lipids of these membranes also showed no significant differences in fluorescent polarization. However, the fluorescent polarization of the membranes was significantly higher than that of the respective total lipids, indicating a possible involvement of protein in controlling the fluidity of the membrane.

Kinetic studies of desaturases in microsomes from starved-refed rats using [14C]20:3-CoA as substrate showed a rapid incorporation of [14C]acyl groups into phospholipids and concomitant extensive hydrolysis (>50%) of the acyl-CoA substrate. After 30 min, 50% of the [^{14}C]acyl groups in the phospholipids was associated with the 20:4 acid, whereas only 10% of the [^{14}C]20:3-CoA substrate had been converted to the 20:4 acid. These results suggest that a considerable proportion of the arachidonic acid present in liver microsomes is formed by the action of the phospholipid desaturase and that this enzyme system may play an important role in control of membrane fluidity.

INTRODUCTION

Interest in phospholipid desaturases in our laboratory derives from studies undertaken at the National Research Council in Ottawa (Kates and Baxter, 1962) on the changes in fatty acid composition during the growth cycle of the mesophilic yeast *Candida lipolytica*. It was found that striking reciprocal changes in proportions of oleic (18:1) and linoleic (18:2) acids occurred during the growth cycle of this organism. In the very early phase of growth, the 18:2 and 18:1 acids showed a marked increase and decrease, respectively, to maximum and minimum values at the beginning of active growth; thereafter the proportions of these acids were restored to their initial values at the end of active growth. In subsequent studies these changes in levels of 18:1 and 18:2 acids were shown to occur on the level of the phospholipids, mainly phosphatidylcholine and phosphatidylethanolamine (Kates and Paradis, 1973) as shown in Fig. 1. The question then arose as to the

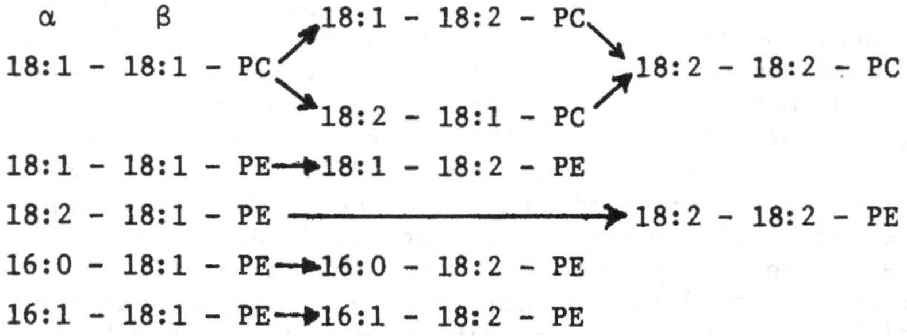

α β

18:1 - 18:1 - PC → 18:1 - 18:2 - PC → 18:2 - 18:2 - PC

18:1 - 18:1 - PC → 18:2 - 18:1 - PC → 18:2 - 18:2 - PC

18:1 - 18:1 - PE → 18:1 - 18:2 - PE

18:2 - 18:1 - PE → 18:2 - 18:2 - PE

16:0 - 18:1 - PE → 16:0 - 18:2 - PE

16:1 - 18:1 - PE → 16:1 - 18:2 - PE

Fig. 1. Changes in molecular species of phosphatidylcholine (PC) and phosphatidylethanolamine (PE) during growth of *Candida lipolytica*. Taken from Kates and Paradis (1973).

mechanism of these changes for which two possibilities existed: (1) desaturation of 18:1-CoA to 18:2-CoA followed by incorporation of the 18:2 chain into phospholipid by transacylation; and/or (2) direct desaturation of an 18:1-containing phospholipid to 18:2-phospholipid.

In regard to mechanism (1), desaturation of acyl chains (e.g. 18:0 and 18:1) in the form of either CoA or ACP derivatives has been established in baker's yeast (Bloomfield and Bloch, 1960; Yuan and Bloch, 1961) and transfer of 18:1 and 18:2 acids newly formed by the action of desaturases to endogenous acceptor has been demonstrated in seeds (Vijay and Stumpf, 1971, 1972). However, evidence in favour of the concept that phospholipids may serve as substrates for the desaturation of oleate to linoleate [mechanism (2)] has been provided by Gurr *et al*. (1969) using *Chlorella* chloroplasts and by Baker and Lynen (1971) using *Neurospora crassa* microsomes.

Direct desaturation of phospholipids by mechanism (2) has now been demonstrated conclusively in *Candida lipolytica* (Pugh and Kates, 1973, 1975b), in *Torulopsis utilis* (Talamo *et al*. 1973) and more recently in rat liver (Pugh and Kates, 1977). We will present here a review of recent work done in our laboratory on desaturation of phospholipids (see also Pugh and Kates, 1979).

PHOSPHOLIPID DESATURASE OF *CANDIDA LIPOLYTICA*

In our early experiments with the phospholipid desaturase system in *C. lipolytica* it was necessary to demonstrate first that hydrolytic breakdown of the substrate and resynthesis by transacylation was negligible. For this purpose we used the doubly-labeled substrates [^{14}C,^{32}P] phosphatidylcholine or [^{14}C,^{32}P]phosphatidylethanolamine which were prepared from cells of *C. lipolytica* grown in the presence of [^{14}C]acetate and [^{32}P]orthophosphate (Pugh and Kates, 1973). On incubation of these substrates with microsomes of *C. lipolytica* in presence of reduced pyridine nucleotides and oxygen, both substrates showed an increased proportion (5-10%) of 18:2 acid but no significant change in $^{14}C/^{32}P$ ratio, indicating that direct desaturation of 18:1 $\rightarrow$ 18:2 acid had occurred without breakdown and resynthesis of the phospholipid substrate. More extensive phospholipid desaturation was demonstrated later with synthetic 1,2-di[^{14}C]oleoyl-PC (Pugh and Kates, 1975a) and 1-acyl-2-oleoyl-PC as substrates in the presence of optimal concentrations (0.1%) of Triton X-100 (Pugh and Kates,

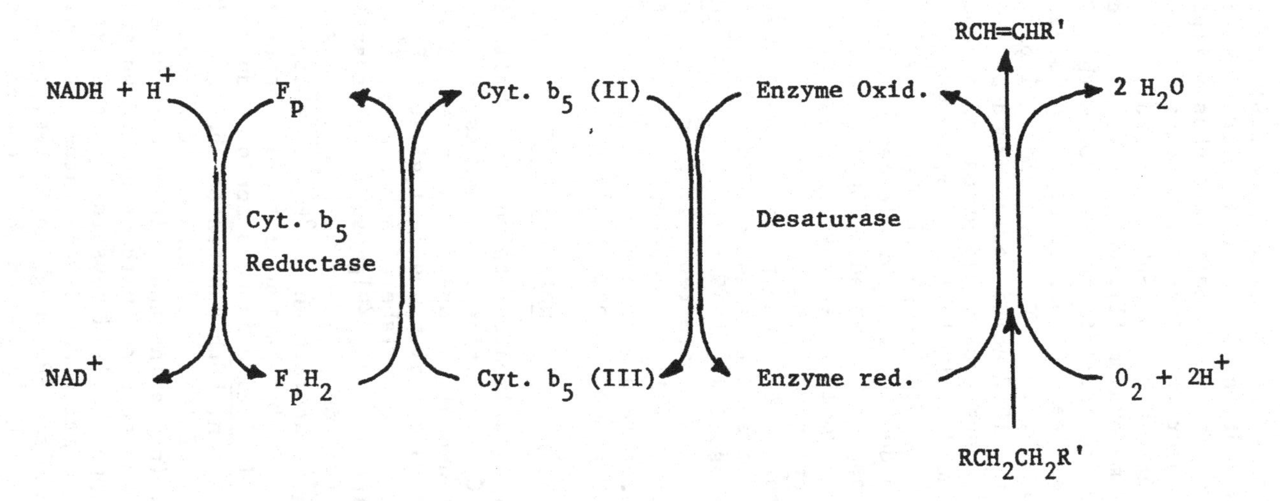

Fig. 2. Pathway of microsomal electron transport coupled to desaturation of acyl chains.

1975b). 1-[^{14}C]Stearoyl-2-acyl-PC or 1,2-di[^{14}C]elaidoyl-PC were not desaturated, indicating that the desaturase was probably specific for the conversion of oleate to linoleate.

The phospholipid desaturase system required molecular oxygen and a reduced pyridine nucleotide (NADH or NADPH) and was inhibited by cyanide but not by carbon monoxide (Pugh and Kates, 1975b). These results indicate that the phospholipid desaturase is coupled to a cytochrome b_5-linked microsomal electron transport system similar to that established for acyl-CoA desaturase systems (Strittmatter *et al.* 1974), as shown in Fig. 2.

The phospholipid desaturase of *C. lipolytica* was shown to act on oleoyl chains in both the C_1- and C_2-positions of lecithin, but the reaction was more rapid at the C_2-position. The reactions catalyzed by the desaturase were thus as shown in Fig. 3.

We have previously shown that the lipids of *C. lipolytica* contain lower proportions of oleic and higher proportions of linoleic acids when the organism is grown at 10°C than when this mesophilic organism is grown at 25°C (Kates and Baxter, 1962; Kates and Paradis, 1973). Microsomal membranes prepared from cells of *C. lipolytica* grown at 10° and 25°C differed similarly in the proportions of oleic and linoleic acids present. Membranes prepared from cells grown to mid-log phase at 25°C contained 50% linoleic and 30% oleic acids (molar ratio 18:2/18:1, 1.7), whereas those grown at 25°C contained 40% linoleic and 40% oleic acids (molar ratio 18:2/18:1, 1.0).

When phospholipid desaturase activity was measured in the two membrane fractions (Pugh and Kates, 1975b) the activity of the 10°C membrane (20-25 pmol/min/mg) was lower than that of the 25°C membrane (50-80 pmol/min/mg) at the standard assay temperature of 25°C. The lower phospholipid desaturase activity in the 10°C membrane was due neither to an unusual temperature sensitivity of the enzyme (see below) nor to an altered apparent K_m for the lecithin substrate (Table I).

Arrhenius plots of phospholipid desaturase activity for the two membrane preparations showed that the activity of the 10°C membrane was lower than that of the 25°C membrane at all temperatures studied over the range 10-37°C. The Arrhenius plots appeared to be linear with both membrane preparations and no clear-cut evidence of a change in slope or phase transition was observed. The apparent lack of any phase transitions was not due to the presence of Triton X-100 in the reaction mixture, since identical

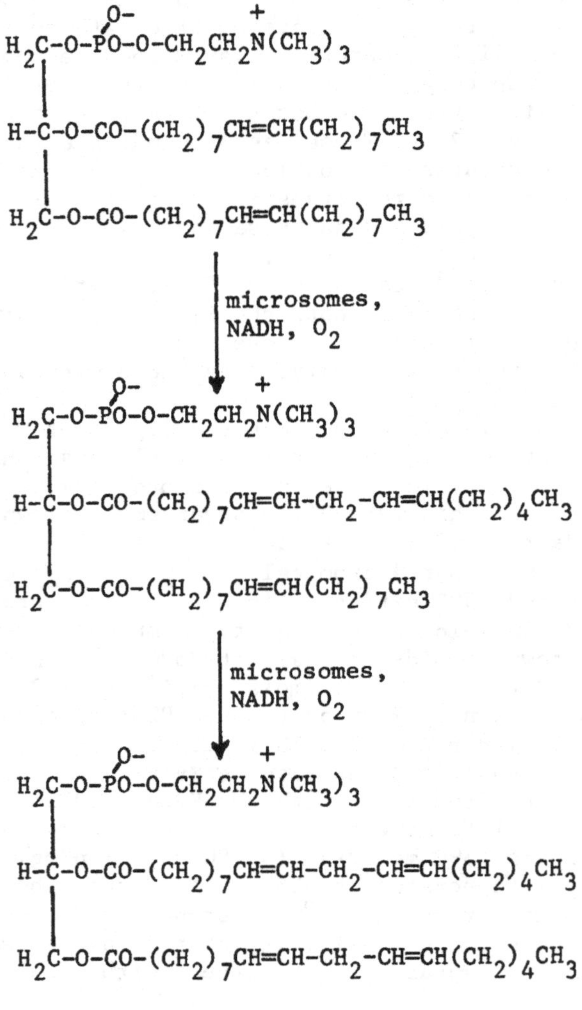

$$H_2C\text{-}O\text{-}\overset{O^-}{P}O\text{-}O\text{-}CH_2CH_2\overset{+}{N}(CH_3)_3$$
$$H\text{-}C\text{-}O\text{-}CO\text{-}(CH_2)_7CH{=}CH(CH_2)_7CH_3$$
$$H_2C\text{-}O\text{-}CO\text{-}(CH_2)_7CH{=}CH(CH_2)_7CH_3$$

microsomes, NADH, O_2

$$H_2C\text{-}O\text{-}\overset{O^-}{P}O\text{-}O\text{-}CH_2CH_2\overset{+}{N}(CH_3)_3$$
$$H\text{-}C\text{-}O\text{-}CO\text{-}(CH_2)_7CH{=}CH\text{-}CH_2\text{-}CH{=}CH(CH_2)_4CH_3$$
$$H_2C\text{-}O\text{-}CO\text{-}(CH_2)_7CH{=}CH(CH_2)_7CH_3$$

microsomes, NADH, O_2

$$H_2C\text{-}O\text{-}\overset{O^-}{P}O\text{-}O\text{-}CH_2CH_2\overset{+}{N}(CH_3)_3$$
$$H\text{-}C\text{-}O\text{-}CO\text{-}(CH_2)_7CH{=}CH\text{-}CH_2\text{-}CH{=}CH(CH_2)_4CH_3$$
$$H_2C\text{-}O\text{-}CO\text{-}(CH_2)_7CH{=}CH\text{-}CH_2\text{-}CH{=}CH(CH_2)_4CH_3$$

Fig. 3. Scheme for desaturation of dioleoyl lecithin by microsomal-bound yeast (Candida lipolytica) desaturase system.

TABLE I. Apparent Kinetic Constants of Microsomal-bound Lecithin Desaturase from C. lipolytica grown at 25°C and 10°C [a]

Growth Tempera-ture °C	18:2/18:1 mole ratio in micro-somes	K_m(app) M x 10^4	V pmol/min/mg	Activa-tion Energy kJ/mol
25°	1.0	2.5	1300	76
10°	1.7	2.5	710	50

[a] Data from Pugh and Kates (1975b). Activity was measured in reaction mixture (1ml) containing 10mM phosphate buffer (pH 7.2), 100mM sucrose, 10mM NADH, 0.1% Triton X-100, 1-2 mg microsomal protein and varying amounts of 1-acyl-2-[^{14}C]oleoyl-PC for 15 min at 25°C.

slopes were obtained in the absence of this detergent. Activation energies calculated from the slopes showed a lower value for the 10°C membrane preparation (Table I).

The effect of temperature on the oleoyl-CoA desaturase activity of the membrane preparations from 10°C and 25°C grown cells was also determined. In contrast to the results obtained with the phospholipid desaturase, the oleoyl-CoA desaturase was found to be more active in cells grown at 10°C than in those grown at 25°C. Arrhenius plots of oleoyl-CoA desaturase activity showed that the activity of the 10°C membrane was higher than that of the 25°C membrane at all temperatures tested from 10-37°C.

The results would suggest that the activity of the phospholipid desaturase is inversely related to the 18:2/18:1 mole ratio of the membrane preparation, or presumably, to the fluidity of the membrane. A similar conclusion has been reached by Thompson (1979) for the palmitoyl-CoA desaturase of Tetrahymena and by Riordan (1979) for the ATPase of rat liver plasma membrane. Thus the activity of the membrane-bound phospholipid desaturase system might be considered as self-regulating if one assumes that interaction and/or conformation of the three proteins were sub-

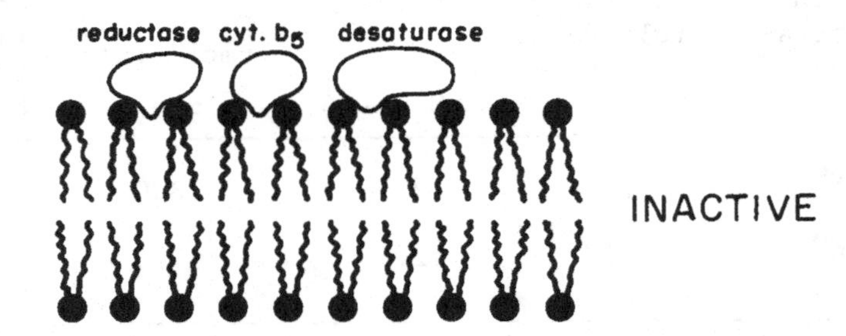

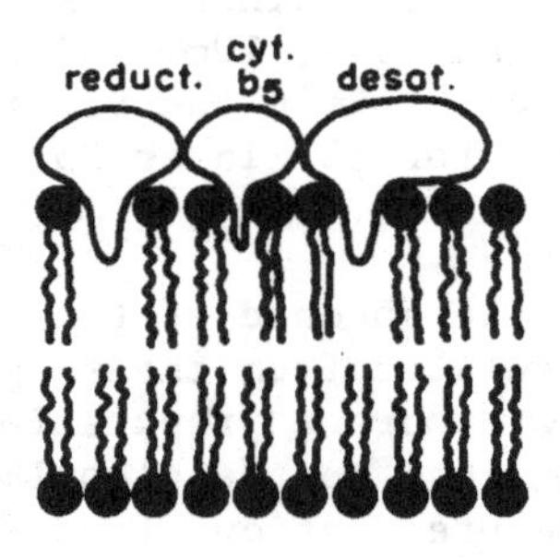

Fig. 4. Hypothetical model for self-regulation of phospholipid desaturase activity by membrane fluidity.

optimal in a membrane of high 18:2/18:1 ratio or high fluidity and optimal in a membrane of low 18:2/18:1 ratio or low fluidity, as illustrated in Fig. 4.

PHOSPHOLIPID DESATURASE OF RAT LIVER MICROSOMES

In animal systems, conversion of oleate to α-linoleate does not occur, so one would not expect to find a phospholipid desaturase capable of converting oleoyl-PC to

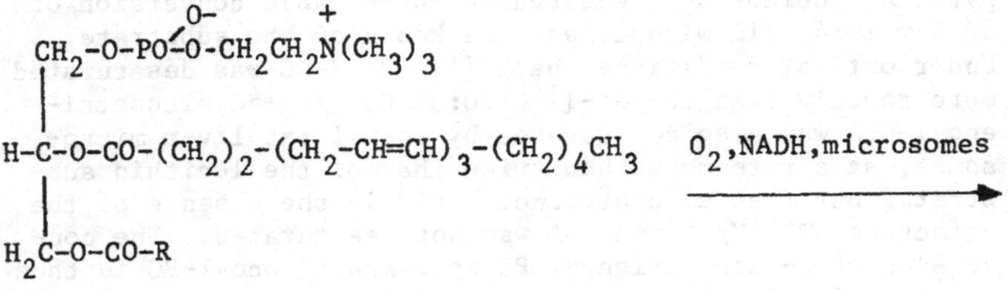

$$\begin{array}{l} \quad\quad\quad O^- \quad\quad + \\ CH_2\text{-}O\text{-}PO\text{-}O\text{-}CH_2CH_2N(CH_3)_3 \\ | \\ H\text{-}C\text{-}O\text{-}CO\text{-}(CH_2)_2\text{-}(CH_2\text{-}CH{=}CH)_3\text{-}(CH_2)_4CH_3 \\ | \\ H_2C\text{-}O\text{-}CO\text{-}R \end{array} \xrightarrow{O_2,\text{NADH, microsomes}}$$

1-Acyl-2-eicosatrienoyl-sn-glycero-3-phosphorylcholine
(20:3-PC)

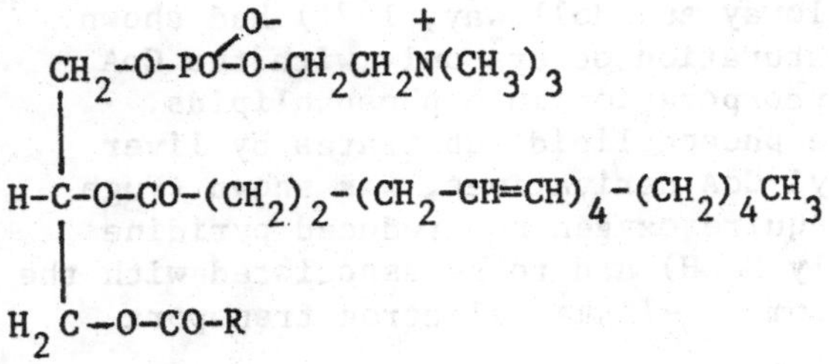

$$\begin{array}{l} \quad\quad\quad O^- \quad\quad + \\ CH_2\text{-}O\text{-}PO\text{-}O\text{-}CH_2CH_2N(CH_3)_3 \\ | \\ H\text{-}C\text{-}O\text{-}CO\text{-}(CH_2)_2\text{-}(CH_2\text{-}CH{=}CH)_4\text{-}(CH_2)_4CH_3 \\ | \\ H_2C\text{-}O\text{-}CO\text{-}R \end{array}$$

1-acyl-2-eicosatetraenoyl-sn-glycerophosphorylcholine
(20:4-PC)

Fig. 5. Scheme for desaturation of eicosatrienoyl lecithin by rat liver microsomes.

α-linoleoyl-PC. However, it was of interest to see whether the more highly unsaturated fatty acids found in animals, e.g. arachidonic acid, could arise by desaturation of phospholipid-linked fatty acids, as well as by desaturation of the acyl-CoA derivatives. We therefore undertook a study of the desaturation of biosynthetically prepared 1-acyl-2-[^{14}C]eicosatrienoyl-glycerophosphorylcholine (2-[^{14}C]20:3-PC), as well as the chemically synthesized 1,2-di-[^{14}C]-eicosatrienoyl-glycerophosphorylcholine (di-[^{14}C]20:3-PC) (Pugh and Kates, 1975a, 1977). It was found that incubation of either substrate with liver microsomes from rats fed a normal diet, in the presence of oxygen and reduced pyridine nucleotides, resulted in appreciable conversion of 20:3 → 20:4 acid without any breakdown of the substrate. Under optimal conditions the 2-[^{14}C]20:3-PC was desaturated more rapidly than the di-[^{14}C]20:3-PC. [1-^{14}C]eicosatrienoyl-CoA was also desaturated by normal rat liver microsomes, at a rate more than twice that of the lecithin substrate, but free eicosatrienoic acid in the absence of the cofactors ATP, Mg^{++} and CoA was not desaturated. The conversion of 2-eicosatrienoyl-PC to 2-arachidonoyl-PC is thus as shown in Fig. 5.

Several lecithin species other than eicosatrienoyl-PC were tested as substrates for the phospholipid desaturase from rat liver. Although dioleoyl lecithin was desaturated in this system (presumably to the cis-6,9-octadecadienoyl-PC, 1-[^{14}C]stearoyl-2-acyl- and 1-acyl-2[^{14}C]linoleoyl-PC did not serve as substrates. Previous studies with hen liver microsomes (Holloway and Holloway, 1974) had shown that stearic acid desaturation occurs only with the CoA ester without prior incorporation into phospholipids.

Desaturation of the phospholipid substrates by liver microsomes, as for acyl-CoA derivatives, was shown (Pugh and Kates, 1977) to require oxygen and reduced pyridine nucleotides (preferably NADH) and to be associated with the multicomponent cytochrome b_5-linked electron transport chain (see Fig. 2).

Desaturation of the lecithin substrate in the rat liver system, as in the yeast system proceeded without a lag period, was linear with time for about 15 min, and was proportional to microsomal protein concentration up to 2 mg/ml. The desaturation appeared to follow Michaelis-Menten kinetics and an apparent K_m for 20:3-PC was calculated to be about 3.6 x 10^{-4}M and with a V of 260 pmoles/min/mg (cf. corresponding values for 18:1-PC desaturase of *C. lipolytica* Table I).

The lecithin desaturase activity was stimulated by addition of detergents such as deoxycholate or Triton X-100. Under optimal conditions (0.1% deoxycholate or 0.2% Triton X-100), the presence of detergent increased desaturation of the lecithin substrate about 3- to 8-fold, respectively. By contrast, desaturation of eicosatrienoyl-CoA was only slightly stimulated by low concentrations of detergent and was partially inhibited by concentrations which were optimal for lecithin desaturation (Pugh and Kates, 1977).

The activity of the phospholipid and acyl-CoA desaturases can be altered by changes in the dietary regimen (Pugh and Kates, 1977). For example, starving animals and

TABLE II. Influence of Diet on Desaturase Activities of Rat Liver Microsomes [a]

Substrate	Desaturation Rate [a] pmol/min/mg Normal	Starved	Starved-Refed
[1-^{14}C]eicosa-trienoyl-CoA	39, 42	n.d. [b]	88, 120
1-Acyl-2-[^{14}C]-Eicosatrienoyl-GPC	13, 19	n.d. [b]	28, 42

[a] Results are given for two independent assays on 2 groups of five animals. Assays were performed as described elsewhere (Pugh and Kates, 1977); incubations were for 30 min at 37°.

[b] n.d., not detected.

re-feeding a fat-free diet, which has been shown to deplete the microsomal membrane of polyenoic fatty acids (Allman *et al*., 1965), results in increased synthesis of arachidonic acid, suggesting that the level of the Δ-5 desaturase activity is closely related to changes in properties of the lipid environment, as has been demonstrated for the stearoyl-CoA desaturase system (Holloway and Holloway, 1977; Lippiello *et al*., 1979).

To probe further the relationship of changes in the Δ-5 desaturase activities to alterations in the lipid environment of the enzyme, we have compared liver microsomes of rats maintained on different dietary regimens with respect to eicosatrienoyl-CoA desaturase, eicosatrienoyl-PC desaturase, fatty acid composition and membrane fluidity as measured by fluorescence polarization with diphenylhexatriene as probe (Shinitzky and Barenholz, 1974).

Influence of Diet on Desaturase Activities of Rat Liver Microsomes.

The dietary regimen had a profound effect on the desaturase activity of rat liver microsomes towards both [1-^{14}C] eicosatrienoyl-CoA and 1-acyl-2-[^{14}C]eicosatrienoyl-PC by rat liver microsomes (Table II). Desaturation of both substrates was increased 2-3 fold in rats starved and refed a fat-free diet compared with the normal controls, as observed previously (Pugh and Kates, 1977). In contrast, neither desaturase activity could be detected in livers of starved animals (Table II).

Fatty Acid Composition of Lipids from Liver Microsomes of Normal, Starved and Starved-Refed Rats.

Fatty acid composition of microsomal membranes prepared from livers of rats maintained on different dietary regimens is shown in Table III. The proportion of saturated fatty acids (palmitic and stearic acids) was about the same in the three membrane preparations. However, the membranes differ in proportions of unsaturated fatty acids. Membranes from rats starved and refed a fat-free diet have increased levels of oleic acid (18:1) and decreased amounts of long-chain polyenoic fatty acids (largely 20:4 and 22:6 acids) over the normal controls, as reported by others (Allman *et al.*, 1965; Peluffo *et al.*, 1976). On the other hand, membranes from starved rats had increased levels of polyunsaturated fatty acids as compared to normal rats.

Changes in fatty acid composition have been related to membrane fluidity by the double bond index: saturated fatty acid mole ratio, defined as total moles of double bonds/mol of saturated fatty acid (Farias *et al.*, 1975; Lippiello *et al.*, 1979). A decrease in this parameter would be expected to correlate with decreased fluidity (i.e. increased fluorescence polarization). The data in Table III demonstrate that the double bond index: saturated fatty acid mole ratio was essentially similar for the three membranes in spite of the quantitative differences in their fatty acid composition.

Fluorescence Polarization of Microsomal Membranes of Varying Fatty Acid Composition and their Total Lipids.

Fluorescence polarization of the lipid-soluble probe 1,6-diphenylhexatriene was measured in the three microsomal membranes of varying fatty acid composition described above. The fluorescence polarization values of the microsomal membranes from starved and starved-refed animals did not vary significantly from that of normal controls over the temperature range 10-37°C despite the quantitative change in fatty acid composition (Table IV). This is in

TABLE III. Fatty Acid Composition of Lipids from Liver Microsomes of Normal, Starved and Starved-Refed Rats [a]

Fatty Acid	% of Total Fatty Acids		
	Normal	Starved	Starved-Refed
16:0	20, 20	22, 21	24, 24
16:1	2, 1	n.d.	6, 6
18:0	20, 21	22, 23	14, 14
18:1	11, 9	5, 5	29, 28
18:2	18, 18	12, 15	6, 7
20:4	23, 22	26, 24	15, 15
20:5	1, 2	n.d.	1, n.d.
22:6	5, 8	13, 12	5, 6
Double bond index [b]	1.8, 1.9	2.1, 2.0	1.4, 1.4
Double bond index: mol saturated fatty acid	4.5, 4.6	4.8, 4.6	3.7, 3.7

[a] Values given are for independent analyses on two groups of five animals with probable error of ± 5%. Fatty acid methyl esters were prepared from total microsomal lipid and analyzed by gas-liquid chromatography.

[b] Number of double bonds per mol fatty acid (Δ/mol).

agreement with the rather constant values of the double bond index: saturated fatty acid ratio for the microsomal membranes which suggest that the saturated and unsaturated fatty acids are adjusted so as to maintain a fairly constant membrane fluidity despite quantitative shifts in fatty acid composition caused by dietary challenges. Plots of the fluorescence polarization vs. temperature for the three membranes appeared to be linear over the temperature range 10-37°C and no clear-cut changes in slope indicating phase-transitions were observed over the temperature range studied.

Fluorescence polarization was also investigated in the total lipids isolated from the microsomal membranes of normal, starved and starved-refed rats. Polarization values were very similar in the three membranes (Table IV)

TABLE IV. Changes in Fluorescence Polarization of Microsomal Membranes and Total Lipids from Normal, Starved, and Starved-Refed Rats as a Function of Temperature.

Temp.	Polarization [a]		
°C	Normal	Starved	Starved-Refed
		Membranes	
37	0.24 ± 0.02	0.24 ± 0.02	0.24 ± 0.02
21	0.26 ± 0.02	0.28 ± 0.02	0.28 ± 0.02
11.5	0.28 ± 0.02	0.29 ± 0.02	0.29 ± 0.02
		Total Lipids	
43	0.174	0.211	0.191
38.2	0.175 ± 0.002	0.191 ± 0.003	0.178 ± 0.002
24.5	0.191 ± 0.002	0.212 ± 0.002	0.206 ± 0.004
13	0.183 ± 0.001	0.202 ± 0.004	0.181 ± 0.002
7	0.161 ± 0.004	0.187 ± 0.002	0.171 ± 0.002

[a] Determined by fluorescence polarization with 1,6-diphenylhexatriene as probe.

and no changes in slope or phase transitions were observed over the temperature range 10-37°C. Fluorescence polarization values for the total lipids, however, were significantly lower than for the intact microsomal membranes (Table IV), indicating that the isolated lipids are in a more fluid state than those imbedded in the membrane and suggesting that microsomal protein may play a role in controlling membrane fluidity.

Lipid Involvement in Eicosatrienoyl-CoA Desaturation

To investigate the possibility that eicosatrienoyl-CoA was being desaturated after transacylation into lipids, we have studied the rate of desaturation of this substrate by liver microsomes from starved-refed rats. In these experiments the incubation mixtures were extracted with the Bligh-Dyer system to separate the methanol-water soluble acyl-CoA fraction from the chloroform-soluble lipids; the latter were subsequently separated by silicic acid column chromatography into neutral and phospholipid fractions. Distri-

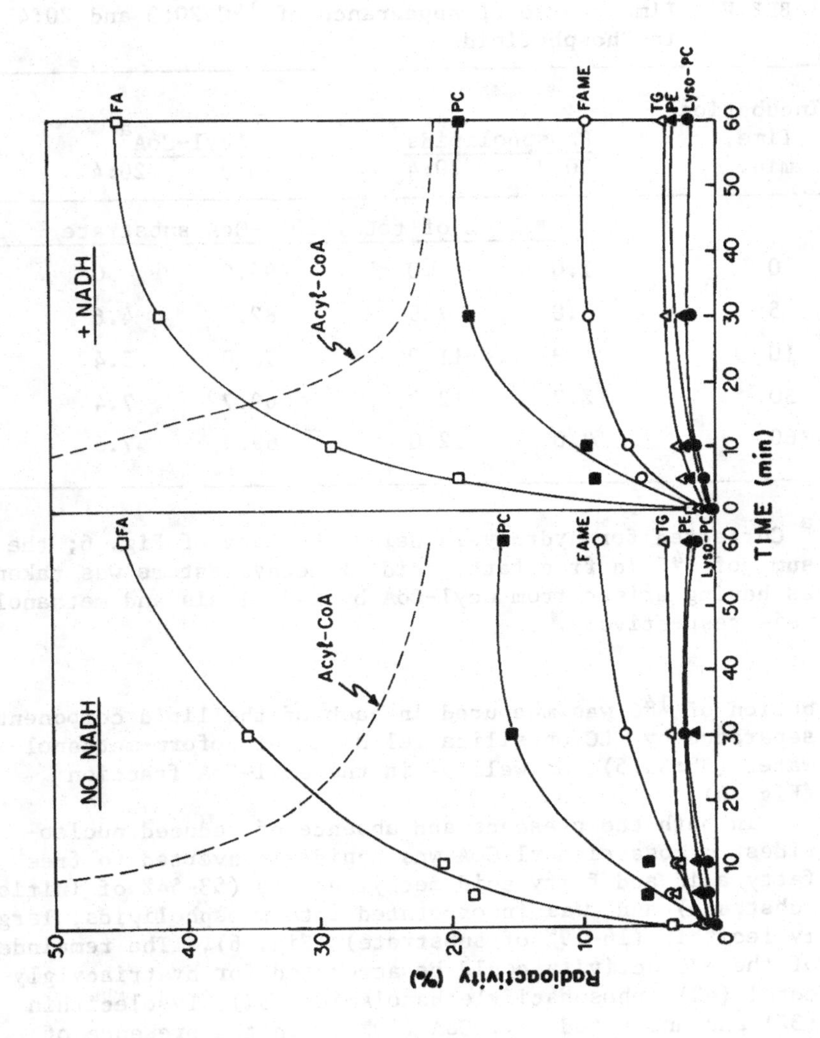

Fig. 6. Distribution of ^{14}C among acyl CoA and lipids during incubation of [1-^{14}C]eicosatrienoyl-CoA with microsomes of starved-refed rats. Assays were performed in the absence and presence of NADH; lipid components were separated by TLC (see text) and counted for ^{14}C. FA, free fatty acid; PC, phosphatidylcholine; FAME, fatty acid methyl ester; TG, triglyceride; PE, phosphatidylethanolamine; lyso-PC, lysolecithin.

TABLE V. Time course of Appearance of ^{14}C 20:3 and 20:4 in Phospholipids

Incubation Time, min.	Phospholipids		Acyl-CoA[a]	
	20:3	20:4	20:3	20:4
	% of total 20:3-CoA substrate			
0	1.0	0	96.0	0
5	3.8	7.9	82.0	4.6
10	3.9	11.0	76.0	5.4
30	12.3	12.0	69.7	7.4
60	12.0	12.0	69.0	7.6

[a] Corrected for hydrolysis using the data of Fig. 6; the sum of ^{14}C in free fatty acids + methyl esters was taken as having arisen from acyl-CoA by hydrolysis and methanolysis respectively.

bution of ^{14}C was measured in each of the lipid components separated by TLC on silica gel H in chloroform-methanol-water (65:35:5), as well as in the acyl-CoA fraction (Fig. 6).

In both the presence and absence of reduced nucleotides, eicosatrienoyl-CoA was rapidly converted to free fatty acid and fatty acid methyl esters (53-54% of initial substrate) and was incorporated into phospholipids, largely lecithin (16-19% of substrate) (Fig. 6). The remainder of the ^{14}C-activity could be accounted for by triacylglycerol (4%), phosphatidylethanolamine (3%), lysolecithin (3%) and unreacted acyl-CoA (20%). In the presence of NADH, desaturation of 20:3 → 20:4 acid occurred and analysis of the total phospholipid and total neutral lipid fractions revealed that the phospholipids were labeled with ^{14}C-arachidonate during the first 30 min. at a faster rate than was the acyl-CoA fraction (Table V). After 30 min. 50% of the [^{14}C]acyl groups in the phospholipids was associated with arachidonic acid, whereas only 10% of the [^{14}C] 20:3-CoA had been converted to the [^{14}C]20:4-CoA.

These findings indicate that the 20:3-CoA (apart from

being hydrolyzed by a thiolase) is rapidly transacylated to phospholipid in rat liver microsomes, and the 20:3-phospholipid is desaturated to 20:4-phospholipid at a greater rate than desaturation of 20:3-CoA. Similar results have been reported recently for the action of safflower seed microsomes on 18:1-CoA (Stymne and Appelqvist, 1978). The present results thus suggest that a considerable amount of the arachidonic acid present in liver microsomes is formed by the action of the phospholipid desaturase and that this enzyme system may play an important role in control of membrane fluidity.

ACKNOWLEDGMENT

This work was supported by a grant from the Ontario Heart Foundation. The authors are grateful to Dr. A.G. Szabo, National Research Council of Canada for help with the fluorescence polarization measurements.

REFERENCES

Allman, D.W., D.D. Hubbard and D.M. Gibson (1965). J. Lipid Res. 6, 63.

Baker, N. and F. Lynen (1971). Eur. J. Biochem. 19, 200.

Bloomfield, D.K. and K. Bloch (1960). J. Biol. Chem. 235, 337.

Farias, R.N., B. Bloj, R.D. Moreno, F. Sineriz and R.E. Trucco (1975). Biochim. Biophys. Acta 415, 231.

Gurr, M.I., M.P. Robinson and A.T. James (1969). Eur. J. Biochem. 9, 70.

Holloway, C.T. and P.W. Holloway (1974). Lipids, 9, 196.

Holloway, C.T. and P.W. Holloway (1977). Lipids, 12, 1025.

Kates, M. and R.M. Baxter (1962). Can. J. Biochem. Physiol. 40, 1213.

Kates, M. and M. Paradis (1973). Can. J. Biochem. 51, 184.

Lippiello, P.M., C.T. Holloway, S.A. Garfield and P.W. Holloway (1979). J. Biol. Chem. 254, 2004.

Peluffo, R.O., A.M. Nervi and R.R. Brenner (1976). Biochim. Biophys. Acta 441, 25.

Pugh, E.L. and M. Kates (1973). Biochim. Biophys. Acta 316, 305.

Pugh, E.L. and M. Kates (1975a). J. Lipid Res. 16, 392.

Pugh, E.L. and M. Kates (1975b). Biochim. Biophys. Acta 380, 442.

Pugh, E.L. and M. Kates (1977). J. Biol. Chem. 252, 68.

Pugh, E.L. and M. Kates (1979). Lipids 14, 159.

Riordan, J.R. (1979). This volume.

Stymne, S. and L.A. Appelqvist (1978). Eur. J. Biochem. 90, 223.

Shinitzky, M. and Y. Barenholz (1974). J. Biol. Chem. 249, 2652.

Strittmatter, P., L. Spatz, D. Corcoran, M.J. Rogers, B. Setlow and R. Redline (1974). Proc. Natl. Acad. Sci. USA 71, 4565.

Talamo, B., N. Chang and K. Bloch (1973). J. Biol. Chem. 248, 2738.

Thompson, Guy, Jr. (1979). This volume.

Vijay, I.K. and P.K. Stumpf (1971). J. Biol. Chem. 246, 2910.

Vijay, I.K. and P.K. Stumpf (1972). J. Biol. Chem. 247, 360.

Yuan, C. and K. Bloch (1961). J. Biol. Chem. 236, 1277.

MEMBRANE FLUIDITY AND THE ACTIVITY OF BOVINE BRAIN PHOSPHOLIPID EXCHANGE PROTEIN

George M. Helmkamp, Jr.

Department of Biochemistry
University of Kansas Medical Center
Kansas City, Kansas 66103 U.S.A.

INTRODUCTION

The major phospholipid exchange protein from bovine brain catalyzes the transfer of phosphatidylinositol (PtdIns) and, to a lesser extent, phosphatidylcholine (PtdCho) into and out of a variety of membrane surfaces, including rat liver mitochondria and microsomes, phospholipid bilayer liposomes, and phospholipid monolayers (Harvey et al., 1973; Helmkamp et al., 1974; Demel et al., 1977). The exchange protein is isolated from the cytosol fraction of cerebral cortex, can be purified to homogeneity, and has a molecular weight of 29,000. Previous experiments demonstrated that the rate of phospholipid transfer was sensitive to both the concentration and the composition of the participating membranes. Systematic changes in the concentration of microsomes (donor membrane) and sonicated liposomes (acceptor membrane) and the isolation of a stoichiometric phospholipid-protein complex yielded results which were consistent with a ping-pong kinetic mechanism for the protein-catalyzed, intermembrane movement of phospholipid molecules (Helmkamp et al., 1976). Transfers of PtdIns and PtdCho were essentially unaffected by the incorporation into egg PtdCho vesicles of phosphatidic acid, phosphatidylserine, or phosphatidylglycerol (5-20 mol%) but were strongly depressed by the incorporation of stearylamine (10-40 mol%). Marked stimulation of transfer activity was observed into vesicles containing phosphatidylethanolamine (2-40 mol%) or sphingomyelin (2-10 mol%); inhibition, however, occurred at higher levels of sphingomyelin (up to 40 mol%). Compared to the egg PtdCho vesicles, the magnitude of K_m tended to increase for vesicles which depressed phospholipid transfer and to decrease for those which stimulated; little change was observed in the values of V_{max} (Helmkamp, 1979). The one exception to these results was PtdIns. Increasing proportions of PtdIns in mixed PtdCho-PtdIns vesicles led to decreased transfer of phospholipids from microsomes to liposomes, and this was accompanied by decreasing K_m's for the interaction between phospholipid exchange protein and the single bilayer vesicle (Helmkamp et al., 1976). Thus, the exchange protein exhibits not only a preference toward PtdIns as an exchangeable substrate but a strong affinity toward membrane surfaces which contain PtdIns.

The current investigation describes variations in the fatty acid composition of single bilayer vesicles and the subsequent effects on phospholipid exchange. Measured are the rates of PtdIns transfer from microsomes to liposomes, values of K_m and V_{max} for different liposome populations, and the relative fluidity of the membranes.

RESULTS AND DISCUSSION

Effect of Dioleoyl and Dielaidoyl PtdCho in Acceptor Liposomes

Since the principal unsaturated fatty acid in egg PtdCho was oleic acid, it was appropriate to examine the acceptor efficiency of liposomes prepared from phospholipids containing monounsaturated fatty acyl residues. Two isomeric forms were considered: oleic (9-*cis*-octadecenoic acid) and elaidic (9-*trans*-octadecenoic acid). With the standard assay system of 1 μmol liposomal phospholipid, transfer activity was 15.2 nmol h^{-1} to dioleoyl PtdCho liposomes, 9.6 nmol h^{-1} to dielaidoyl PtdCho liposomes, and by way of comparison 15.4 nmol h^{-1} to egg PtdCho liposomes (Table I).

To more completely describe the interaction between phospholipid exchange protein and a particular membrane population, transfer activity was measured as a function of liposome phospholipid concentration. Representative results, using dioleoyl PtdCho, are shown in Fig. 1. The observed hyperbolic relationships were indicative of a saturable process, and transposition of the data to double reciprocal

TABLE I

ACTIVITY PARAMETERS OF PHOSPHATIDYLCHOLINE LIPOSOMES OF VARIOUS FATTY ACID COMPOSITIONS

Liposomal Phosphatidylcholine	T_c	Fluorescence Polarization[a]	Transfer Activity[b]	K_m	V_{max}
	°C		nmol h^{-1}	mM	nmol h^{-1}
Egg	-15 to -5	0.102	15.4	0.09	17.2
Dioleoyl	-22	0.100	15.2	0.11	20.0
Dielaidoyl	11	0.137	9.6	0.93	26.3
Dimyristoyl	21	0.155	1.6	n.d.[c]	n.d.[c]

[a]The fluorescence polarization of 1,6-diphenyl-1,3,5-hexatriene was determined at 37° at a probe/phospholipid ratio of 1:500.
[b]Transfer activity was measured at 37° in the presence of 1 µmol of liposomal phospholipid as the acceptor membrane.
[c]n.d., not determined.

plots produced linear curves from which the kinetic parameters K_m and V_{max} were calculated (Table I). V_{max} remained relatively constant for the various liposomes, in a range from 17.2 to 26.3 nmol h^{-1}. On the other hand, significant differences in K_m were found: 0.09 mM and 0.11 mM for egg PtdCho and dioleoyl PtdCho, respectively, and an increase to 0.93 mM for dielaidoyl PtdCho. It is clear that both *cis* and *trans* unsaturated phospholipids function in the intermembrane movement of phospholipid molecules. Yet the geometry of the *cis* species, present in both dioleoyl PtdCho and egg PtdCho, permits a more rapid incorporation of PtdIns.

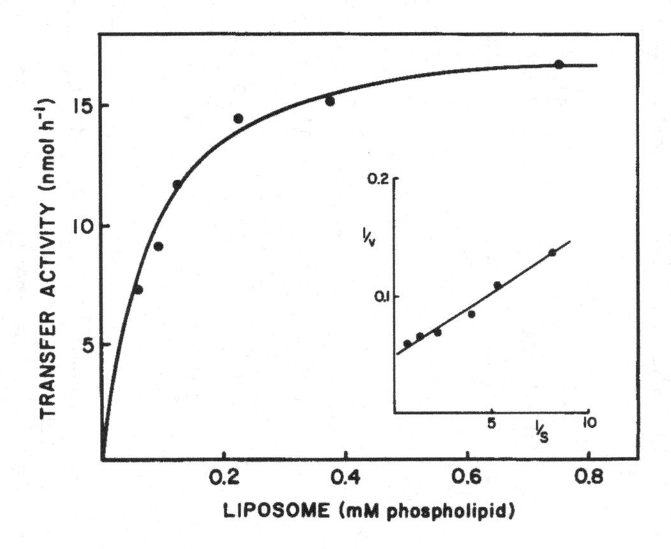

Figure 1. Phospholipid transfer activity in the presence of different liposome concentrations. Transfer activity was measured at 37° for 30 min in the following system: 1 mg rat liver microsomes labelled specifically with 40 nmol phosphatidyl [2-^{3}H] inositol, the indicated quantity of sonicated phospholipid liposomes containing 98 mol% dioleoyl PtdCho, 2 mol% egg phosphatidic acid, and 0.3 mol% cholesteryl [1-^{14}C] oleate, and 1.5 µg bovine brain phospholipid exchange protein in a total volume of 2.5 ml 10 mM HEPES, 50 mM NaCl, 1 mM Na_2EDTA (pH 7.4). Control incubations were separated by acidification to pH 5.0 and centrifugation at 10,000*g* for 20 min. Activity represents the quantity of PtdIns transferred from microsomes to liposomes. The inset depicts the data in a double reciprocal plot.

Effect of Dimyristoyl PtdCho in Acceptor Liposomes

Preliminary experiments indicated that liposomes containing 98 mol% dimyristoyl PtdCho were extremely poor acceptors of PtdIns transferred from rat liver microsomes (Table I). Therefore, a series of liposomes containing varying proportions of egg PtdCho and dimyristoyl PtdCho was prepared and evaluated in the microsome-liposome assay system. As seen in Fig. 2, the highest transfer activity occurred when only egg PtdCho was present in the sonicated liposomes. Upon substituting the saturated PtdCho, the activity gradually decreased, such that dimyristoyl PtdCho liposomes supported less than one-tenth the activity of egg PtdCho liposomes. Thus, liposomes composed of this saturated PtdCho species do not participate in the protein-catalyzed exchange of phospholipids, even at a temperature well above the gel to liquid crystalline phase transition temperature of 21°C. Using nuclear magnetic resonance spetroscopy, de Kruijff _et al_. (1976) reported that the structures of single bilayer vesicles prepared from egg, dioleoyl, dielaidoyl, and dimyristoyl PtdCho were similar in terms of membrane thickness, vesicle outer radius, and the molar ratio of lipid distributed between the outer and inner surfaces of the bilayer. The observed differences in the activity of these vesicles cannot be attributed to the size, surface area, or surface chemistry of the membrane particles.

Transfer Activity and Liposomal Membrane Fluidity

The ability of a liposome to participate in intermembrane phospholipid transfer has been, thus far, characterized by the rate at which it accepted PtdIns from rat liver microsomes and by the magnitude of the Michaelis constant for the interaction between the liposomes and phospholipid exchange protein. Because of the nature of the fatty acyl residues of these phospholipids, it was decided to explore the possibility of differences in the liquid crystalline state of these molecules in bilayer membranes. The degree of fluorescence polarization of 1,6-diphenyl-1,3,5-hexatriene (DPH), which partitions to the hydrophobic interior of the bilayer, permits an approximation of membrane fluidity (Shinitzky and Barenholz, 1978). Increased polarization is associated with decreased rotational mobility and, indirectly, with decreased membrane fluidity. As summarized in Table I, DPH in liposomes prepared from egg PtdCho and dioleoyl PtdCho gave polarizations of 0.102 and 0.100, respectively. Polarization increased to 0.137 for

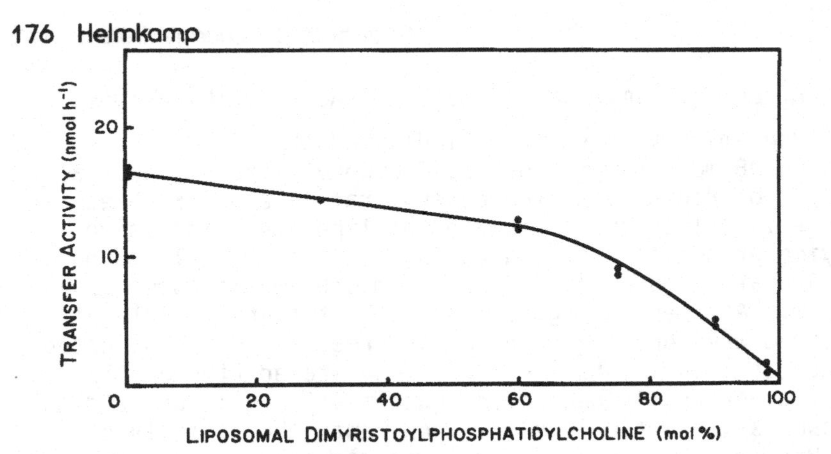

Figure 2. Effect of dimyristoyl PtdCho on phospholipid transfer activity. Measurements of transfer activity were carried out, as described in Fig. 1, using 1 μmol egg PtdCho liposomes which contained the indicated molar proportion of dimyristoyl PtdCho.

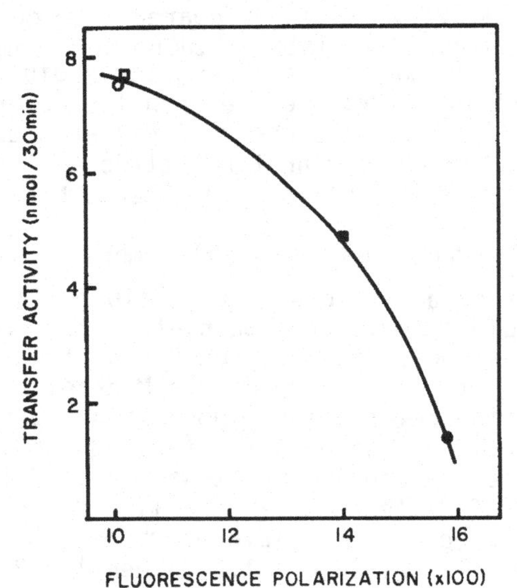

Figure 3. Relationship between phospholipid transfer activity and membrane fluidity. The following PtdCho liposomes were investigated: ○ , dioleoyl; □ , egg; ■ , dielaidoyl; and ● , dimyristoyl. Fluorescence polarization of DPH was determined at 37° at a DPH/phospholipid ratio of 1:500.

dielaidoyl PtdCho and to 0.155 for dimyristoyl PtdCho. Clearly, DPH encounters a spectrum of physical environments in these various lipid bilayers and exhibits a more hindered rotational behavior in those liposomes prepared from phospholipids containing saturated or *trans* unsaturated fatty acids. There was a reasonable correlation between transfer activity with a specific liposome and the fluorescence polarization of DPH in that liposome (Fig. 3). That is to say, the more fluid a bilayer interior the more rapid was the transfer of PtdIns into that bilayer. Maximum and minimum rates of PtdIns transfer occurred over a range of DPH fluorescence polarization that was much narrower than the range accompanying a typical gel to liquid crystalline phase transition (Andrich and Vanderkooi, 1976). This fact suggests that whatever physical event within the bilayer is responsible for altering phospholipid exchange activity must involve changes in fatty acyl chain mobility and organization different from those described for a thermotropic phase transition (Luzzati and Tardieu, 1974).

To examine the relationship between membrane fluidity and exchange activity in terms of protein-lipid interactions, several discrete events may be considered: (a) penetration of the phospholipid exchange protein into and subsequent explusion from the lipid bilayer; (b) insertion of a PtdIns molecule into and extraction of a PtdCho molecule from the bilayer; or (c) a combination of the above. These events share a common determinant in that phospholipid bilayer presents a formidable barrier to molecular movements into, out of, and through the membrane. It follows that such movements may be facilitated by a more fluid environment or a more loosely packed array of lipid molecules. The present study has demonstrated significant differences in fluidity between saturated and unsaturated phospholipid vesicles, as well as between *cis* and *trans* unsaturated vesicles. Surface pressure-molecular area measurements have likewise pointed toward a more expanded monolayer as the PtdCho fatty acid composition varied from stearate to elaidate to oleate (Phillips and Chapman, 1968).

In conclusion, the present series of *cis* unsaturated, *trans* unsaturated, and saturated phospholipid vesicles may be characterized by decreasing activity as an acceptor of microsomal PtdIns, increasing magnitude of K_m in a phospholipid exchange system, decreasing membrane hydrocarbon fluidity, and decreasing surface area per lipid molecule. (This work has been supported by grant GM24035 from the US National Institutes of Health.)

REFERENCES

Andrich, M.P., and Vanderkooi, J.M. (1976). Biochemistry 15, 1257-1261.

de Kruijff, B., Cullis, P.R., and Radda, G.K. (1976). Biochim. Biophys. Acta, 436, 729-740.

Demel, R.A., Kalsbeek, R., Wirtz, K.W.A., and van Deenen, L.L.M. (1977). Biochim. Biophys. Acta 466, 10-22.

Harvey, M.S., Wirtz, K.W.A., Kamp, H.H., Zegers, B.J.M., and van Deenen, L.L.M. (1973). Biochim. Biophys. Acta 323, 234-239.

Helmkamp, G.M., Jr., (1980). Biochim. Biophys. Acta, in press.

Helmkamp, G.M., Jr., Harvey, M.S., Wirtz, K.W.A., and van Deenen, L.L.M. (1974). J. Biol. Chem. 249, 6382-6389.

Helmkamp, G.M., Jr., Wirtz, K.W.A., and van Deenen, L.L.M. (1976). Arch. Biochem. Biophys. 174, 592-602.

Luzzati, V., and Tardieu, A. (1974). Ann. Rev. Phys. Chem. 25, 79-94.

Phillips, M.C., and Chapman, D. (1968). Biochim. Biophys. Acta 163, 301-313.

Shinitzky, M., and Barenholz, Y. (1978). Biochim. Biophys. Acta 515, 367-394.

ABSTRACT

The interaction of bovine brain phospholipid exchange protein with membranes has been investigated as a function of membrane phospholipid fatty acid composition. Single bilayer liposomes were prepared by sonication, centrifugation, and molecular sieve chromatography and were used as acceptor membranes in the exchange protein-catalyzed transfer of phosphatidylinositol from rat liver microsome donor membranes. For the series egg, dioleoyl-, dielaidoyl-, and dimyristoylphosphatidylcholine, initial rates of phosphatidylinositol transfer were highest with the two *cis*-unsaturated species and lowest with the saturated species, the *trans*-unsaturated species being intermediate. A progressive decrease in transfer rate was noted with liposomes containing a mixture of egg and dimyristoylphosphatidylcholine as the molar proportion of the latter phospholipid increased. Michaelis constants for the interaction between protein and different liposome preparations decreased in the order: saturated > *trans*-unsaturated > *cis*-unsaturated. Maximum velocities were independent of fatty acid composition. The fluorescence polarization of diphenylhexatriene in liposome preparations also decreased in the same order, under conditions well above the thermotropic gel to liquid crystalline phase transition of all phospholipids studied. These results suggest that the fatty acid composition, the degree of unsaturation, and in particular, the hydrocarbon fluidity of the membrane are important determinants in the activity of bovine brain phospholipid exchange protein.

The regulation of bacterial membrane fluidity by modification of phospholipid fatty acyl chain length

N.J. Russell and S.P. Sandercock

Department of Biochemistry, University College,

P.O. Box 78, Cardiff, CF1 1XL, U.K.

INTRODUCTION

Microorganisms are particularly useful for studying the control of membrane fluidity for two main reasons. Firstly, they respond readily to changes in their environment, and, secondly, it is relatively easy to isolate mutants, which may provide a great deal of information that it would be very difficult otherwise to obtain. Bacteria also have the advantage that usually they do not store fat, and with few exceptions do not synthesise polyunsaturated fatty acids. Consequently, growth temperature-dependent changes in fatty acid composition can be related more easily to membrane lipid function. The most commonly observed growth temperature-dependent alteration in lipid acyl composition is that of the degree of fatty acyl unsaturation. In addition, fatty acyl chain length may change, either independently or together with changes in unsaturation. In gram positive bacteria, which contain iso- and anteiso-branched chain fatty acids, as well as n-chain, odd-numbered fatty acids, the ratio of the branched chain isomers may change with growth temperature (e.g. in Bacilli, Kaneda, 1977). This article concentrates on fatty acyl chain length changes and discusses possible mechanisms for such changes including their control by temperature.

BACKGROUND

The gram negative psychrophilic bacterium *Micrococcus cryophilus* provides an example of a bacterium in which there is a particularly marked temperature-dependent fatty acyl chain length change. This bacterium contains essentially palmitoleic and oleic acids only in its membrane phospholipids, and the ratio of these fatty acids is modified in response to changes in growth temperature; a lowering of the growth temperature from 20° to 0°C results in a decrease of the C18/C16 ratio from 3.7 to 0.8, but there is no change in the percentage of phospholipid unsaturated fatty acids, which remains at 95-97% throughout the growth temperature range (Russell, 1971; 1978a). The unsaturated fatty acids are produced by the action of a Δ9 desaturase (Russell, 1977; 1978b; Foot & Russell, 1978), whose activity is modified by temperature so that the desaturation of palmitate is energetically preferable to that of stearate at lower temperatures (Foot, M., Jeffcoat, R. & Russell, N.J., unpublished results). However, this difference may be an artifact of the assay procedure, and in any case is much too small to account for the large change in C18/C16 ratio that occurs.

What are the possible mechanisms to account for the control of acyl chain length by growth temperature? Fatty acid elongation has been well studied in plants (e.g. the palmitoyl-ACP elongase, Jaworski *et al*., 1974) and animals (e.g. mammary gland fatty acid synthetase, Carey, 1977; Libertini & Smith, 1978), but there are few studies using microorganisms, and little is known about the control of acyl chain length in them. *Acholeplasma laidlawii* B is capable of elongating C6-C15 fatty acids, which may well be a property of fatty acid synthetase (Saito *et al*., 1978). Bloch and coworkers have characterised the multienzyme complex (Type I) fatty acid synthetase of *Mycobacterium smegmatis*, which differs from other Type I fatty acid synthetases in that it produces a bimodal pattern of fatty acyl chain lengths, with peaks at C16/C18 and C24/C26. The proportion of these fatty acids can be controlled *in vitro* by varying the acetyl CoA/malonyl CoA ratio or by adding compounds such as mycobacterial polysaccharides that sequester end products,or *in vivo* by temperature, but the temperature-sensitive enzyme has not yet been identified (Bloch, 1977). Odriozola & Bloch (1977) have also proposed that membrane phospholipids might have a role in regulating acyl chain length *in vivo*. *M. smegmatis* also contains a second fatty acid synthetase (ACP-dependent, Type II), which forms very long chain fatty acids (Odriozola *et al.*, 1977). Bloch has pointed out that it is unusual for a fatty acid synthetase to produce such a wide range of acyl chain lengths. In this context it is worth mentioning the mammary gland fatty acid synthetase in which the large fluctuations in acyl chain length are regulated by a thioesterase (Libertini & Smith, 1978). It is more usual, however, when a wide range of fatty acyl chain lengths are required for there to be a separate system for elongating the products (usually C16 + C18) of fatty acid synthetase such as the microsomal (Guchhait *et al.*, 1966; Ayala *et al.*, 1973) and mitochondrial (Harlan & Wakil, 1963; Bond & Pynadath, 1976) elongation systems in mammals. Bearing in mind the narrow acyl chain length range (2C only) of the membrane phospholipids in *Micrococcus cryophilus*, it might be argued that in this microorganism it is more likely that fatty acid synthetase produced a mixture of C16 + C18 fatty acids; their relative amounts might be controlled either directly by temperature affecting synthetase activity (e.g. by the relative activities and specificities of the β-keto acyl synthetase and acyl thioester hydrolase - cf Sumper *et al.*, 1969), or indirectly by a protein whose induction was regulated by temperature (cf thioesterase of mammary gland). The presence of two distinct fatty acid synthetases, such as occur in *Mycobacterium smegmatis*, is considered unlikely due to the narrow chain length range of the fatty acids in *Micrococcus cryophilus*. However, it is noteworthy that in plants two enzymes exist for the production of stearate; fatty acid synthetase produces palmitate, which is elongated to stearate by a specific palmitoyl-ACP elongase (Jaworski *et al.*, 1974). Brain microsomes also probably contain a specific elongase that synthesises stearoyl-CoA using palmitoyl-CoA as substrate (Pollet *et al.*, 1973; Goldberg *et al.*, 1973).

In order to determine the enzymatic basis of the temperature-dependent acyl chain length change in *M. cryophilus* we have investigated the effects of temperature on the synthesis of fatty acids from radioactive precursors, in particular the elongation of exogenous [1-^{14}C]-fatty acids. A primary aim of this work was to establish whether or not a mechanism for elongating fatty acids, distinct from fatty acid synthetase, exists. The results in this paper provide

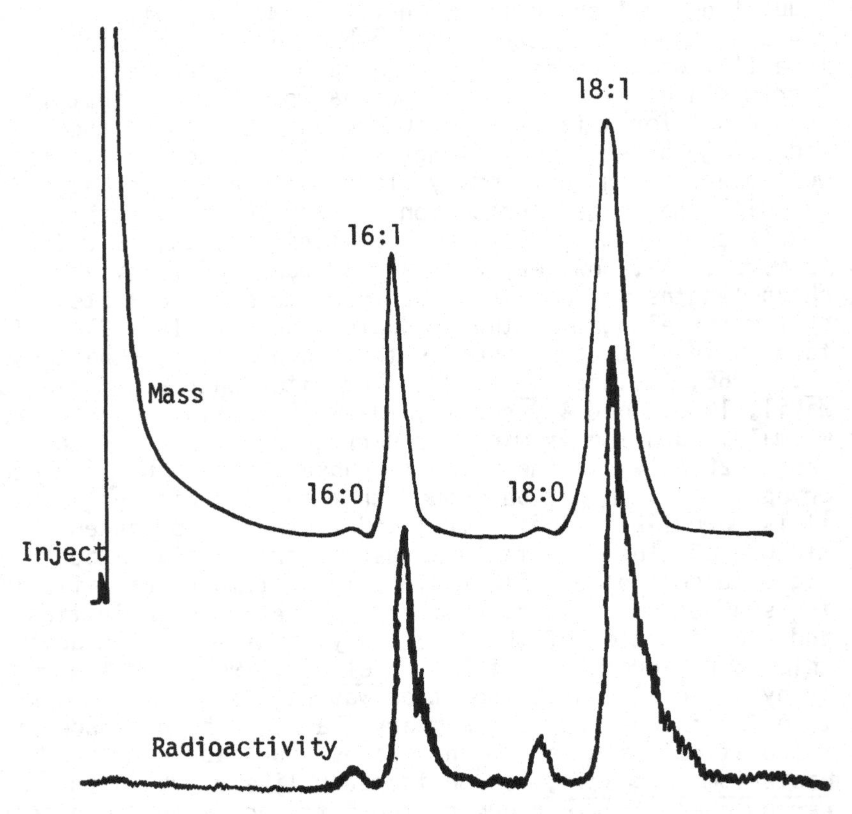

Fig. 1. Radio-gas liquid chromatogram of phospholipid acyl chains isolated from Micrococcus cryophilus incubated at 20°C for 2h with Na[U-^{14}C]acetate.

evidence for such an elongation system.

ELONGATION STUDIES USING MICROCOCCUS CRYOPHILUS

The strict control of phospholipid acyl chain length in M. cryophilus is readily seen when cultures are incubated with sodium [U-^{14}C] acetate; radioactivity is recovered in C16 and C18 acyl chains only (Fig. 1) and even on over-loaded chromatograms radioactivity is not detected in longer or shorter acyl chain lengths. The

Table 1. Distribution of radioactivity in phospholipid acyl chains in Micrococcus cryophilus incubated at 20°C for 2h with sodium [U-^{14}C] acetate or [1-^{14}C] fatty acids.

Radioactive precursor	Percentage distribution of radioactivity in acyl chains					
	14:0	14:1	16:0	16:1	18:0	18:1
Na[U-^{14}C]acetate	n.d.*	n.d.	1.9	13.0	3.7	81.4
[1-^{14}C] 12:0	n.d.	n.d.	0.7	10.8	2.4	86.0
[1-^{14}C] 14:0	3.3	5.4	1.5	11.3	1.8	76.7
[1-^{14}C] 16:0	n.d.	n.d.	3.4	36.2	3.8	56.6
[1-^{14}C] 18:0	n.d.	n.d.	2.3	7.9	2.7	87.1

* Not detected

same pattern of labelling is observed when [1-^{14}C] lauric acid is used as the precursor (Table 1), although the incorporation of radioactivity is low. In contrast, the incorporation of [1-^{14}C]myristic acid is greater and the pattern of radioactive labelling significantly different to that of laurate incubations, since small amounts of radioactivity are detected in C14 acyl chains (Table 1). Most radioactivity, however, is present in C16 and C18 acyl chains, and chemical α-oxidation studies show that >95% radioactivity is present in C14-C18 oxidation fragments. This is evidence for direct elongation, but does not distinguish between fatty acid synthetase activity and a putative elongase. Since M. cryophilus possesses a system for the strict control of fatty acyl chain length, it might be expected that levels of exogenous myristate incoporated into phospholipids would be kept at a low level by elongation so as to maintain correct membrane fluidity. Exogenous [1-^{14}C]palmitic and stearic acids are incorporated into phospholipids to higher levels than lauric or myristic acids, which may reflect the specificity of the (trans)acylating enzymes. Palmitate is elongated to C18 acids (Table 1), but radioactivity is not detected in acyl chains other than C16 and C18. The latter is true also for stearate, a proportion of which (10-32%) is converted to C16 acids ('retroconversion') (Table 1). Chemical α-oxidation studies of fatty acyl chains synthesised during palmitate or stearate incubations show that >95% radioactivity is present in C16-C18 oxidation fragments. This result is readily explained for palmitate by postulating direct elongation, but is more difficult to explain for stearate. Presumably [1-^{14}C] acetyl-CoA produced by (partial) β-oxidation of [1-^{14}C]stearate normally would enter the acetyl-CoA 'pool' with a consequent dilution of the radioactivity. It is unlikely, therefore, that the results of the chemical α-oxidation experiments would be obtained if there was β-oxidation of stearate and resynthesis to palmitate. Instead it is proposed that the [1-^{14}C]acetyl-CoA is removed by an 'elongase' enzyme, and that the radioactive acetyl-CoA remains bound either within the membrane or directly to the enzyme: the elongase interconverts C16 and C18 acyl chains via C14 and C16 intermediates. In this way radioactivity is not diluted by the cellular pool of acetyl-CoA. Such a mechanism explains, also, the efficient elongation of exogenous [1-^{14}C]myristate as this would be a substrate, presumably as an activated thioester, for the elongase.

These experiments show that a strict system for the control of acyl chain length exists in *M. cryophilus*, and that it is capable of elongating directly exogenous fatty acids. However, the results do not distinguish whether this is a property of fatty acid synthetase or whether a separate elongation system exists. In an attempt to clarify this point the effect of sodium arsenite on fatty acid elongation was studied, since it is known to be an inhibitor of palmitate elongation in plants (Harwood & Stumpf, 1971). The effect of arsenite is to reduce the amount of radioactivity in C18 acyl chains, whether acetate or fatty acids are used as the radioactive precursor (Table 2). The C18/C16 ratios of radioactivity obtained from bacteria incubated with acetate, laurate or myristate are very similar for a particular growth/incubation temperature. The ratio obtained from palmitate or stearate incubations appeared to reflect the nature of the precursor in that higher values are obtained with stearate than with palmitate. In all cases the C18/C16 ratios of radioactivity obtained in 20°C

Table 2. Effect of 10mM sodium arsenite on the ratio of radioactivity in C18/C16 phospholipid acyl chains in *Micrococcus cryophilus* incubated for 2h with sodium [U-^{14}C]acetate or [1-^{14}C] fatty acids.

Radioactive precursor	Bacterial growth and incubation temperature			
	0°C		20°C	
	Control	+$NaAsO_2$	Control	+$NaASO_2$
Na[U-^{14}C]acetate	0.9	0.6	5.7	0.6
[1-^{14}C] 12:0	1.0	0.5	7.7	0.6
[1-^{14}C] 14:0	1.0	0.4	6.1	0.4
[1-^{14}C] 16:0	0.6	0.1	1.5	0.1
[1-^{14}C] 18:0	2.1	5.5	8.8	9.7

incubations are higher than those obtained for 0°C incubations. With one exception (0°C, stearate incubation) arsenite reduces the C18/C16 ratio of radioactivity. The effect of arsenite on acetate, laurate and myristate incubations is to virtually abolish the difference between 0°C and 20°C ratios (Table 2). Despite the similar C18/C16 ratios of radioactivity for laurate and myristate incubations there is a major difference in the effect of arsenite on the two precursors; arsenite causes the accumulation of C14 acyl chains when myristate is the precursor (Table 3), but there is no accumulation of C12 or C14 acyl chains when laurate is the precursor. It is assumed that laurate incorporation is via β-oxidation and resynthesis, which accounts for the low incorporation of radioactivity from laurate, compared with other fatty acids. The accumulation of C14 acyl chains in myristate incubations in the presence of arsenite (Table 3) supports the view that myristate is incorporated directly into phospholipids. Arsenite causes a sharp decline in the incorporation into phospholipid acyl chains of radioactivity

Table 3. Effect of 10mM sodium arsenite on the distribution of radioactivity in phospholipid acyl chains in *Micrococcus cryophilus* incubated at 20°C with {1-^{14}C}myristic acid.

Acyl constituent	Percentage distribution of radioactivity in acyl chains	
	Control	+$NaAsO_2$
14:0	3.3	43.5
14:1	5.4	23.7
16:0	1.5	12.1
16:1	11.3	12.6
18:0	1.8	2.8
18:1	76.7	6.4

from acetate, which is interpreted as an inhibition of fatty acid synthetase. We believe, however, that the effect of arsenite on the C18/C16 ratio of radioactivity is not merely a result of fatty acid synthetase inhibition, since this would not explain the difference in the effects of arsenite on laurate and myristate incubations. Arsenite is an inhibitor of lipoic acid - containing enzymes such as pyruvate dehydrogenase and α-ketoglutarate dehydrogenase. Thus the inhibition of chain elongation could be a consequence of reduced pyridine nucleotide or ATP starvation. This is considered unlikely since fatty acid synthetase has a greater demand for reducing equivalents than does elongation, but the point was checked by carrying out control experiments using sodium malonate to inhibit succinate dehydrogenase, and therefore Kreb's cycle activity, and potassium cyanide to inhibit respiratory chain activity. Under conditions when 97% of the succinate dehydrogenase activity is inhibited by malonate the elongation of palmitate is inhibited by less than 40%.

CONCLUSIONS

There probably exists in M. cryophilus a fatty acid elongation system, distinct from fatty acid synthetase, whose function is to regulate the proportion of C16 and C18 acyl chains. Both fatty acid synthesis and elongation are membrane-bound. One means of confirming the above results will be to isolate the two systems and for example determine the chain length of the products of fatty acid synthetase. As a working hypothesis we propose that elongation and retroconversion are performed by a membrane-bound elongation system. The phospholipid acyl C18/C16 ratio is regulated by the balance of elongation and retroconversion, which are controlled by membrane fluidity, and which in turn is affected by changes in growth temperature. Retroconversion of stearate to palmitate at 0°C is greater than at 20°C (32% compared with 10%), which is consistent with such a hypothesis, since the C18/C16 ratio is less at the lower growth temperature. In addition, the fact that arsenite abolishes the sevenfold difference between the C18/C16 ratio of radioactivity obtained for 20°C and 0°C acetate incubations suggests that the control of acyl chain length by temperature is mediated at the level of fatty acid elongation rather than fatty acid synthesis, in which case the C18/C16 ratio of radioactivity obtained in the presence of arsenite (0.5-1.0) may represent the fatty acid synthetase product ratio. Such a mechanism for elongation would be capable of modifying fatty acid composition in the absence of net phospholipid synthesis, and possibly growth; it could mediate, therefore, rapid compositional changes and be responsible for membrane lipid turnover.

REFERENCES

Ayala, S., G. Gaspar, R.R. Brenner, R.O. Peluffo and W. Kunau (1973). J. Lipid Res. 14, 296

Bloch, K. (1977) Adv. in Enzymol. 45, 1

Bond, L.W. and T.I. Pynadath (1976). Biochim. et Biophys. Acta 450, 8

Carey, E.M. (1977). Biochim. et Biophys. Acta 486, 91

Foot, M. and N.J. Russell (1978). Soc. for Gen. Microbiol. Quart. 6, 21

Goldberg, I., I. Shechter and K. Bloch (1973). Science 182, 497

Guchhait, R.B., G.R. Putz and J.W. Porter (1966). Arch. Biochem. Biophys. 117, 541

Harlan, W.R. Jr. and S.J. Wakil (1963). J. Biol. Chem. 238, 3216

Harwood, J.L. and P.K. Stumpf (1971). Arch. Biochem. Biophys. 142, 281

Jaworski, J.G., E.E. Goldschmidt and P.K. Stumpf (1974). Arch. Biochem. Biophys. 163, 769

Kaneda, T. (1977). Bacteriol. Rev. 41, 391

Libertini, L.J. and S. Smith (1978). J. Biol. Chem., 253, 1393

Odriozola, J.M. and K. Bloch (1977). Biochim. et Biophys. Acta 488, 198

Odriozola, J.M., J.A. Ramos and K. Bloch (1977). Biochim. et Biophys. Acta 488, 207

Pollet, S., J-M. Bourre, G. Chaix, O. Daudu and N. Bauman (1973). Biochimie 55, 333

Russell, N.J. (1971). Biochim. et Biophys. Acta 231, 254

Russell, N.J. (1977). Biochem. Soc. Trans. 5, 1492

Russell, N.J. (1978a). FEMS Microbiol. Lett. 4, 335

Russell, N.J. (1978b). Biochim. et Biophys. Acta 531, 179

Saito, Y., J.R. Silvius and R.N. McElhaney (1978). J. Bacteriol. 133, 66

Sumper, M., D. Oesterhelt, C. Riepertinger and F. Lynen (1969). Eur. J. Biochem. 10, 377

THE QUESTION OF MEMBRANE FLUIDITY IN AN ANAEROBIC GENERAL FATTY ACID AUXOTROPH

G.P. Hazlewood, R.M.C. Dawson and *H. Hauser,
ARC Institute of Animal Physiology, Babraham,
Cambridge and *ETH, Zurich CH-8092, Switzerland.

ABSTRACT

Butyrivibrio S2, a general fatty acid auxotroph, grows vigorously in the presence of a single saturated fatty acid such as palmitic acid. We have analysed lipid composition and used spin-labelled probes to determine whether membrane fluidity is maintained in the absence of acyl chain unsaturation.

INTRODUCTION

Much of the recent progress made in understanding the relationship between the physical properties of component lipids and the biological function of cell membranes has been achieved through the study of microbial systems, in particular those where induced mutation has provided a means of manipulating lipid composition (Cronan and Gelmann, 1976; Silbert, 1976) or alternatively where alteration in lipid composition has been produced by the use of antilipogenic substances which suppress normal fatty acid biosynthesis (Silvius and McElhaney, 1978). As a result, homeoviscous adaptation, whereby the fatty acid composition of membrane lipids may be adjusted to maintain so-called membrane fluidity irrespective of growth temperature, is a well-demonstrated phenomenon in Escherichia coli (Sinensky, 1974), Acholeplasma laidlawii (Esser and Souza, 1976) and Bacillus stearothermophilus (Esser and Souza, 1974). Furthermore, it has been established for all microbial systems studied that at the prescribed growth temperature a certain minimum amount of fluid lipid is essential for maintenance of membrane function and cell viability (Jackson and Cronan, 1978; McElhaney, 1974). Hence, it is to be expected in organisms where suppression or genotypic absence of normal biosynthetic mechanisms has produced a reliance on exogenous lipid hydrocarbon chain precursors, that the temperature range over which growth can occur may be determined by the physical state of the complex membrane lipids synthesized from the available fatty acid precursors (McElhaney, 1976).

METHODS

For experimental details, the reader is referred to previous publications (Clarke *et al*, 1976; Hazlewood and Dawson, 1979; Hauser *et al*, 1979).

RESULTS

Culture of *Butyrivibrio* S2 with a Single Saturated Straight-Chain Fatty Acid

In studying the pregastric digestion of the lipid components of herbage which occurs in the ovine rumen, we have isolated from rumen contents by the use of a selective medium, an anaerobic gram negative curved rod of the genus *Butyrivibrio* which is strongly lipolytic and catabolises chloroplast membrane lipids by identical pathways to those which operate *in vivo* (Hazlewood and Dawson, 1979). The organism (strain designation S2) is a general fatty acid auxotroph and will not grow in the absence of long-chain fatty acids. Fatty acids may be provided in the growth medium esterified in a variety of plant lipids or the non-ionic Tween detergents, or complexed with commercial preparations of bovine serum albumin or in the free form may be dispersed by the use of taurocholate or by hydrating alcoholic solutions of the fatty acid. In addition to differing from previously described bacterial fatty acid auxotrophs by virtue of its inability to synthesise any long-chain fatty acid either from acetate or by elongation, *Butyrivibrio* S2 proved unique in being able to grow vigorously when supplied with a single species of high-melting saturated straight-chain fatty acid such as palmitic or stearic acid.

Examination of the range and type of fatty acids which supported growth at 39° (the environmental temperature within the rumen) revealed that straight-chain saturated fatty acids such as myristic, pentadecanoic, palmitic or margaric acids added to a fatty acid-free but otherwise complete medium promoted vigorous growth of Butyrivibrio S2 with little or no lag phase. Lauric acid did not support growth and stearic acid promoted good growth at 45° but not at 39° unless a culture subcultured several times at 45° was used to provide the inoculum. S2 grew well in the presence of *trans*-11-octadecenoic acid (vaccenic acid) which has physicochemical properties rather akin to the saturated fatty acids, but with *cis*-monoenoic acids such as palmitoleic and oleic acids growth only occurred after a long lag phase. Linolenic and linoleic acids supported growth but were rapidly isomerised and hydrogenated to *trans*-11-octadecenoic acid which then presumably served as the growth-promoting fatty acid.

Incorporation of Saturated Fatty Acids into Plasmalogen and a Long-Chain Dicarboxylic Acid

Each fatty acid species which supported growth of S2 was incorporated into a large number of membrane lipids without any chain shortening, chain elongation or desaturation occurring. Methanolysates of total lipids examined by GLC revealed that when any single fatty acid (tridecanoic, myristic, pentadecanoic, palmitic, margaric, stearic, oleic or vaccenic acid) was used as the sole exogenous fatty acid species for promoting growth of S2, each was incorporated into unusual membrane lipids of the plasmalogen type exclusively as unchanged fatty acid (acyl ester) or the corresponding fatty aldehyde (alkenyl ether). An additional involatile hydrophobic component was subsequently discovered by TLC of the methanolysis products, and we conclude from evidence summarised below that this new substance is a long-chain terminal dicarboxylic acid with vicinal methyl groups located at the centre of the chain (Klein *et al*, 1980). Formation of this molecule probably involves a (ω-1) reductive condensation reaction between two molecules of fatty acid, and we have isolated from the organism an homologous series of these dicarboxylic acids depending on the nature of the fatty acid which has been used to promote growth. Thus stimulation of growth of Butyrivibrio S2 by palmitic acid results in the production of 15,16-dimethyltriacontan-1,30-dioic acid (diabolic acid) (Klein *et al*, 1980). The infra red spectrum of the dimethyl ester of this dicarboxylic acid had a carbonyl absorption at 1744 cm^{-1} and peaks at 1175 cm^{-1} and 1250 cm^{-1} corresponding to stretching of the C-O bond in $\overset{O}{\overset{\|}{C}}-O-R$; there was no evidence of unsaturation or of oxygen in ether or hydroxy linkage. The 70 eV mass spectrum of the same compound gave a molecular ion of formula $C_{34}H_{66}O_4$, which did not change on prior treatment with acetyl chloride or trimethylsilyl imidazole. Esterification of the free acid in the presence of CD_3OH instead of CH_3OH produced an increase of 6 mass units to the molecular ion indicating the presence of two carboxylic acid groupings. In the mass spectrum of the dimethyl ester intense ions occurred at m/e 74 and 87 as well as a series of ions of the general type $CH_3.OCO(CH_2)n$, suggesting a saturated methylester. This series of ions continued up to and included two particularly intense clusters of ions 28 mass units apart centred at m/e 269 (n=15) and m/e 297 (n=17) indicative of the presence of a fragmentation directing functional group such as a branching point. When S2 was grown in the presence of fatty acids other than palmitic acid, mass spectrometry of the dimethyl esters of the dicarboxylic acids isolated from total membrane lipids showed homologous behaviour predictable on the basis of a reductive condensation of two fatty acid molecules. The dimethyl ester of the dicarboxylic acid prepared from cells grown with palmitic acid was converted into its parent 32-carbon hydrocarbon which was compared with a 26-carbon hydrocarbon having mid-chain vicinal dimethyl branching and synthesised from 2-bromotridecane; the 70 eV mass spectra of the two hydrocarbons showed similar fragmentation patterns dominated by cleavage and rearrangement associated with the tertiary carbons. Proton and ^{13}C NMR spectroscopy yielded results consistent with the structure we have proposed for this series of dicarboxylic acids, and in addition the dimethyl ester of the C22 homologue has been synthesised by an unambiguous route and has been shown to have equivalent properties to the homologues produced by S2.

ESR Studies of Butyrivibrio S2 Grown with a Single Saturated Straight-Chain Fatty Acid

The exclusive incorporation of *n*-saturated hydrocarbon chains into membrane lipids of *Butyrivibrio* S2 did not appear consistent with accepted concepts regarding the physicochemical state of membrane lipids since a degree of hydrocarbon chain fluidity has been found to be essential for correct membrane function in the microbial systems so far studied. We therefore undertook a preliminary ESR study of membrane fluidity in Butyrivibrio S2 using TEMPO and 5-doxylstearic acid probes.

TEMPO (2×10^{-6}M) was equilibrated with lipids obtained from S2 grown in the presence of myristic or trans-11-octadecenoic acids and examined by ESR spectroscopy at temperatures up to 70°C. Even at this high temperature, there was no evidence of TEMPO partitioning into fluid hydrocarbon regions; instead as the {lipid}:{label} ratio was decreased from 4.175×10^4 to 8.35×10^2, a 3-line spectrum characteristic of the isotropic motion of free TEMPO in an aqueous medium appeared, suggesting that the mixed lipids in some way adsorb TEMPO up to a saturating concentration.

At all temperatures examined, the spectrum of 5-doxylstearic acid equilibrated with whole cells or extracted lipids of S2 cultured in the presence of palmitic acid was typical for anisotropic motion of the probe, and the high values of the maximum hyperfine splitting ($2A_{\parallel}$) were reminiscent of those observed when the same spin label was bound to the rigid specialised purple membrane of Halobacterium sp. (Chignell and Chignell, 1975). At the growth temperature of 39° $2A_{\parallel}$ for 5-doxylstearic acid bound to S2 cells was 55.5 gauss (Hauser et al, 1979), which is similar to the value obtained when the same spin label is incorporated into multilayer vesicles of dipalmitoyllecithin in the gel phase when the hydrocarbon chains of this phospholipid are packed in an orderly crystalline lattice (Janiak et al, 1976).

The relationship between temperature and the maximum hyperfine splitting ($2A_{\parallel}$) of the spectrum of 5-doxylstearic acid bound to S2 cells grown in the presence of either myristic, palmitic or stearic acid is shown in Fig.1.

From the plots there was no evidence that varying the length of the exogenously-provided hydrocarbon chain had a significant or predictable effect on the fluidity of the cell membranes as determined by evaluation of $2A_{\parallel}$. However, for each of the fatty acids examined there was a discontinuity in the plot which we interpret as corresponding to a phase transition or structural reorganisation, so that at temperatures above the point at which it occurs, a greater degree of disorder exists in the hydrocarbon chains of membrane lipids. It is probable, considering the high values of $2A_{\parallel}$ at which these transitions occurred, that they should not be regarded as indicating complete "melting" of the hydrocarbon chains but instead a more limited or subtle release from the crystalline gel state. Discontinuities or

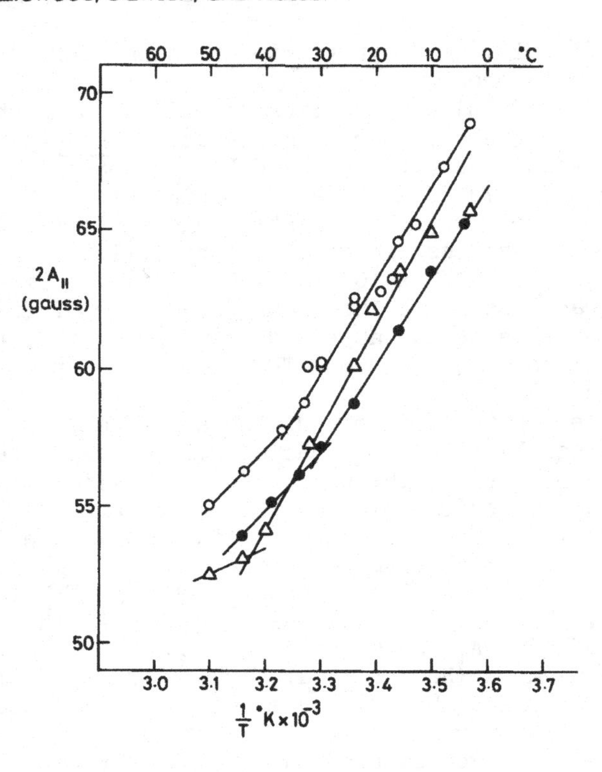

Fig.1. Relationship between maximum hyperfine splitting ($2A_{\parallel}$) and temperature for 5-doxylstearic acid bound to Butyrivibrio S2 cells grown in the presence of myristic O, palmitic ● or stearic Δ acid.

transitions occurring at a different temperature for each of the three fatty acids which have been used as growth supplement were similarly observed in plots showing the temperature dependence of $2A_{\parallel}$ for 5-doxylstearic acid equilibrated with total lipids extracted from cells which had been cultured in the presence of the different fatty acids. Within the limits of experimental error, for each fatty acid examined it would appear that the transition temperature approximated to the lowest temperature supporting significant growth (Table 1). However, it is clear from more recent observations that this is not the only factor determining the lowest temperatures at which the organism can multiply. Thus pentadecanoic acid only supports limited growth of the organism at 32° whereas the phase transition observed when 5-doxylstearic acid was mixed with the lipids extracted from the C15:0 grown bacteria, was considerably below this value.

Table 1. Transition temperatures, as determined by ESR spectroscopy of the 5-doxylstearic acid probe, of membrane lipids in whole cells of Butyrivibrio S2 cultured with saturated fatty acids of different chain lengths, compared with the minimum temperature supporting growth in a medium containing each of the fatty acids.

Fatty Acid	Transition temperature observed in $2A_{\parallel}$ plot	Minimum temperature supporting significant growth* of organism
Myristic	35°	34°
Palmitic	31°	33°
Stearic	42°	41°

*In 20h.

Whatever the nature of the physical change in the lipids producing this increased molecular motion, it seems that this minimum degree of disorder is essential for the growth of the organism. Nevertheless the high absolute values obtained for $2A_{\parallel}$ at the transition temperature would suggest that in Butyrivibrio S2 a more limited degree of hydrocarbon chain motion is required for membrane function than has been observed in other bacteria. This does not however contradict the postulate that the minimum growth temperature of a microorganism may be determined in certain circumstances by the lower boundary of the membrane lipid phase transition (McElhaney, 1976).

Possible Mechanism by which Membrane Fluidity is Maintained

What then is the nature of the temperature-dependent transition observed in membrane lipids which as far as our analyses can tell contain no hydrocarbon chain unsaturation? We have now purified the main complex lipids produced by the organism grown in the presence of a single saturated fatty acid (palmitic acid). They are either galactolipids or phospholipids or galactose-containing phospholipids, and are characterized by a very high content of plasmalogen. Many also contain the long-chain dicarboxylic acid previously discussed (i.e. 15,16 dimethyltriacontan-1,30 dioic acid) as an integral part of their structure with the two carboxylic acid groups esterified to the glycerol moieties of either two phosphoglycerides or a phosphoglyceride and a galacto-glyceride. With lipids of such a unique structure it is at present not possible to say whether the very long hydro-carbon chain spans the lipid bilayer existing within the organism's plasma membrane, reducing cooperative lipid motion and producing an unusually rigid membrane. In addition, many of the complex lipids contained within the organism S2 are characterized by the presence of esterified butyric acid, an acid which is produced by sugar fermentation and is also incorporated into the lipids of other species within the same genus.

Apart from the unusual characteristics mentioned above, there is nothing obvious in the lipid composition of *Butyrivibrio* S2 which could account for the apparent transitions observed in the ESR experiments. There is little evidence that the vinyl ether unsaturation contained in normal plasmalogen structures can increase the fluidity of lipid bilayers when substituted for acyl chains (Khuller and Goldfine, 1974; Roots and Johnston, 1968). Although this has not been investigated with plasmalogen groupings in the present type of structures, it seems likely that the results would be essentially as predicted. On the other hand, the presence of butyroyl groups and the centrally-located methyl branches of the long-chain dicarboxylic acid could both produce a perturbation in the orderly packing of the hydro-carbon chains in the lipid bilayer. We have already shown (Hauser *et al*, 1979) that the presence of mid-chain vicinal dimethyl branching lowers the capillary melting point of a long-chain dicarboxylic acid to a value well below that of its straight-chain counterpart. Physicochemically, this may be compared with the situation existing for monocarboxylic acids where those fatty acids with dimethyl substituents on a single carbon atom (Deuel, 1951) or anteiso methyl-branched acids, with a methyl group present on the ante-penultimate carbon atom, pack less readily and therefore have a much lower melting point than their straight-chain or iso-branched counterparts. Also, if the short hydrocarbon chain of the esterified butyric acid in the complex lipids is orientated, as seems likely, within the lipid hydrophobic environment of the bilayer, it could exert a fluidising effect by preventing the close and orderly packing of the much longer hydrocarbon chains of the growth-promoting fatty acid incorporated into the membrane complex lipids.

In its natural environment (the rumen), _Butyrivibrio_ S2 has a plentiful supply of saturated fatty acids (palmitic and stearic acids) and _trans_-octadecenoic acids formed by the rapid hydrogenation of dietary polyunsaturated fatty acids. If such fatty acids were incorporated directly into the complex lipids usually present in bacteria, they would produce a very rigid membrane at the environmental temperature (39°). Presumably the continued survival of S2 as a naturally-occurring general fatty acid auxotroph has depended on its ability to incorporate these fatty acids into lipid structures which can modify the close-packed organisation of a rigid bilayer. The introduction of butyroyl groups and vicinal dimethyl branching in the hydrocarbon chains may perhaps therefore be considered equivalent to the evolutionary adaptation of psychrophilic bacteria. When compared to their mesophilic and thermophilic counterparts, such organisms have adapted to low growth temperatures not only by increasing their content of unsaturated fatty acids but also by increasing the proportion of anteiso methyl branched fatty acyl chains and reducing the average length of the hydrocarbon chains present in their membrane lipids (McElhaney, 1976).

REFERENCES

Chignell, C.F. and Chignell, D.A. (1975) Biochem.Biophys.Res. Commun. 62, 136-143.

Clarke, N.G., Hazlewood, G.P. and Dawson, R.M.C. (1976) Chem.Phys.Lipids 17, 222-232.

Cronan, J.E. and Gelmann, E.P. (1976) Bact.Rev. 39, 232-256.

Deuel, H.J. (1951) In The Lipids; their Chemistry and Biochemistry Vol.1. Interscience, N.Y. p.55.

Esser, A.F. and Souza, K.A. (1974) Proc.natn.Acad.Sci., U.S.A. 71, 4111-4115.

Esser, A.F. and Souza, K.A. (1976) In M.R. Heinrich (ed.), Extreme Environments: Mechanisms of Microbial Adaptation. Academic Press, New York, San Francisco & London, pp.283-294.

Hauser, H., Hazlewood, G.P. and Dawson, R.M.C. (1979) Nature 279, 536-538.

Hazlewood, G.P. and Dawson, R.M.C. (1979) J.gen.Microbiol. 112, 15-27.

Huang, L., Lorch, S.K., Smith, G.G. and Haug, A. (1974) FEBS Lett. 43, 1-5.

Jackson, M.B. and Cronan, J.E. (1978) Biochim.biophys.Acta 512, 472-479.

Janiak, M.J., Small, D.M. and Shipley, G.G. (1976) Biochemistry 15, 4575-4580.

Khuller, G.K. and Goldfine, H. (1974) J.Lipid.Res. 15, 500-507.

Klein, R.A., Hazlewood, G.P., Kemp, P. and Dawson, R.M.C. (1980) in press.

McElhaney, R.N. (1974) J.molec.Biol. 84, 145-157.

McElhaney, R.N. (1976) In M.R. Heinrich (ed.), Extreme Environments: Mechanisms of Microbial Adaptation. Academic Press, New York, San Francisco & London. pp.255-281.

Roots, B.I. and Johnston, P.V. (1968) Comp.Biochem.Physiol. 26, 553-560.

Silbert, D.F. (1976) In E.D. Korn (ed.), Methods in Membrane Biology vol.6. Plenum Press, New York & London, pp. 151-182.

Silvius, J.R. and McElhaney, R.N. (1978) Can.J.Biochem. 56, 462-469.

Sinensky, M. (1974) Proc.natn.Acad.Sci., U.S.A. 71, 522-525.

THERMAL CONTROL OF FATTY ACID SYNTHETASE FROM *Brevibacterium ammoniagenes*

A.Kawaguchi*, Y.Seyama**, K.Sasaki*, S.Okuda* and T.Yamakawa** *Inst. of Applied Microbiol. and **Dept. of Biochem., Faculty of Medicine, The Univ. of Tokyo, Bunkyo-ku, Tokyo, Japan

ABSTRACT

The fatty acid composition of *Brevibacterium ammoniagenes* was affected by the temperature of growth. As the growth temperature was lowered, the proportion of unsaturated fatty acids increased. The fatty acid synthetase obtained from *B. ammoniagenes* produced oleic acid as well as saturated fatty acids. The ratio of unsaturated to saturated fatty acids synthesized by this enzyme *in vitro* was dependent on the temperature of the enzyme reaction but not on the growth temperature of *B. ammoniagenes* from which the enzyme was prepared. These results suggest that the changes of composition in cellular fatty acids reflect the temperature dependence of the fatty acid synthetase.

INTRODUCTION

Various organisms, ranging from bacteria to higher plants and animals, adjust the fatty acid composition of their membrane lipids in response to their environmental temperature. The mechanisms responsible for temperature-induced fatty acid alteration seem to operate at the level of both phosphatidic acid synthesis and fatty acid synthesis. We have studied the control of fatty acid synthetase from *Brevibacterium ammoniagenes*. Fatty acid synthetase from *B. ammoniagenes* is a multienzyme complex. The novel

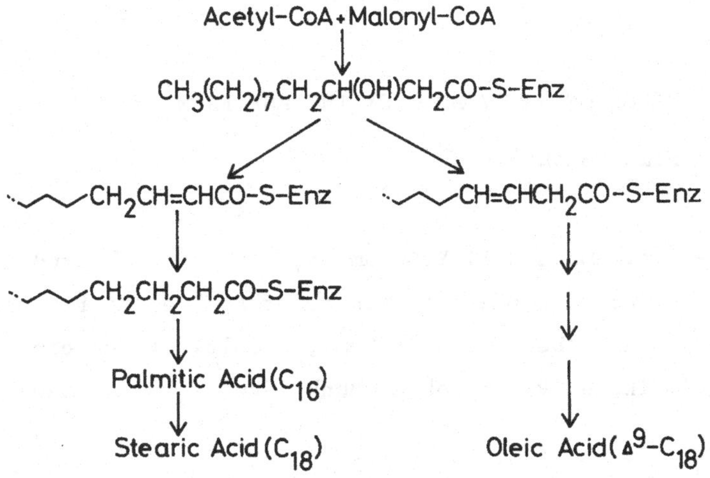

Acetyl-CoA + Malonyl-CoA

↓

$CH_3(CH_2)_7CH_2CH(OH)CH_2CO\text{-}S\text{-}Enz$

↙ ↘

$\sim CH_2CH{=}CHCO\text{-}S\text{-}Enz$ | $\sim CH{=}CHCH_2CO\text{-}S\text{-}Enz$

↓ ↓

$\sim CH_2CH_2CH_2CO\text{-}S\text{-}Enz$

↓ ↓

Palmitic Acid (C_{16})

↓ ↓

Stearic Acid (C_{18}) | Oleic Acid (Δ^9-C_{18})

Fig. 1. Mechanism of fatty acid synthesis in *B. ammoniagenes*

feature of this enzyme complex is its ability to synthesize oleic acid as well as saturated fatty acids (Kawaguchi and Okuda, 1977). This enzyme synthesizes oleic acid by an anaerobic process involving β,γ-dehydration of β-hydroxy-acyl thioester intermediates and subsequent chain elongation of β,γ-enoate without reduction of the double bond. The enzyme component responsible for this dehydration is tightly associated with the various activities for carbon chain elongation as an integral part of the fatty acid synthetase complex (Fig. 1) (Kawaguchi and Okuda, 1977). The present paper reports that the ratio of unsaturated to saturated fatty acids synthesized by the fatty acid synthetase *in vitro* depends on the temperature of the enzyme reaction but not on the growth temperature of the bacteria.

EXPERIMENTAL PROCEDURE

In order to measure fatty acids synthesized by the enzyme reaction, we devised a new assay method using mass

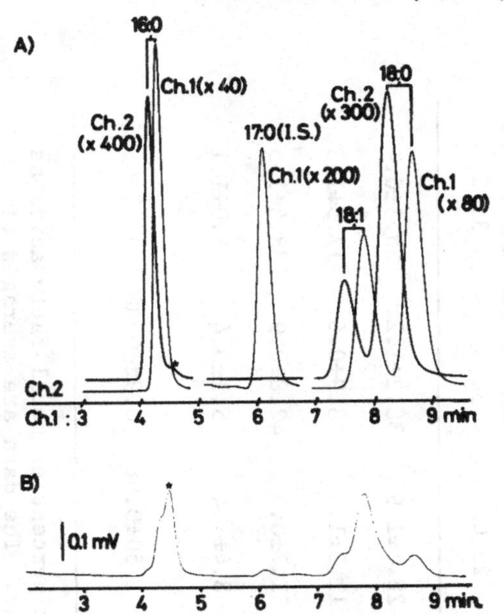

Fig. 2. Mass fragmentogram (A) and gas chromatographic recording (B) of fatty acids synthesized by the fatty acid synthetase. Channels 1 and 2 were set for m/e 74 and 77, respectively.

fragmentography (Seyama *et al.*, 1978). The amounts of fatty acids synthesized in deuterated water are determined by mass fragmentography by monitoring the intensities of m/e 77 and m/e 74 fragment ions using heptadecanoic acid as an internal standard (Fig. 2).

RESULTS

Effect of Growth Temperature on the Fatty Acid Composition of *B. ammoniagenes*

The principal fatty acids of *B. ammoniagenes* were identified as palmitic, stearic, oleic and 10-methylocta-decanoic acids (Kawaguchi *et al.*, 1979). The amounts of these fatty acids in cells grown at various temperatures were calculated from the areas of the peaks on gas

Table I. Fatty acid composition of *Brevibacterium ammoniagenes* grown at various temperatures

Fatty acid	Fatty acid composition (%)[a]			
	20°C	25°C	30°C	35°C
Palmitic	24.9±0.5	28.7±1.9	38.2±1.2	38.3±0.4
Stearic	11.2±0.9	11.3±1.4	13.9±0.6	17.5±1.3
Oleic	57.3±1.0	51.7±0.9	42.8±0.9	38.3±1.5
10-Methyloctadecanoic	6.4±0.4	8.4±1.4	5.1±0.4	6.0±1.1
$\frac{C_{18:1}+C_{19:0}}{C_{16:0}+C_{18:0}}$	1.76±0.06	1.50±0.08	0.92±0.08	0.79±0.05

[a] Results are expressed as the weight percent of total fatty acids as determined by gas-liquid chromatography. The data are averages of 5 separate experiments.

chromatograms(Table I). The percentage of oleic acid decreases with increasing growth temperature. On the contrary, the proportion of saturated fatty acids (palmitic + stearic acids) increased as the growth temperature was raised. When the culture temperature was constant, an increase in 10-methyloctadecanoic acid and a decrease in oleic acid occurred at the transition from exponential to stationary phases. The increase in 10-methyloctadecanoic acid was accompanied by a concomitant stoichiometric loss of oleic acid. Therefore, the $C_{18:1}+C_{19:0}/C_{16:0}+C_{18:0}$ ratio is almost constant throughout growth at a definite culture temperature. However, this ratio is affected by changes of the culture temperature, as shown in Table I.

Effect of Enzyme Reaction Temperature on Fatty Acid Synthetase Activity

The fatty acid synthetase from *B. ammoniagenes* catalyzes the synthesis of oleic, palmitic and stearic acids

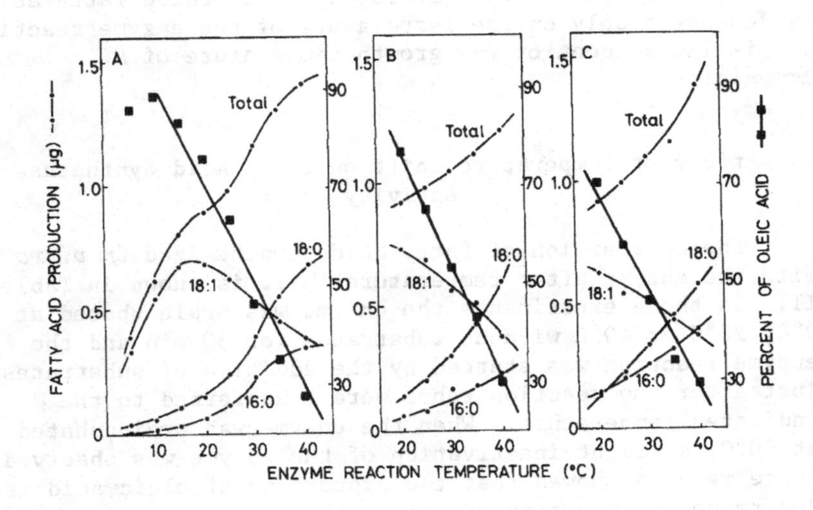

Fig. 3. Temperature-sensitive synthesis of fatty acid species by fatty acid synthetase. The enzyme was prepared from cells which were grown at 30°C (A), 20°C (B) and 37°C (C), respectively.

(Kawaguchi and Okuda, 1977). The ratio of unsaturated to saturated fatty acids depends on the temperature of the enzyme reaction (Fig. 3A). Crude enzyme preparations were used for the experiments in Fig. 3 but the temperature dependence of the products was essentially the same with the purified fatty acid synthetase. The amounts of newly synthesized palmitic and stearic acids increased continuously with increasing temperature of the enzyme reaction over the temperature range of 5°C to 40°C. On the other hand, the amount of oleic acid became maximum at 20°C and decreased at lower or higher temperatures. The proportion of oleic acid decreased with increasing temperature of the enzyme reaction.

In another series of experiments shown in Fig. 3, we measured the amounts of saturated and unsaturated fatty acids synthesized with crude enzyme preparations from cells grown at 20°C (Fig.3B) and at 37°C (Fig. 3C). The enzyme preparations from cells grown at 20°C, 30°C and 37°C showed almost the same profiles of temperature-dependent synthesis of palmitic, stearic and oleic acids. These results suggest that the ratio of unsaturated to saturated fatty acids is dependent only on the temperature of the enzyme reaction and is independent **of** the growth temperature of *B. ammoniagenes*.

Effect of Temperature Shift on Fatty Acid Synthetase Activity

The composition of fatty acids synthesized *in vitro* with the enzyme after temperature shift is shown in Table II. In these experiments the enzyme was preincubated at 0°C, 20°C or 40°C without substrates for 30 min and the enzyme reaction was started by the addition of substrates just after the reaction tubes were transferred to the indicated temperature. When the enzyme was preincubated at 40°C, a slight inactivation of the enzyme was observed. These results showed that the proportion of oleic acid is determined by the temperature of the enzyme reaction, not by preincubation temperature. From these results, we can rule out the possibility that the decrease of oleic acid production at higher temperature is a result of inactivation of the enzyme component responsible for oleic acid synthesis.

Table II. Fatty acid species synthesized by fatty acid synthetase after temperature shifts

Preincubation temperature (°C)	Reaction temperature (°C)	Palmitic acid (ng)	Stearic acid (ng)	Oleic acid (ng)	Percent of oleic acid (%)
0	40	160	730	455	33.8
20	40	220	456	464	40.7
40	40	164	480	374	36.7
0	20	40	80	332	73.5
20	20	25	51	272	78.2
40	20	16	37	195	78.6

DISCUSSION

Variation in the growth temperature altered the fatty acid composition in the lipids of *B. ammoniagenes*. As the growth temperature was lowered, the proportion of unsaturated fatty acids increased. It has been noted by a large number of workers (for review see Cronan and Vagelos, 1972) that *Escherichia coli* also adjusts the fatty acid composition of its phospholipids in response to growth temperature. The mechanisms regulating this alteration have been extensively studied by many investigators (Sinensky, 1971; Kito *et al.*, 1975; Cronan, 1975; Cronan and Gelmann, 1975; Okuyama *et al.*, 1977), and seem to operate at the level of both phosphatidic acid synthesis and fatty acid synthesis. The results of Sinensky (1971) and Kito *et al.* (1975) indicate that the change in cellular fatty acids reflects the temperature dependence of acyltransferase specificity. Using several strains of *E. coli*, Cronan (1974, 1975) suggested that the level of β-hydroxydecanoyl thioester dehydrase, which introduces the double bond into the hydrocarbon chain, is also important for the thermal regulation of fatty acid composition. Recently, Okuyama *et al.* (1977) reported that the ratio of unsaturated to saturated fatty acids synthesized by a fatty acid synthetase system of *E. coli* was affected by both the temperature of the enzyme reaction and the growth temperature. The data in the present paper indicate that fatty acid synthetase of *B. ammoniagenes* is temperature dependent. The ratio of unsaturated to saturated fatty acids synthesized by the enzyme was affected by the temperature of the enzyme reaction, but not by the growth temperature of the cells. Thermal changes in the composition of cellular fatty acids seem to reflect the temperature dependence of fatty acid synthetase. The nature of the fatty acid synthetase appears to be unchanged at any growth temperature. The quantity and quality of the enzyme are also the same at any growth temperature (Fig. 3). The ratio of unsaturated to saturated fatty acids is determined solely by the enzyme reaction temperature. There are some differences in the ratio of palmitic to stearic acids in the fatty acid composition of cells (Table I) compared with fatty acids synthesized by the enzyme (Fig. 3). However, the percentages of oleic plus 10-methyloctadecanoic acids in the fatty acid composition of cells at various growth temperatures are well correlated with the percentages of oleic acid in enzyme reaction products at corresponding reaction temperatures (Table I and Fig. 3). This close correlation makes it attractive to consider that the fatty acid composition of *B. ammoniagenes* might be determined by the temperature dependence of the fatty acid synthetase itself.

The changes in the ratio of unsaturated to saturated fatty acids with temperature may be ascribed to conformational changes of the fatty acid synthetase. These conformational changes are reversible judging from the data in Table II. Oleic acid is synthesized by β,γ-dehydration of β-hydroxyacyl thioester and this dehydration activity is tightly associated with activities for chain elongation (Fig. 1). As a result of conformational changes of the synthetase, the enzyme component responsible for oleate formation becomes more active at lower temperatures.

REFERENCES

Cronan, J. E. (1974). Proc. Natl. Acad. Sci. USA 71, 3758.

Cronan, J. E. (1975). J. Biol. Chem. 250, 7074.

Cronan, J. E., and Gelmann, E. P. (1975). Bacteriol. Rev. 39, 232.

Cronan, J. E., and Vagelos, P. R. (1972). Biochim. Biophys. Acta 265, 25.

Kawaguchi, A., and Okuda, S. (1977). Proc. Natl. Acad. Sci. USA 74, 3180.

Kawaguchi, A., Seyama, Y., Sasaki, K., Okuda, S., and Yamakawa, T. (1979). J. Biochem. 85, 865.

Kito, M., Ishinaga, M., Nishihara, M., Kato, M.,Sawada, S., and Hata, T. (1975). Eur. J. Biochem. 54, 55.

Okuyama, H., Yamada, K., Kameyama, Y., Ikezawa, H., Akamatsu, Y., and Nojima, S. (1977). Biochemistry 16, 2668.

Seyama, Y., Kawaguchi, A., Okuda, S., and Yamakawa, T. (1978). J. Biochem. 84, 1309.

Sinensky, M. (1971). J. Bacteriol. 106, 449.

WHY DO PROKARYOTES REGULATE MEMBRANE FLUIDITY?

John R. Silvius, Nanette Mak and Ronald N. McElhaney

Department of Biochemistry, University of Alberta, Edmonton, Alberta, Canada T6G 2H7

ABSTRACT

While adaptation of microbial membrane lipid composition is widely observed in response to changes in temperature, and in some cases to the presence of membrane-fluidizing substances, the physiological importance of such adaptations remains largely unexplored. To shed some light on this question we have used our ability to extensively vary the membrane fatty acid composition of Acholeplasma laidlawii B to study a variety of membrane functions in membranes whose lipids contain a single fatty acyl species whose structure and physical properties can be widely varied. Our results suggest that control of the membrane lipid phase state is considerably more important to the proper functioning of the organism than is control of the 'fluidity' of liquid-crystalline lipids.

A. laidlawii B can be grown to normal or near-normal yields when cultured with avidin plus any of a large number of individual fatty acids, with structures ranging from a fourteen-carbon anteisobranched species to an eighteen-carbon monounsaturated species to an eighteen-carbon isobranched species. Cells will not grow, however, if their membrane lipids are largely in the gel state. The temperature dependence of a key membrane enzyme, the osmoregulatory (Na^{+},Mg^{2+})-ATPase, is independent of the lipid fatty acid composition in the membranes of such 'fatty acid-homogeneous' cells, so long as the lipids remain liquid-crystalline.

The thermostability of this enzyme is likewise independent of fatty acid composition, and the absolute ATPase activity shows no consistent variation with composition that could be attributed to 'fluidity' effects. However, when the membrane lipids enter the gel state, the ATPase is entirely inactivated. Studies of cellular permeability to ions and polyols indicate that while more 'fluid' membranes are more permeable to these compounds, membranes in which gel and liquid-crystalline lipids coexist are more permeable still. Therefore, the phase state of the membrane lipids affects certain key physiological properties (including growth) much more dramatically than do variations in the 'fluidity' of liquid-crystalline membrane lipids, and we suggest that regulation of the membrane phase state is the primary function of 'homeoviscous adaptation' in prokaryotic organisms.

INTRODUCTION

It is now well established that many organisms alter their membrane lipid compositions in response to changes in temperature or to other environmental perturbations that would be expected to alter the properties of the membrane lipid phase. These findings have led to the proposal that organisms seek to maintain an 'optimal fluidity' of the membrane lipid phase under varying environmental conditions by means of such lipid compositional adjustments (Sinensky, 1974). It has in fact been shown that very drastic changes in the physical properties of membrane lipids can be deleterious to the normal function of the organism (McElhaney, 1974; Davis and Silbert, 1975; Saito and Silbert, 1979; Baldassare et al., 1979). Beyond these findings, however, two questions remain to be answered. First, how finely must lipid fluidity be regulated in order to maintain normal membrane function in a living system? Secondly, what membrane functions are most likely to be impaired by changes in fluidity from a presumed 'optimal fluidity', and by what mechanism do the altered lipid properties cause such impairment? Answers to these questions are obviously of considerable importance in furthering our overall understanding of the effects of membrane lipid physical properties on the function of biological membranes.

We have previously shown (Silvius and McElhaney, 1978) that when the simple, cell wall-less prokaryote *Acholeplasma laidlawii* B is grown in the presence of avidin, the organism becomes wholly unable to synthesize or modify the structure of medium- and long-chain fatty acids. Cells grown with avidin readily incorporate exogenously supplied fatty acids into their membrane lipids to very high levels (>97% of total lipid acyl chains), a fact which allows us to vary the membrane lipid fatty acid composition and the membrane lipid physical properties in this organism far more widely than would normally occur in a living system. We have studied the effects of such drastic changes in the membrane lipid properties on the growth and various membrane functions in A. laidlawii B in an effort to obtain some answers to the two questions posed above.

RESULTS AND DISCUSSION

Our initial studies focused on the growth of A. laidlawii B when various single fatty acids were added to the culture medium along with avidin. As can be seen in Table 1, fatty acids of widely varying structure can support good

Table 1

Growth Yields, Membrane Lipid Enrichment in Exogenous Fatty Acid Species, and Lipid Phase Transition Temperatures for Cultures of A. laidlawii B Grown with Avidin Plus Various Exogenous Fatty Acids

Fatty Acid	% Total Lipid Acyl Chains	Tc (°C)	Maximal Turbidity (% of Control)
14:0*ai*	95	-14.5	61
18:1*c*Δ^{11}	96	-8.3	75
16:1*t*Δ^{9}	98	7.6	87
17:0*ai*	96	8.2	95
14:0*i*	96	10.1	103
18:1*t*Δ^{11}	99	20.0	99
17:0*i*	98	28.8	98
Control[a]	-	30.8	-
14:0/16:0	49/50	32.9	91

[a]Grown in the absence of avidin or fatty acid; the major membrane fatty acyl chains are ∿45% palmitate and ∿30-35% myristate, plus lesser amounts of lauric, stearic and oleic acids.

cell growth (and presumably proper membrane function) when they constitute essentially all of the acyl chains of the membrane glycerolipids. The broad range of lipid phase transition midpoint temperatures (determined by differential thermal analysis) which are observed in various 'fatty acid-homogeneous' membranes strongly suggests that membrane lipid 'fluidity' can vary appreciably without impairing essential membrane functions in this organism. More extensive studies (Silvius and McElhaney, 1978) have shown that fatty acids, such as linoleic acid, whose diacylglycerolipid derivatives give extremely fluid bilayers, do not support growth of *A. laidlawii* B in the presence of avidin, nor do long-chain saturated fatty acids which would give rise to glycerolipids with transition temperatures well above the growth temperature. Therefore, it seems that there are fairly broad limits to the range of membrane 'fluidity' that is compatible with normal cell growth.

The observed tolerance of *A. laidlawii* B to wide variations in its membrane lipid composition and physical properties could be explainable in either of two ways. First, the key membrane functions in this organism could be quite insensitive to the physical properties of the membrane lipids. Secondly, some key functions could be influenced by the lipid properties, but they might not be altered to such an extent that a significant loss in cell viability would result. To investigate this question further, we examined certain membrane functional properties which were most likely to be affected by the membrane lipid properties if the latter explanation was the correct one. These functional properties were as follows: the absolute activity, apparent activation enthalpy and heat capacity, and thermostability of the membrane (Na^+,Mg^{2+})-ATPase, which appears to function as a transmembrane cation pump (Jinks *et al.*, 1978); the permeability of the membrane to nonelectrolytes; and the leakiness of the membrane to monovalent cations. Our results in each of these areas are briefly outlined below.

In Table 2, we have tabulated the absolute (Na^+,Mg^{2+})-ATPase activities in isolated membranes (Silvius *et al.*, 1978) of various fatty acid compositions (listed in order of increasing lipid phase transition midpoint temperature). While different membrane samples exhibit substantial variations in the specific activity of the ATPase, such variations cannot be systematically related to differences in the lipid physical properties (e.g., the transition temperature)

Table 2

Dependence of the Membrane (Na^{+},Mg^{2+})-ATPase Activity on the Membrane Lipid Fatty Acyl Composition of Avidin-Grown A. laidlawii B

Fatty Acid	ATPase Activity (μmoles/mg membrane protein/min)
14:0*ai*	930
18:1*c*Δ^{11}	420
16:1*t*Δ^{9}	270
17:0*ai*	540
14:0*i*	740
18:1*t*Δ^{11}	440
17:0*i*	810
Control[a]	520
14:0/16:0	730

[a]No avidin or fatty acid was added to the culture medium.

of various membrane preparations. Arrhenius plots of the ATPase activity in membranes of different fatty acid composition (Figure 1) show that the enthalpy and heat capacity of activation for the ATPase reaction are independent of the lipid fatty acyl composition so long as the lipids are in the liquid-crystalline state. For example, in Figure 1 it may be noted that the Arrhenius plots for cis-vaccenate-, isomyristate-, and isopalmitate-homogeneous membranes are superimposable (using a small scaling constant to allow for absolute activity differences) at all temperatures above their respective upper phase transition boundaries (which are indicated on the graph by 'T_{ℓ}' for the latter two membranes, that for cis-vaccenate-homogeneous membranes lying below 0°C). A similar result has been obtained for the mediated uptake of glucose by this organism (Read and McElhaney, 1975). It seems likely that the absolute activity differences seen in various membranes are attributable in both cases to differences in the number of active enzyme or carrier molecules rather than to a selective effect of the lipids on the entropy of activation, $\Delta S^{\ddagger}$. However, the latter possibility, while somewhat unlikely in view of the total lack of a lipid compositional effect on $\Delta H^{\ddagger}$ or $\Delta Cp^{\ddagger}$, cannot be absolutely ruled out. We have also found (data

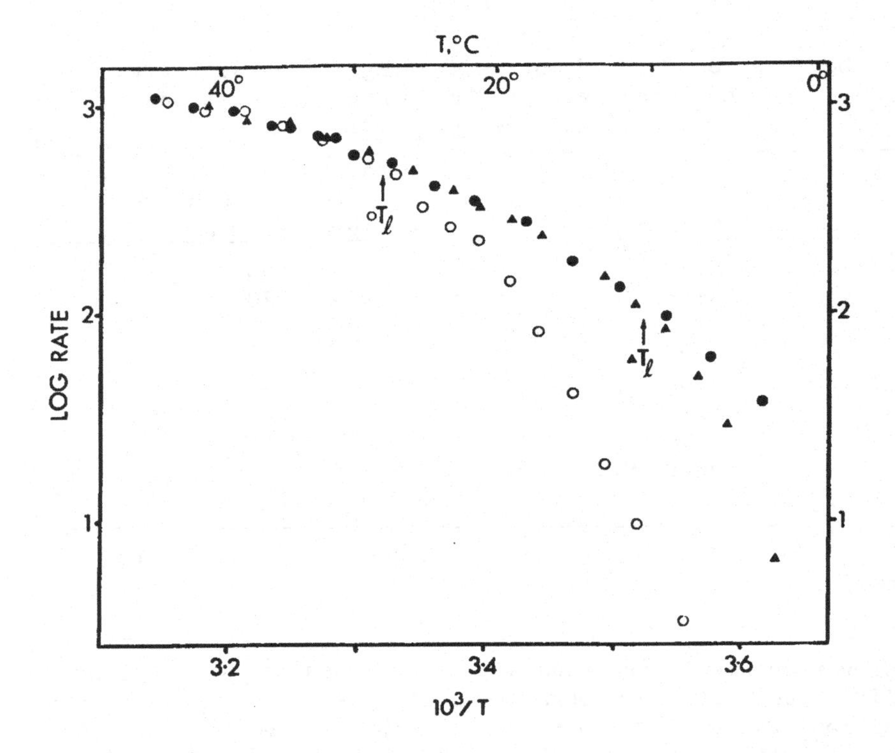

Figure 1. Superimposed Arrhenius plots for the (Na^{+},Mg^{2+})-ATPase activity (expressed in nmoles P_i released per mg membrane protein per minute) in membranes whose lipids contain exclusively cis-vaccenoyl (●), isomyristoyl (▲) or isopalmitoyl (o) chains. The upper boundaries for the thermotropic lipid phase transitions in the latter two types of membranes are indicated by arrows and the symbol 'T_ℓ' on the graph.

not shown) that the lipid fatty acyl composition has no effect on the thermostability of the (Na^{+},Mg^{2+})-ATPase activity. We have, however, detected one important effect of the membrane lipid physical properties on the ATPase activity. Theoretical analyses of Arrhenius plots for the ATPase activity in membranes exhibiting a lipid phase transition over the temperature range studied indicate that the enzyme becomes inactivated when the lipids surrounding it enter the gel (or, more properly when one is discussing boundary lipids, a 'gel-like') state. Therefore, while variations in the composition (and presumably in the fluidity) of the liquid-crystalline lipid phase do not affect the enzyme's activity or stability in any systematic manner, the phase state (pure liquid-crystalline, pure gel or mixed) of the lipids strongly modulates the ATPase activity.

In addition to influencing the activities of lipid-associated membrane enzymes, a second obvious way in which lipid physical properties could affect membrane (and hence cellular) function is by altering the nonmediated permeability of the membrane to various solutes. To evaluate this latter effect of lipid composition on membrane function, we measured the glycerol and erythritol permeability of A. laidlawii B cells of varying membrane fatty acid composition, using a swelling assay as previously described (DeGier et al., 1971). As well, we evaluated the leakiness of the membrane to monovalent cations by turbidometrically measuring the rate of swelling in NaCl or KCl buffers of energy-depleted cells (which are apparently unable to carry out the normal extrusion of cations as they leak into the cell (Jinks et al., 1978)). Representative data are given in Table 3 for cell swelling in glycerol (200 mM) or NaCl/

Table 3

Swelling Rate Constants for Avidin-Grown A. laidlawii B of Various Membrane Lipid Fatty Acid Compositions in Glycerol or NaCl/NaF Buffers

Fatty Acid	Glycerol Swelling Rate Constant ($hsec^{-1}$)	NaCl Swelling Rate Constant (hr^{-1})
14:0 *ai*	12.7	0.46
18:1 $c\Delta^{11}$	16.0	0.32
16:1 $t\Delta^{9}$	21.9	0.23
17:0 *ai*	6.6	0.28
14:0 *i*	10.2	0.25
18:1 $t\Delta^{11}$	10.2	0.20
17:0 *i*	5.9	0.28
Control[a]	8.8	0.32
14:0/16:0	25.3	1.02

[a] No avidin or fatty acid added to the growth medium.

NaF (125 mM/25 mM) solutions buffered with 5 mM Tris (pH 8.0). In both types of experiments, as well as in experiments with erythritol or KCl solutions, the same basic trends are evident: while membranes with very low lipid phase transition midpoint temperatures (Tc's) tend to be more permeable than are membranes with intermediate Tc's, the permeabilities do not decrease further as Tc approaches the assay temperature of 37° but instead tend to rise again. Our findings to date suggest that this last effect may be a consequence of an enhanced permeability of the lipid bilayer within the temperature range of the phase transition, as has been observed with pure lipid/water systems (Marsh *et al.*, 1976; Kanehisa and Tsong, 1978). The data shown in Table 3 thus suggest that very large changes in the nature and properties of the membrane lipid acyl chains (e.g., the replacement of fourteen-carbon anteiso-branched acyl chains by eighteen-carbon *trans*-unsaturated acyl chains, a substitution which increases the transition temperature by 35° in the *A. laidlawii* membrane and by 47° in the pure phosphatidylcholine/water system (Silvius and McElhaney, 1979a,b)) lead to moderate changes in membrane permeability if the lipids are in the liquid-crystalline state, while a change in the phase state of the lipids (from all liquid-crystalline to mixed gel and liquid-crystalline, for example) can lead to permeability changes of equal or even greater magnitude.

The results discussed above generally suggest that membrane 'fluidity' need not be tightly regulated, in *A. laidlawii* B at least, in order to preserve proper membrane function, although there are limits to the range of membrane lipid fluidity variations that the organism can tolerate. Our present results, and previous studies on the effect of temperature and lipid properties on the growth of *A. laidlawii* B (McElhaney, 1974), suggest that changes in the lipid phase state (e.g., a shift from all liquid-crystalline lipids to a gel-liquid-crystalline mixture) can affect certain membrane functions at least as strongly as do even fairly major changes in the lipid composition of membranes (so long, of course, as the compositional changes do not themselves lead to a change in the phase state). Our results to date do not provide an unambiguous answer to the question of what membrane functions suffer most readily when lipid fluidity falls outside of the range that *A. laidlawii* B can tolerate while remaining viable, but the evidence currently available suggests that the barrier function of the membrane may be among the first vital functions to be impaired when very highly fluidizing lipid species (e.g., dilinoleoyl lipids) are present in the membrane. Since the fluidity of normal membrane lipids in the liquid-crystalline state would not be expected to ever approach such extremes, we suggest that the major function of 'homeoviscous adaptation' in prokaryotic organisms is not to maintain an absolutely constant membrane fluidity but is rather to adjust the lipid phase transition temperature range, thereby maintaining a proper lipid phase state (all liquid-crystalline, or a mixture of gel and liquid-crystalline lipids) for optimal membrane functioning at the growth temperature.

ACKNOWLEDGEMENTS

J.R.S. gratefully acknolwedges the financial support of the Medical Research Council in the form of a Studentship award during the course of this work. These studies were supported by Research Grant MT-4261 from the Medical Research Council of Canada.

REFERENCES

Baldassare, J.J., Y. Saito, and D.F. Silbert (1979). J. Biol. Chem. 254, 1108.

Davis, M.T., and D.F. Silbert (1975). Biochim. Biophys. Acta 363, 1.

DeGier, J., J.G. Mandersloot, J.V. Hupkes, R.N. McElhaney, and W.P. Van Beek (1971). Biochim. Biophys Acta 233, 610.

Jinks, D.C., J.R. Silvius, and R.N. McElhaney (1978). J. Bacteriol. 136, 1027.

Kanehisa, M.I. and T.Y. Tsong (1978). J. Am. Chem. Soc. 100, 424.

Marsh, D., A. Watts, and P.F. Knowles (1976). Biochemistry 15, 3570.

McElhaney, R.N. (1974). J. Mol. Biol. 84, 145.

Read, B.D., and R.N. McElhaney (1975). J. Bacteriol. 123, 47.

Saito, Y., and D.F. Silbert (1979). J. Biol. Chem. 254, 1102.

Silvius, J.R., and R.N. McElhaney (1978). Can. J. Biochem. 56, 462.

Silvius, J.R., B.D. Read, and R.N. McElhaney (1978). Sci-

ence 199, 902.
Silvius, J.R., and R.N. McElhaney (1979). *Chem. Phys. Lipids*, in press
Silvius, J.R., and R.N. McElhaney (1979). *Chem. Phys. Lipids*, submitted for publication.
Sinensky, M. (1974). *Proc. Nat. Acad. Sci. U.S.A.* 71, 522.

DOCOSAHEXAENOYL CHAINS ARE INTRODUCED IN PHOSPHATIDIC ACID DURING DE NOVO SYNTHESIS IN RETINAL MICROSOMES

NICOLAS G.BAZAN and NORMA M.GIUSTO

Instituto de Investigaciones Bioquímicas
Universidad Nacional del Sur - Consejo Nacional
de Investigaciones Científicas y Técnicas
8000 Bahía Blanca, Argentina

ABSTRACT

Phosphatidic acid from microsomal membranes of bovine retina contains 21% of docosahexaenoate. If entire retinas are incubated during short-periods of time with dl-propranolol about a fourfold increase takes place in the content of microsomal phosphatidic acid. Moreover docosahexaenoate as well as other acyl chains are increased. It is suggested that a significative proportion of docosahexaenoyl groups of other phospholipids is introduced into the glycerolipids during the de novo biosynthesis of phosphatidic acid in the endoplasmic reticulum. Diacylglycerol of the toad retina is highly enriched in docosahexaenoate and has been implicated to have a biosynthetic origin. However, bovine retina diacylglycerols contain relatively low proportions of this fatty acid. This may indicate either that there is a mixture of several diacylglycerols with different fatty acid profiles or that the docosahexaenoate enriched phosphatidic acid is metabolized without conversion into diacylglycerols. The alternative pathway investigated at present is docosahexaenoate containing phosphatidylserine synthesis from phosphatidic acid without involving base exchange reaction. When using 2-^{3}H-glycerol as a marker of the *de novo* synthesis we obtained evidence of a rapid phosphatidic acid formation. A route introducing docosahexaenoate during phosphatidic acid synthesis may actively participate in controlling membrane function by changing membrane fluidity.

INTRODUCTION

The metabolic steps regulating membrane fluidity through modifications in the fatty acid composition of lipids involve a) the metabolism of phosphatidic acid (PA); b) the metabolism of other glycerolipids; c) intermembrane distribution of phospholipids; and d) transbilayer mobility and methylation reactions that modify the net charge in the polar head of phospholipids (Hirata and Axelrod, 1978). The major pathways depicted in Figure 1 are the Lands (1965) cycle of deacylation-acylation (steps 2 and 3), the desaturation reactions of unsaturated acyl chains of phospholipids (step 4, Kates and Pugh, 1979) and the synthesis of PA (step 1). Although PA comprises a very small pool both the biosynthetic flux (Akesson, 1970; Bazán and Bazán, 1976, Bazán et al., 1976) as well as its turnover (Bleasdale and Hawthorne, 1975; Schacht and Agranoff, 1974) are among the highest of those observed in cellular glycerolipids. Thus the acyl groups channeled through PA likely influence in a significative manner the fatty acid composition of membrane lipids. Moreover it is of interest to determine which fatty acids are introduced in the de novo pathway.

The fatty acyl groups of retinal phospholipids are characterized by a high content of long-chain polyunsaturated components (e.g. 20:4 ω6, 22:4 ω6, 22:5 ω3, 22:5 ω6 and 22:6 ω3). These fatty acyl groups are present mainly in mitochondria , outer plasma membrane of neural cells and in photoreceptor membranes from visual cells. It is not known the role that PA plays in the synthesis of polyenoic containing phospholipids, at which rate these steps operate and how they are regulated. In **addition,** to approach these problems information about the pool size and fatty acid composition of the intrinsic PA pool in microsomes is required. Such experiments have been previously hampered by the lack of suitable techniques for the rapid isolation of PA on a preparative scale. There were only a few reports on the fatty acid composition of PA in entire organs such as in liver (e.g. Possmayer _et al_, 1969; Akesson _et al_, 1970), brain (Baker and Thompson, 1972), bovine and toad retina (Aveldaño and Bazán, 1977) and even fewer in microsomal membranes (Su and Sun, 1978).

However, to the best of our knowledge, no one has either shown a high content of 22:6 in microsomal PA or has demonstrated increased content of microsomal PA, including

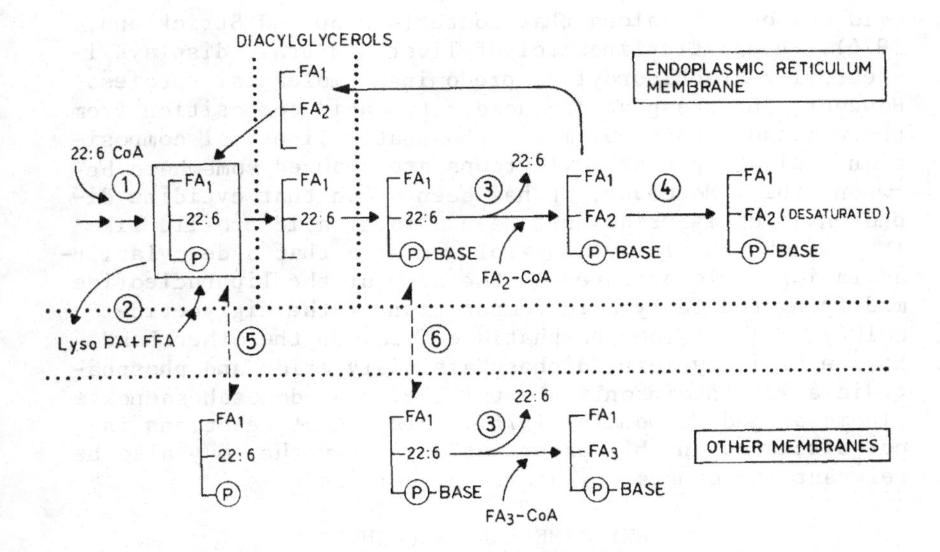

Fig. 1 Examples of reactions involved in acyl group metabolism of membrane lipids: (1), acyltransferases in the de novo synthesis of PA, the present chapter provides data indicating that docosahexaenoyl-CoA is introduced during this step; (2), phospholipase A_2 acting on PA and acylating enzymic system; (3) deacylation-acylation cycle; (4) desaturation of esterified fatty acid; (5) PA distribution to other membranes by exchange; (6) phospholipid exchange. This outline also includes formation of diacylglycerols from phospholipids and the step catalyzed by diglyceride kinase that converts diglyceride into phosphatidic acid. For simplicity and for reasons given in the Introduction the step towards cytidine diphosphate diglyceride and other lipids is not included.

the hexaenoic fatty acid, in any experimental condition. This chapter summarizes recent studies from this laboratory devised to answer several of these questions using the retina as an experimental model. A recent communication from our laboratory reported that the microsomal PA contains about 21% of docosahexaenoate (Giusto and Bazán, 1979b).

The pathway leading towards phosphatidylinositol formation has not been included in Fig. 1 because available evidence indicates that there is a modification in the fatty acid composition along that route (Bishop and Strickland, 1976). Phosphatidylinositol of liver and brain displays 1-stearoyl-2-arachidonoyl as predominant molecular species. However, the phosphatidic acid fatty acid composition from those tissues does not match phosphatidylinositol composition indicating that acyl groups are changed somewhere between them. Moreover, it has been shown that cytidine diphosphate diglyceride does have a fatty acid profile like that of phosphatididylinositol and also that a deacylation-acylation cycle operates at the step of the liponucleotide modifying the fatty acid composition of the diglyceride moiety derived from phosphatidic acid. On the other hand, bovine brain cytidine diphosphate diglyceride and phosphatidic acid contain only about 1.1-1.4% of docosahexaenoate (Thompson and Mc Donald, 1976). This set of reactions is not discussed in this paper but obviously they may also be relevant for membrane fluidity properties.

EXPERIMENTAL PROCEDURES

Eyes were kept packed in crush ice from the time of killing the cattle until dissecting and incubating the retinas (usually about 2 hours). Incubation was in an ionic medium (Ames and Hastings, 1956) pH 7.4 containing 2 mg/ml of glucose during 40 min. Other additions are described along with the pertinent experiment. Retinas were then homogenized with 0.32 M sucrose pH 7.4 containing 50 mM Tris HCl and 10^{-4} M EDTA by means of a motor driven teflon pestle Potter-Elvejhem homogenizer. Microsomes were obtained by spinning 50 min at 140,000 x g the post-mitochondrial supernatant. Each sample was prepared with about 2 g of tissue wet weight comprised of 4 retinas. Lipid extraction was performed as described elsewhere (Giusto and Bazân, 1973, 1979a) and phosphatidic acid isolated by the oblique spotting technique on thin-layer chromatography (Rodrîguez de Turco and Bazân, 1977). Methanolysis was carried out with 14% BF_3-MeOH without prior elution from Silica gel scrapings (Bazân and Bazân, 1975). Fatty acyl methyl esters were separated and quantified in a 5700 model Varian Aerograph gas-liquid chromatograph using methyl 21:0 as internal standard. And other conditions were essentially as previously described (Bazân and Bazân, 1975).

RESULTS AND DISCUSSION

Phosphatidic acid of microsomes from bovine retina contains about 21% of docosahexaenoate (Table I). This represents to the best of our knowledge, the highest content of 22:6 ω6 reported in PA of a subcellular fraction. In whole rat brain only 1.2% of the acyl groups of PA is 22:6 (Baker and Thompson, 1972) and in brain microsomes about 10% of 22:6 was found to be esterified to PA (Su and Sun, 1978). Retinal diacylglycerols in the toad are also rich in 22:6 like phospholipids of that tissue (Aveldaño and Bazán, 1972, 1973, 1974). Thus it was suggested that the docosahexaenoyl-DG may be linked to a neobiosynthetic route derived from PA and leading to phospholipids rather than an accumulated product of polar lipid catabolism (Aveldaño and Bazán, 1972, 1973, 1974). However at that time procedures for the rapid

TABLE I

ACYL CHAINS OF PHOSPHATIDIC ACID, DIACYLGLYCEROLS AND PHOSPHOLIPIDS IN BOVINE RETINA

Percent values ± S.E.M. are given. Microsomal PA are from ten individual determinations

Acyl chains	Microsomes (from Giusto and Bazán, 1979b)	Entire retina (from Aveldaño and Bazán, 1977)	
	Phosphatidic acid	Diacylglycerols	Total Phospholipids
16:0	15.3 ± 1.4	18.2 ± 1.2	22.0 ± 1.4
16:1	1.5 ± 1.2	1.3 ± 0.5	1.2 ± 0.2
18:0	28.2 ± 3.9	25.7 ± 1.6	22.5 ± 0.7
18:1	16.9 ± 2.5	8.9 ± 0.8	14.3 ± 0.5
18:2	1.6 ± 0.6	1.2 ± 0.1	1.0 ± 0.5
20:4 ω6	6.0 ± 0.9	26.6 ± 1.8	6.1 ± 0.2
22:4 ω6	0.4 ± 0.2	0.3 ± 0.1	0.2 ± 0.1
22:5 ω3 + ω6	1.9 ± 0.2	1.1 ± 0.2	2.4 ± 0.1
22:6 ω3	21.1 ± 2.6	7.1 ± 1.3	24.1 ± 0.2
Unsaturation Degree	188	168	199

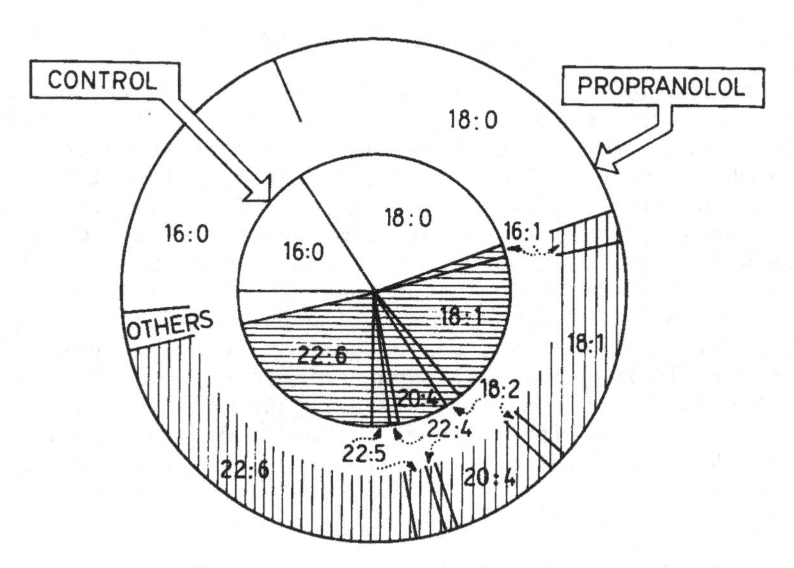

Fig. 2 Increased content of individual acyl chains of phosphatidic acid of endoplasmic reticulum from dl-propranolol treated retinas (outer circle) as compared with controls (inner circle). Areas are proportional to concentration. Hatched areas indicate unsaturated acyl chains. Drug-treated retinas were incubated during 40 min with 500 µM dl-propranolol.

isolation of intrinsic PA of membranes at a preparative scale were not available. Such method was developed by Rodríguez de Turco and Bazán (1977) and allowed the quantitative recoveryof PA preserving highly unsaturated fatty acids such as 22:6, due to its rapidity and other precautions followed against peroxidation (Bazán and Bazán, 1975).

Docosahexaenoate-containing phospholipids are mainly concentrated in the outer segments of visual cells and in plasma membranes from the neural portion of the retina. Microsomal fractions obtained from the retina contain membranes derived from the endoplasmic reticulum of the cells located in the neural retina. In addition there is endoplasmic reticulum in the inner segment of the visual cells. If during homogenization inner segments are broken down microsomes derived from them will also be present in the microsomal fraction. Our studies were made on microsomes prepared

TABLE II

LABELING OF PHOSPHATIDIC ACID FROM ^{3}H-GLYCEROL AND RATES OF ACCUMULATION IN ENDOPLASMIC RETICULUM FROM DL-PROPRANOLOL-TREATED RETINAS

	Labeling ^{a} % above controls	Rates of accumulation^b nmol h^{-1} mg of protein^{-1}
Phosphatidic acid	560	13.26
Docosahexaenoyl containing PA	-	4.92
Phosphatidylinositol	630	-

a: Labeling was measured by incubating bovine retinas with 5 µCi of ^{3}H-glycerol (specific activity 9.52 Ci/mmol) during 30 min as described by Giusto and Bazán, 1979a.

b: Rates were determined after 40 min of incubation. In both cases 500 µM dl propranolol were used. Other details as in Experimental Procedures.

from entire retinas. Microsomal PA is a complex fraction composed of several pools involved in the biosynthesis of diacylglycerides or CDP-diacylglycerides. In addition the contribution of phospholipid degradation through diacylglycerol kinase or through phospholipase D remains to be evaluated in the retina.

Further experiments are needed to clarify several of these questions. In addition to a high content of docosahexaenoate in microsomal PA of retina (Table I) a drug-induced accumulation of this lipid was observed. Short-term incubation of bovine retinas with dl-propranolol give rise to a four-fold enlargement of the PA content in microsomes. Moreover when the content of acyl chains is determined an increased level of all of them was found (Fig.2). This represents an augmented rate of PA formation including the 22:6 containing PA. This conclusion is supported by experiments carried out under similar conditions with 2-^{3}H-glycerol as a marker of biosynthesis (Bazán and Bazán, 1976).

We have previously shown that dl-propranolol as well as phentolamine exert several effects on lipid metabolism in the retina, mainly an increase in the rate of biosynthesis of acidic lipids such as PA, phosphatidylinositol and phosphatidylserine (Bazán et al., 1976, 1977). In Table II the increased labeling of PI and PA from 2-^{3}H-glycerol in microsomes from propranolol-treated retinas is shown. In addition, the rates of accumulation of docosahexaenoyl-PA represent a significant proportion of the rates measured for total PA. An enhancement in the labeling of PA can be seen in microsomal membranes from propranolol-treated retinas at 5 min of incubation. However, a decreased percent incorporation takes place thereafter (Fig.3). The stimulatory effect of phosphatidylinositol and the diminished uptake in diacylglycerols can also be observed in the microsomes. Moreover, no detectable change became apparent in phosphatidylethanolamine and there was a drastic inhibition in triacylglycerols and in phosphatidylcholine (Fig. 3).

CONCLUSIONS

The main feature of the fatty acid composition of PA from retinal microsomes is the relatively high content of docosahexaenoate (Giusto and Bazán, 1979b). Short term incubation of retinas with dl-propranolol yields an enhancement in the net synthesis of microsomal PA (Table II). This was ascertained by measuring the PA pool size through the determination of the concentration of each acyl group. Palmitate, stearate and docosahexaenoate containing PA were mainly responsible for the increment (Fig.2). The accumulation rate of docosahexaenoate containing PA amounted to about 25% of that of total PA. The increase in net synthesis of PA was also assessed by following the labeling by 2-^{3}H-glycerol. In agreement with previous observations, phosphatidylinositol labeling also increased and diacylglycerol decreased during early incubation times with propranolol (Bazán et al. 1976, 1977). Thus the conversion of certain PA pools into diacylglycerol was inhibited and the synthesis of phosphatidylinositol was stimulated (Fig. 4, A).

However, in addition to these effects, the drug is also likely to exert an effect on the de novo synthesis of PA. dl-Propanolol is a β-adrenergic receptor blocker and it elicits also several other effects. This drug is

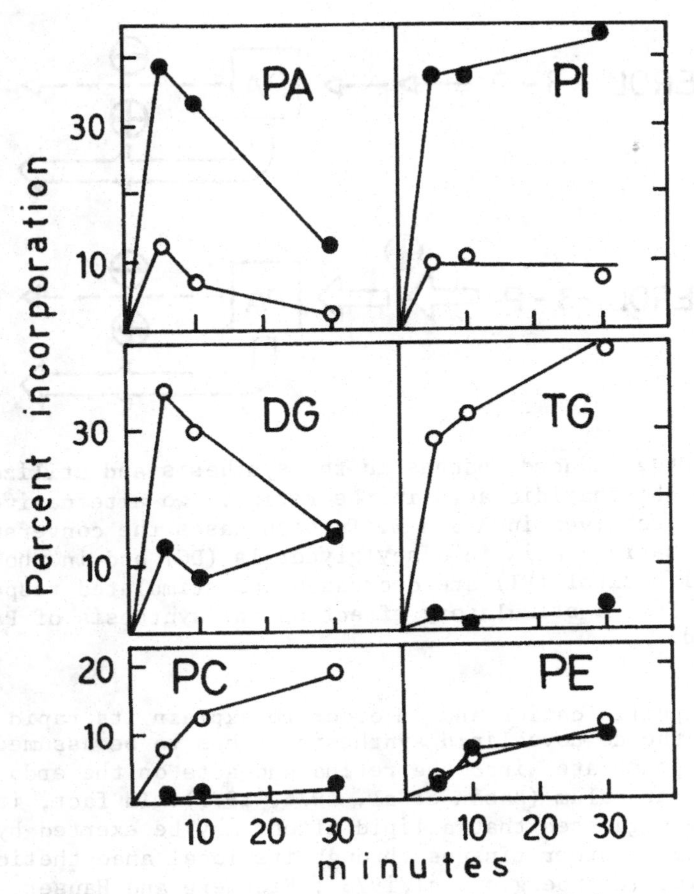

Fig. 3 Comparative effect of propranolol on the synthesis of phosphatidic acid and other glycerolipids in the endoplasmic reticulum of the retina. Fresh bovine retinas were preincubated during 20 min with 500 µM of dl-propranolol. Then 2-^{3}H-glycerol (as in Table II) was added and retinas were sampled after 5, 10 and 30 minutes of further incubation. Details are given elsewhere (Giusto and Bazán, 1979a). Subcellular fractionation followed sampling. Lipid extraction and further analysis of microsomal pellets were performed as described in Experimental Procedures. Closed circles stand for propranolol-treated retinas and open circles for controls. PI: phosphatidylinositol; DG: diacylglycerol; TG: triacylglycerol; PC: phosphatidylcholine and PE: phosphatidylethanolamine.

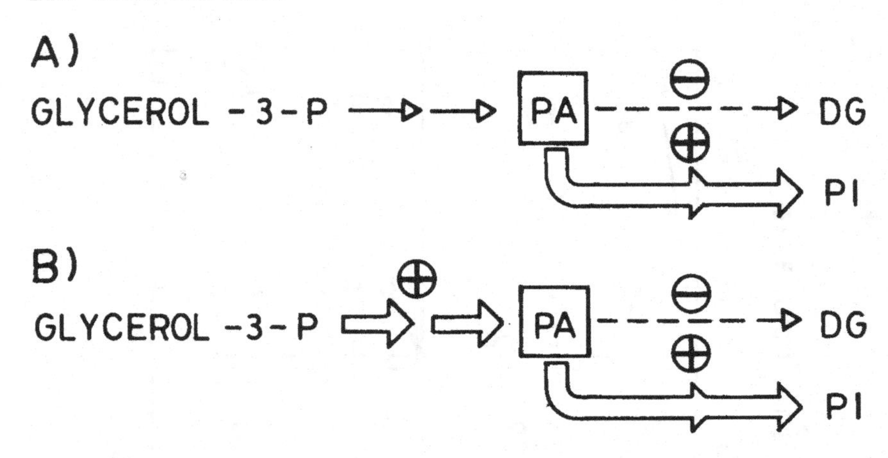

Fig. 4 Drug-induced changes in the synthesis and utilization of phosphatidic acid in the retina. Two alternative effects are given in A and B. In both cases the conversions of phosphatidic acid in diacylglycerols (DG) and in phosphatidylinositol (PI) are decreased and stimulated respectively. In B a stimulatory effect on the synthesis of PA is depicted.

an amphiphilic cation and in order to explain its rapid action on the de novo lipid synthesis it has to be assumed that it penetrates into the retina and acts on the endoplasmic reticulum (Bazán et al. 1976, 1977). In fact, it has been suggested that a lipid effect may be exerted by this drug in other tissues through its local anaesthetic properties (Eichberg et al, 1973 ; Eichberg and Hauser, 1974, Abdel-Latif, 1976). During the past years (Bazán et al, 1976, 1977) we have identified several lipid effects on the retina by dl-propranolol or phentolamine, unrelated to their adrenergic blocker properties. The effect currently described on the synthesis of PA points also to an interesting set of reactions. This implies either that the biosynthetic flux through PA is exceedingly high under basal conditions or that propranolol also stimulates the synthesis of PA from 2-^{3}H-glycerol (Fig.4B). The data on the synthesis of different microsomal membrane lipids from propranolol-treated retinas further support the later alternative. If similar actions are exerted by physiological substances, then, membrane fluidity might be controled, at least in part, by regulating the rate of de novo biosynthesis of PA containing highly unsaturated fatty acids such as docosahexaenoate. The straight-forward experimental outline followed avoided pitfalls often encountered in the study of acylation reactions of PA in systems widely differing from the in vivo situation.

The free fatty acid pool of the retina is a highly labile fraction (Aveldaño and Bazán, 1974; Bazán et al. 1976) similar to that described in the brain (Bazán, 1970) and containing several long-chain polyenoic components (Aveldaño and Bazán, 1975). Since ischemia actively releases free fatty acids in the neural tissue (Bazán, 1970) every precaution was taken to minimize it. However between killing the animals and packing the eyes in crush-ice there is a small period of time not always fully controlable. At any rate the concentration of free 22:6 in retinal microsomes under our experimental conditions is rather low (unpublished). Free docosahexaenoate might stimulate the synthesis of a PA enriched in 22:6 in a manner analogous to that suggested by MacDonald et al (1975). Moreover, if this is the case, it means that the retina is equiped with the enzymes able to acylate 22:6-CoA into PA precursors. The high content of 22:6 in PA of retina described here is in agreement with the proposal that docosahexaenoyl-DG is mainly the product of PA dephosphorylation. However 22:6 in diacylglycerols from the entire bovine retina represents only about 7% of the total acyl chains (Table I). This may be due to the presence of several different diacylglycerol pools. On the contrary, the high content of 22:6 in the toad retina (Aveldaño and Bazán, 1972, 1973, 1974) indicates the predominance of one molecular species in diacylglycerol. Alternatively, a major part of 22:6-containing PA may be used for the synthesis of 22:6-containing phosphatidylserine, without involving diacylglycerol as an intermediate (Bazán et al, 1976). This suggestion is based upon a) the high content of 22:6 in phosphatidylserine in the retina (Aveldaño and Bazán, 1977), b) the redirecting of the de novo biosynthesis of retinal glycerolipids by drugs (Bazán et al, 1976) and c) other metabolic experiments in the retina using radioactive serine (Giusto et al, 1979). Such alternate route which does not involve the base exchange reaction is being investigated at present in our laboratory.

Possmayer et al (1969) demonstrated that in liver saturated fatty acids were incorporated at the 1-position and that unsaturated fatty acids were acylated to the 2-position of phosphatidic acid. The regulation of acyltransferases acting in phosphatidic acid synthesis may participate in the fate of fatty acids among membranes (Bremer et al, 1976, Heming and Hajra, 1977). Moreover the asymmetry given by these reactions to membrane lipids (Tamai and Lands, 1974, Yamashita et al, 1975, Yamada and Okuyama, 1978), hormonal influences (Schneider, 1972), divalent cations (Giusto and Bazán, 1977) and other regulatory effects (Possmayer, 1974) described in the synthesis of phos phatidic acid points to key metabolic steps at this early stage of the biosynthetic pathways of glycerolipids.

Our data further suggest that in order to explain the several actions of the dl-propranolol on lipid synthesis an unspecific interaction of the drug with the endoplasmic reticulum membrane has to be evaluated (Singer, 1975). The effect of dl-propranolol described in this chapter may be an additional action elicited by an intercalation of the amphiphilic cationic drug on the cytoplasmic leaflet of the microsomal membrane through its hydrofobic moiety. The action of propranolol on sarcoplasmic reticulum Ca++ ATPase provides further support to this suggestion (Noack et al., 1978).

Such physical interaction may explain the various lipid effects exerted by this type of drug (Bazán et al, 1976, 1977). Moreover, the study of this drug action may shed some light on the regulation of glycerolipid synthesis and also on the spatial organization of enzymic proteins involved in this microsomal metabolic process. Interactions of the drug altering the molecular packing of lipid molecules in the membrane might cause changes in membrane fluidity and in turn alterations in enzymic membrane proteins. Further studies of the interactions of drugs such as dl-propranolol with membrane lipids and the resulting metabolic effects will contribute to understand the relationships between bulk lipids, lipid synthesis and membrane functions. This is particularly important in the case of docosahexaenoate enriched membranes such as excitable and visual membranes.

REFERENCES

ABDEL-LATIF, A.A. (1976) In G. Porcellati, L. Amaducci and C. Galli (Eds.) Function and Metabolism of Phospholipids in the Central and Peripheral Nervous Systems. Plenum Publishing Corp. New York. 227 pp.

AKESSON, B. (1970) Biochim. Biophys. Acta, 218, 57.

AKESSON, B., J. ELOVSON and G. ARVIDSON (1970) Biochim. Biophys. Acta, 210, 15.

AMES, A., and B.A. HASTINGS (1956) J. Neurophysiol., 19, 201.

AVELDAÑO, M.I., and N.G. BAZAN (1972) Biochem. Biophys. Res. Commun., 48, 689.

AVELDAÑO, M.I., and N.G. BAZAN (1973) Biochim. Biophys. Acta, 296, 1.

AVELDAÑO, M.I., and N.G. BAZAN (1974) J. Neurochem., 23, 1127.

AVELDAÑO, M.I., and N.G. BAZAN (1974) Febs Letters, 40, 53.

AVELDAÑO de CALDIRONI, M.I., and N.G. BAZAN (1977) In N.G. Bazán, R.R. Brenner and N.M. Giusto (Eds.) Function and Biosynthesis of Lipids. Plenum Publishing Corp. New York. 397 pp.

BAKER, R.R., and W. THOMPSON (1972) Biochim. Biophys. Acta, 270, 489.

BAZAN, N.G. (1970) Biochim. Biophys. Acta, 218, 1.

BAZAN, N.G., and H.E.P. BAZAN (1975) In N. Marks and R. Rodnight (Eds.) Research Methods in Neurochemistry. Plenum Publishing Corp. New York. 309 pp.

BAZAN, H.E.P., and N.G. BAZAN (1976) J. Neurochemistry, 27, 1051.

BAZAN, N.G., M.I. AVELDAÑO, H.E.P. BAZAN and N.M. GIUSTO (1976) In R. Paoletti, G. Porcellati and G. Jacini (Eds.) Lipids. Raven Press. New York. 89 pp.

BAZAN, N.G., M.G. ILINCHETA de BOSCHERO, N.M. GIUSTO and H. E. PASCUAL de BAZAN (1976) In G. Porcellati, L. Amaducci and C. Galli (Eds.) Function and Metabolism of Phospholipids in the Central and Peripheral Nervous Systems. Plenum Publishing Corp. New York. 139 pp.

BAZAN, N.G., M.G. ILINCHETA de BOSCHERO and N.M. GIUSTO (1977) In N.G. Bazán, R.R. Brenner and N.M. Giusto (Eds.) Function and Biosynthesis of Lipids. Plenum Publishing Corp. New York. 377 pp.

BISHOP. H.H., and K.P. STRICKLAND (1976) Can. J. Biochem., 54, 249.

BLEADSDALE, J.E., and J.N. HAWTHORNE (1975) J. Neurochem., 24, 373.

BREMER, J., K.S. BJERVE, B. BORREBAEK and R. CHRISTIANSEN (1976) Molec. Cell. Biochem., 12, 113.
EICHBERG, J., H.M. SHEIN, M. SCHWARTZ, and G. HAUSER (1973) J. Biol. Chem., 248, 3615.
EICHBERG, J., and G. HAUSER (1974) Biochem. Res. Commun. 60, 1460.
GIUSTO, N.M., and N.G. BAZAN (1973) Biochim. Biophys. Res. Commun., 55, 515.
GIUSTO, N.M., and N.G. BAZAN (1977) In N.G. Bazán, R.R. Brenner and N.M. Giusto (Eds.) Function and Biosynthesis of Lipids. Plenum Publishing Corp. New York. 481 pp.
GIUSTO, N.M, H.E.P. de BAZAN, M.G.I. de BOSCHERO, M.M. CAREAGA, and N.G. BAZAN (1979) In Abstract Book, XIth International Congress of Biochemistry, Toronto, Canada.
GIUSTO, N.M., and N.G. BAZAN (1979a) Exp. Eye Res. (In press).
GIUSTO, N.M., and N.G. BAZAN (1979b) Biochim. Biophys. Res. Commun. (In press).
HEMING, P.J., and A.K.A. HAJRA (1977) J. Biol. Chem., 252, 1663.
HIRATA, F., and J. AXELROD (1978) Nature, 275, 219.
KATES, M., and E.L. PUGH. This volume.
LANDS, W.E.M. (1965) Ann. Rev. Biochem., 34, 313.
MAC DONALD, G., R.R. BAKER, and W. THOMPSON (1975) J. Neurochem., 24, 655.
NOACK, E., M. KURZMACK, S. VERJOVSKI-ALMEIDA, and G. INESI (1978) J. Pharmacol. Exp. Ther., 206, 281.
POSSMAYER, F., G.L. SHERPHOF, T.M.A.R. DUBBELMAN, L.M.G. GOLDE and L.L.M. VAN DEENEN (1969) Biochim. Biophys. Acta, 176, 95.
POSSMAYER, F. (1974) Biochem. Biophys. Res. Commun., 61, 1415.
RODRIGUEZ de TURCO, E.B., and N.G. BAZAN (1977). J. Chromatog., 137, 194.
SCHACHT, J., and B.W. AGRANOFF (1974) J. Biol. Chem., 249, 1551.
SCHNEIDER, P.B. (1972) J. Biol. Chem., 247, 7910.
SINGER, M. (1975) In B.R. Fink (Ed.) Molecular Mechanism of Anesthesia. Raven Press. New York. 223 pp.
SU, K.L., and G.Y. SUN (1978) J. Neurochem., 31, 1043.
TAMAI, Y., and W.E.M. LANDS (1974) J. Biochem., 76, 847.
THOMPSON, W. and G. MAC DONALD (1976) Eur. J. Biochem., 65, 107.
YAMADA, K., and H. OKUYAMA (1978) Arch. Biochem. Biophys., 190, 409.
YAMASHITA, S., N. NAKAYA, M. NUMA (1975) Proc. Nat. Acad. Sci., 72, 600.

PART IV

PHOSPHOLIPID CHANGES ACCOMPANYING PHYSIOLOGICAL EVENTS

DYNAMICS OF MEMBRANE FATTY ACIDS DURING LYMPHOCYTE STIMULATION BY MITOGENS

E. Ferber, E. Kröner, B. Schmidt, H. Fischer, B.A. Peskar* and C. Anders

Max-Planck-Institut für Immunbiologie, Freiburg and *Pharmakologisches Institut der Universität, Freiburg D-78 Freiburg, West-Germany

INTRODUCTION

When antigens or mitogens bind to the surface of lymphocytes, a sequence of metabolic events is triggered leading to proliferation of the cells and to the expression of new functions. Several studies showed (Greaves and Bauminger, 1972; Betel and van den Berg, 1972; Andersson et al., 1972) that the binding of the stimulating ligands on the plasma membrane is sufficient to activate the cell. Consequently one can assume that the activation is initiated by processes which occur in the outer cell membrane. However, it is difficult to establish a single sequence of events since a variety of changes obviously occur in the plasma membrane during the early phase of the activation. Thus the permeability for nucleosides (Peters and Hausen, 1971a) sugars (Peters and Hausen, 1971b) and amino acids (Mendelsohn et al., 1971; van den Berg and Betel, 1973) or ions such as K^+ (Quastel and Kaplan, 1971) or Ca^{++} (Whitney and Sutherland, 1972) is increased. Since many of these functional changes are transient events it seems necessary to find mechanisms which lead to stable structural changes of the outer cell membrane. In the past, we have described changes of phospholipids which are connected with the activation process and lead to structural changes of the plasma membrane. In the first section some of this work on the fatty acid metabolism of stimulated lymphocytes will be summarized. Since the stimulation of lymphocytes only occurs when they interact with small numbers of macrophages we will then report on recent findings concerning de-acylation (phospholipase A) and re-acylation processes of bone-marrow macrophages.

Finally, the preparation of membrane areas will be described followed by a discussion of the phospholipid dynamics for membrane structure and function.

LIPID COMPOSITION OF RESTING LYMPHOCYTES

In lymphocytes the major components of phospholipids are phosphatidyl choline and phosphatidyl ethanolamine which represent 70% of the total phospholipids. The sphingomyelin and cholesterol content (cholesterol/phospholipid ratio) was found to be relatively high compared to other cells (e.g. 5.4% for liver ref. Ray et al., 1969). This might reflect the high proportion of the plasma membrane in lymphocytes (Patton, 1970).

In recent years it has been established that the fatty acid moieties of phospholipids determine many membrane functions (de Kruyff et al., 1973; McElhaney et al., 1973; Wilson and Fox, 1971; Schairer and Overath, 1969; Bloj et al., 1973; Goldemberg et al., 1973). Therefore it was important to determine the fatty acid composition of the main phospholipids from lymphocytes of different species (Ferber et al., 1975) and to compare these data with those from rat liver. Some special features of the fatty acid composition of lymphocyte phospholipids are apparent: lecithin contained relatively small amounts of polyunsaturated fatty acids (18:2 and 20:4) and stearic acid (18:0), counterbalanced by a higher content of palmitic acid (16:0) and oleic acid (18:1). A higher content of oleic acid (18:1) and a lower content of stearic acid (18:0) were also present in phosphatidyl ethanolamine, when compared to rat liver. However, in contrast to lecithin, the arachidonic acid (20:4) content was higher than in rat liver, equalized by a low content of palmitic acid (16:0). In addition, some fatty acids were distributed in lymphocytes between positions 1 and 2 of the phospholipid molecule in an unusual pattern. In lecithin, and in phosphatidyl ethanolamine, the content of the saturated fatty acid palmitic acid (16:0) was high in position 2. The unsaturated fatty acid oleic acid was even found preferentially in position 1. These distributions strongly suggest that in lymphocytes, there exist species of phospholipids other than those that occur in rat liver. The lower content of polyunsaturated fatty acids in the main phospholipid lecithin is compatible with relatively rigid membranes occurring in resting lymphocytes.

Interestingly, normal lymphocytes from different species exhibit a very similar phospholipid fatty acid composition. This fact suggests the existence of tissue-specific patterns of individual phospholipids rather than species-specific ones, thus reflecting adaptation to environment and function.

DYNAMICS OF PHOSPHOLIPIDS IN STIMULATED LYMPHOCYTES

Incorporation Studies

On the basis of the relatively high content of saturated fatty acids in resting lymphocytes enzymatically catalyzed exchange reactions which occur in stimulated lymphocytes seem to be of great importance. In contrast to the *de novo* synthesis of phospholipids which is increased not earlier than 6 hours after stimulation enhanced incorporation of unsaturated fatty acids we observed within the first hour of stimulation (Resch and Ferber, 1972; Resch *et al.*, 1972). Moreover, the quantitative comparison of both processes revealed that the incorporation of long chain fatty acids exceeds *de novo*

Table 1: Incorporation of ^{14}C-Oleate into Microsomal Lecithin

Incubation (min)	cpm per 10^9 Lymphocytes: Controls	Lymphocytes Stimulated by: PHA	Lymphocytes Stimulated by: PPD
10	728 $\pm$ 29[a]	2134 $\pm$ 24	nd[b]
30	2563 $\pm$ 71	5540 $\pm$ 117	nd
60	3098 $\pm$ 282	8787 $\pm$ 88	3819 $\pm$ 152
120	4439 $\pm$ 66	15301 $\pm$ 556	nd
180	5887 $\pm$ 251	nd	7337 $\pm$ 37
240	6583 $\pm$ 627	19276 $\pm$ 636	8818 $\pm$ 282

Note: 10^9 lymphocytes were cultured in 20 ml of Hepes-buffered (pH 7.2) Eagles medium containing 200 nmol of ^{14}C-oleate. PHA was added to a final concentration of 40 µg/ml; PPD was added to a final concentration of 50 µg/ml. After incubation, the cells were washed twice with phosphate-buffered saline, resuspended, and disrupted. The microsomes were isolated and washed once with 0.001 M Hepes buffer (pH 7.5) to remove cytoplasmic protein.
Incorporation of ^{14}C-oleate into lecithin was measured in microsomes recovered from 10^9 lymphocytes. The recovery of microsomal protein from 10^9 cells ($\pm$ SD) was 0.77 $\pm$ 0.19 mg in controls and 0.91$\pm$0.13 PHA-stimulated lymphocytes. In PPD-stimulated lymphocytes, the same amount of microsomal protein was recovered as in the controls. Microsomes from lymphocytes consist predominantly of plasma membranes and free ribosomes.

[a]Data are given with ranges from experiments performed in duplicate. [b]Not done.

synthesis by more than one hundredfold. These effects were even more pronounced when we followed the transfer of fatty acids to phospholipids of the plasma membrane (Resch and Ferber, 1975).
Table 1 shows that in non-purified plasma membranes the incorporation of oleic acid into lecithin is enhanced 3 fold already 10 minutes after the addition of PHA.

Thus at least during the early phase of the stimulation process the elevated fatty acid metabolism mainly consists of a preferential transfer of long chain fatty acids from intracellular sources to phospholipids of the plasma membrane. The discrepancy of low de novo synthesis and high rates of fatty acid incorporation can only be explained by a separate turnover of the fatty acid moieties of phospholipids. This separate turnover consists of two reactions which are catalyzed by two specific enzymes: phospholipase A_1 or A_2 liberates from diacylphospholipids, phosphatidyl choline or phosphatidyl ethanolamine, one of the two long chain fatty acids thus generating the corresponding lyso-compounds (Fig. 1).

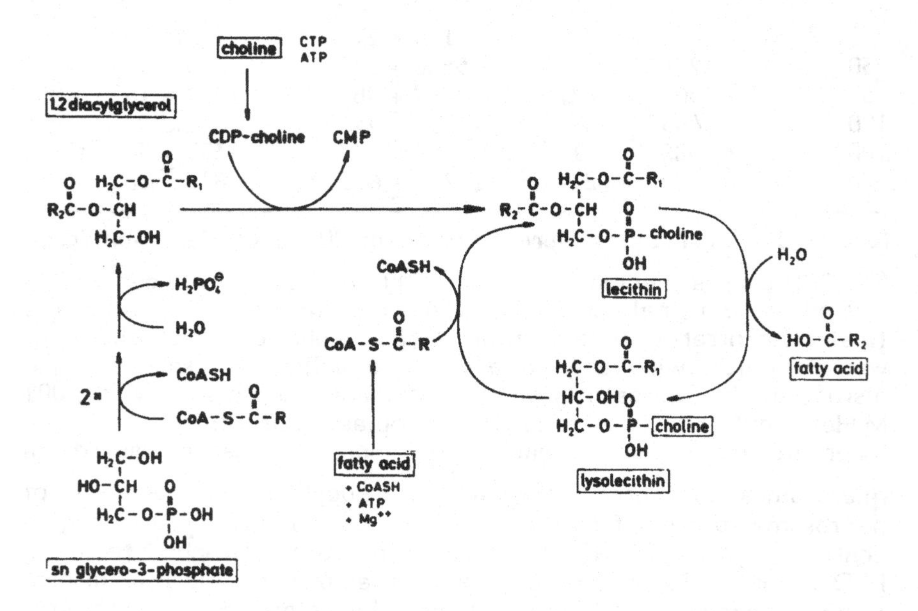

Fig. 1 Pathways of lecithin metabolism

These toxic (cytolytic) substances are either detoxified by further degradation by the action of lysophospholipases or are reutilized by specific acyltransferases. These acyltransferases exhibit a great specificity for both the position of the molecule to which the fatty acid is transferred and the structure of the fatty acid. The acyltransferases only reacylate lysophosphatides and thus form again the original phospholipids. The biological significance of this system is based on the fact that it functions as an exchange cycle for the fatty acid moieties of phospholipids (Ferber, 1973).

Lysolecithin-Acyltransferase of the Lymphocyte Plasma Membrane

The main findings concerning the turnover cycle of fatty acids in stimulated lymphocytes can be summarized as follows: in contrast to other cells in lymphocytes the acyl-CoA: lysolecithin acyltransferase is located predominantly in the plasma membrane. Directly after binding of the stimulant a transferase is activated which exhibits high affinity for polyunsaturated fatty acids (e.g. arachidonic acid $C_{20:4}$) (Ferber and Resch, 1973; Ferber et al., 1976) (Table 2).

Table 2: Substrate Specificity of Microsomal Acyl CoenzymeA: Lysolecithin Acyltransferase of Con A-stimulated Thymocytes

Substrate	Oleoyl CoA		Arachidonoyl CoA	
	V_{max}[a]	K_{m} (M)	V_{max}	K_{m}(M)
Control	3.9	10^{-5}	7.6	6.4×10^{-7}
Con A	7.9	1.2×10^{-5}	31.7	8.5×10^{-7}

Calf thymocytes were incubated for 60 min with or without Con A in Hepes-buffered Eagle's medium. Cell density 5×10^{7} cells/ml; Con A concentration, 20 µg/ml.
Then the cells were disrupted by nitrogen cavitation and the acyltransferase activity was determined in microsomal membranes.
[a] Given as nmol x mg^{-1} x min^{-1}

This finding explains the results obtained from the incorporation studies as described above.
The conclusion that the activation of the transferase occurs immediately is based not only on the comparison of the time course of binding of the stimulating ligand with the enzyme activation but also on the fact that this process needs no metabolic

energy. Stimulation of lymphocytes with concanavalin A at 0°C results in an activation of the same degree (Table 3).

Table 3: Acyl Coenzyme A: Lysolecithin Acyltransferase in Microsomes of Con A-stimulated Thymocytes

	nmol/mg/min	
	0°C	37°C
Control	16.1	15.1
Con A	47.2	38.0

Calf thymocytes were incubated for 60 min with or without Con A in Hepes-buffered Eagle's medium at 0°C and 37°C. Cell density, 5×10^7 cells/ml; Con A concentration, 20 µg/ml. Then the cells were disrupted and the acyltransferase activity was determined in microsomal membranes.

However, this does not mean that the enzyme itself is the receptor for lectins. Membrane fragments which have the same number of binding sites for concanavalin A do not respond with a similar activation of the transferase. Thus we have to assume that the activation is caused by a more complex interaction of specific receptors with the enzyme. Obviously this interaction is impaired when the membrane is fragmented into small vesicles. Kinetic studies of the activation of the acyltransferase revealed further evidence for the underlying mechanism. We were able to show that the activation is caused by a cooparative interaction of the Con A-receptor complex and the enzyme (Ferber et al., 1975).

Changes of the Fatty Acid Composition

The activation of plasma membrane bound acyltransferase occurs immediately and requires no energy metabolism.

The consequence of the activated enzymatic exchange cycle i.e. a changed fatty acid composition can be detected, however, not earlier than after 4 hours of cultivation with the stimulating ligand (Ferber et al., 1975). The 2-acyl-esters of phospholipids usually contain unsaturated fatty acids. Unstimulated lymphocytes, however, have a high percentage of saturated fatty acids in position 2 of the molecule (Table 4). The changes in the fatty acid composi-

tion observed after 4 hours of stimulation with concanavalin A exclusively concern 2-acyl-esters. Predominantly the arachidonic acid and to a lesser extent linoleic acid content is increased. Thus these changes reflect the positional and substrate specificity of the activated lysolecithin acyltransferase.

Table 4: Changes of Fatty Acid Distribution of Phosphatidyl choline and Phosphatidyl ethanolamine in Calf Thymocytes During Stimulation with Con A

		Mol %						Ratio of Polyenoic/Saturated
		16:0	18:0	18:1	18:2	20:4	22:6	
Phosphatidyl choline								
Control	Pos. 1	31.8	24.7	27.7	10.8	-[a]	-	0.191
Con A	Pos. 1	30.9	23.8	29.5	10.9	-	-	0.199
Control	Pos. 2	40.6	1.8	23.3	16.6	6.9	tr[b]	0.554
Con A	Pos. 2	35.2	1.5	23.5	21.1	13.1	tr	0.934
Phosphatidyl ethanolamine								
Control	Pos. 1	17.4	52.6	25.2	4.8	-	-	0.069
Con A	Pos. 1	16.1	51.8	26.9	5.2	-	-	0.077
Control	Pos. 2	28.5	4.5	25.2	15.1	26.8	tr	1.270
Con A	Pos. 2	20.8	3.9	27.9	12.2	35.2	tr	1.919

Note: Con A concentration, 5 μg/ml; cultivation time, 4 hr.

[a]Not detectable. [b]Trace.

Fatty Acid Metabolism of Macrophages

The reactions involved in the resynthesis of phospholipids are fairly well established. On the other hand, much less is known about the preceding steps of phospholipid degradation resulting in the generation of lysocompounds.

Lymphocytes obviously exhibit only trace activities of phospholipases. Therefore, one has to assume that this acceptor substrate is provided from an exogeneous source. It is known that lysophosphatides can be easily exchanged from serum to cells and also between adjacent cells. One possibility is that macrophages generate and exchange lysophosphatides to lymphocytes. Munder (Munder et al., 1966; Munder et al., 1969) showed that these cells, when activated by adjuvants, exhibit high phospholipase A activities which lead to the release of lysolecithin.

On the other hand it is known that even mitogenic stimulation of lymphocytes requires the presence of small numbers of macrophages (Habu and Raff, 1977) and that only those lymphocytes which come in contact with macrophages proliferate (Matthes et al., 1971). Therefore it was necessary to study the fatty acid metabolism of macrophages. Instead of peritoneal macrophages which are a mixture of less defined macrophages, lymphocytes and granulocytes, bone-marrow-derived macrophages cultured according to Meerpohl et al. (1976) were used in these studies. In the experiment shown in Fig. 2 macrophages were prelabelled with 1-^{14}C-arachidonic acid for 60 minutes. Most of the fatty acid was incorporated in phosphatidyl choline (60%) while 25% were incorporated into phosphatidyl ethanolamine and 10% into the neutral lipid fraction. Fig. 2 shows that for up to 6 hours there was a continuous degradation of phospholipids (predominantly of phosphatidyl choline) in macrophages stimulated with zymosan but only when p-chloromercuribenzoate was added. The reason for this is given in Tab. 5 where it is shown that p-chloromercuribenzoate effectively inhibits the acylation (probably the re-acylation of lysolecithin to phosphatidyl choline) but does not impair the phospholipase A. Although there is no doubt that other enzymes are also inhibited for up to 2 hours at least the effect is fully reversed when the cells are washed.

This experiment shows that the amount of liberated arachidonic acid is not only controlled by phospholipase A but even more by the high acyltransferase activities of macrophages.

In accordance with these results are experiments with the Ca^{++}-ionophore A 23187. This compound is known to activate phospholipase A (Pickett et al., 1977). Using this compound phosphatidyl choline was degraded without the addition of p-chloromercuribenzoate and when added it did not enhance the liberation of arachidonic acid (Fig. 3). Here again arachidonic acid only accumu-

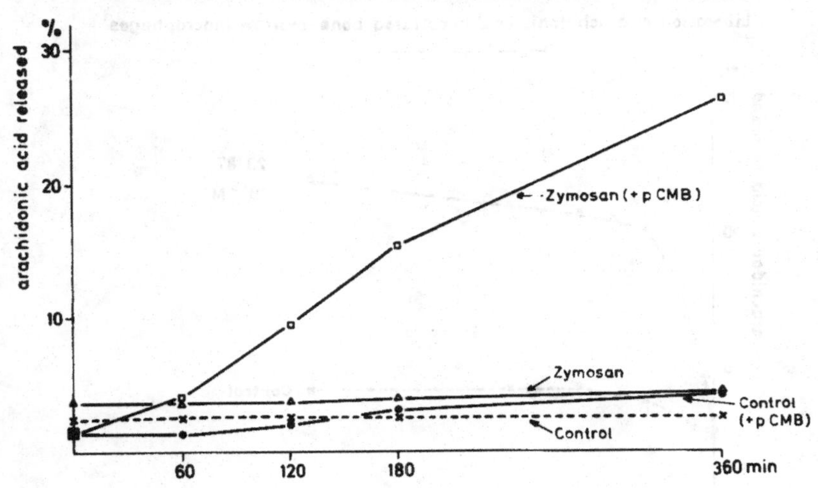

Fig. 2 1×10^6 mouse bone-marrow-derived macrophages were labelled with 0.05 µCi (0.9 nmol) 1-^{14}C-arachidonic acid for 60 min in 1.0 ml. Then the cells were washed and the pre-incubation was continued for 60 min with and without p-chloromercuribenzoate (pCMB). After pre-incubation 1 mg zymosan was added and the cells were incubated up to 6 hours.
About 60% of the arachidonic acid was incorporated into phosphatidyl choline while 25% were incorporated into phosphatidyl ethanolamine and 10% into the neutral lipid fraction.

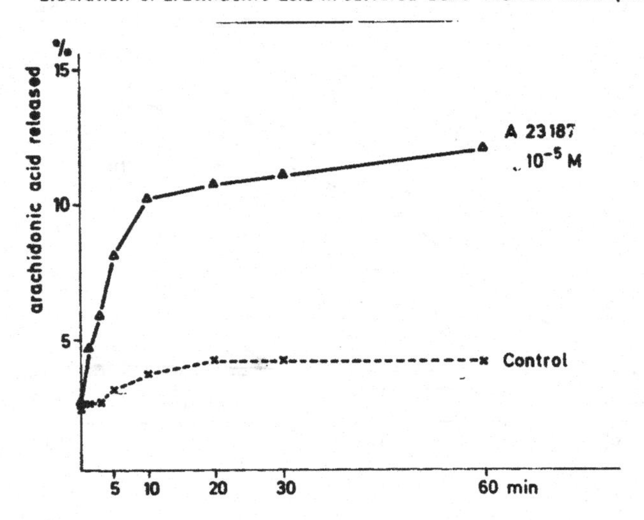

Fig. 3 1×10^6 mouse bone-marrow-derived macrophages were labelled with 0.05 µCi (0.9 nmol) 1-^{14}C-arachidonic acid for 60 min in 1.0 ml. Then the cells were washed, Ca^{++}-ionophore A23187 was added and the cells were incubated up to 60 min. Further details see Fig. 2.

lates when the acyltransferase is inhibited. As shown in Tab. 5 this inhibition is achieved by the ionophore itself*.
In contrast to lymphocytes concanavalin A does not stimulate the fatty acid turnover of macrophages. Neither the acylation (Tab. 5) nor the liberation of arachidonic acid (with and without p-chloromercuribenzoate) was changed by concanavalin A. With respect to the mechanism by which zymosan activates the phospholipase A, it is of interest that phagocytosis itself - which is necessary for the uptake of zymosan particles-does not increase the liberation of arachidonic acid from phospholipids since phagocytosis of inert latex particles does not influence this reaction.
Thus concanavalin A and zymosan can be favorably used in this system to activate distinct pathways of the phospholipid metabolism in distinct cells. Concanavalin A activates the acylation of lysophosphatides in lymphocytes without having any effect on their low phospholipase A activity and without any effect on the de- and re-acylation-reactions of macrophages.
On the other hand zymosan selectively activates the phospholipase A of macrophages without any effect on the acylation reaction (Tab. 5, Fig. 2). If one rationalizes these findings with respect to the stimulation of lymphocytes and the interaction of macrophages with lymphocytes it seems reasonable to expect that the macrophage has the capacity to provide lysolecithin and free arachidonic acid to adjacent cells. The conditions necessary are the activation of the phospholipase A and at least a partial inhibition of the acyltransferase. The control of the steady-state concentrations of free fatty acids by phospholipase A and the lysolecithin-acyltransferase has further implications for the formation of compounds derived from arachidonic acid as the prostaglandins. Since the availability of free arachidonic acid is the rate limiting step during the synthesis of prostaglandins in most studies it was assumed that the phospholipase A is this rate-limiting enzyme (Vogt, 1978). From the results described above it is evident that the amount of free fatty acids at a certain level of phospholipase A activity is controlled much more efficiently by the acyl-CoA: lysolecithin acyltransferase. This is also clearly demonstrated by the determination of these two enzymes in homogenates of macrophages and thymocytes (Tab. 6).

---

*In addition to these effects it is of interest that also indomethacin inhibits effectively the lysolecithin-acyltransferase.

Table 5: Acylation of Lysolecithin

Incorporation of arachidonic acid into phosphatidyl choline of cultured bone-marrow macrophages

compounds tested (content/ml)		Incorporation nmol/ 5 x 10^5 cells/60min	%
Control		1.76	100
p-Chloromercuribenzoate*	100 nmol	0.27	15.4
Ca^{++}-Ionophore A 23187	10 nmol	0.91	51.6
Idomethacin	100 μg	0.33	19.0
Zymosan	1 mg	1.79	101.0
Concanvalin A	100 μg	1.79	101.0

* After 120 min incubation the inhibition caused by p-chloromercuribenzoate could be fully reversed by washing the cells.

Tab. 6:

Phospholipase A_2 and Acyl-CoA: Lysolecithin Acyltransferase Activities in Homogenates of Mouse Macrophages and Mouse Thymocytes

	nmol . mg^{-1} . min^{-1} Macrophages	Thymocytes
Phospholipase A_2	1.1	0.01
Lysolecithin-Acyltransferase	15.6	2.95

Acyl-CoA: lysolecithin acyltransferase activities were determined according to Ferber and Resch (1973). For phospholipase A_2 determination 1.0 samples containing 50 nmol ^{14}C-phosphatidyl choline (fatty acid labelled) in 10 mM HEPES-buffer pH 7.5 containing 2 mM $CaCl_2$ and 0.5% sodium cholate were incubated for 60 min at 37^oC.

RECEPTORS, ENZYMES AND MEMBRANE-AREAS

The studies on the mechanism of the acyltransferase activation revealed indirect evidence for the heterogeneous structure of the lymphocyte plasma membrane. In particular two findings support this notion. Membrane fragments, although they contain acyltransferase activity and ConA-binding sites, do not respond with an activation of the enzyme. On the other hand there was evidence that the activation was caused by the interaction of high affinity binding sites with the enzyme. Therefore it seemed justified to try a subfractionation of plasma membranes in order to enrich these areas which bear the stimulating receptor. Moreover, the preparative separation and the analysis of these membrane fragments should enable studies to find the criteria of a membrane-area competent to transfer the stimulating signal. The idea behind this is that the triggering does not depend on a special structure of the receptor molecule but on the structure of the surrounding membrane.
Two methods which are based on completely different principles have allowed the separation of membrane vesicles from purified plasma membranes i.e. affinity chromatography on ConA-Sepharose (Brunner et al., 1976) and the free-flow electrophoresis (Brunner et al., 1977). Fig. 4 shows the separation of purified thymocyte plasma membranes by free-flow electrophoresis. Three fractions of membrane vesicles which exhibit different enzyme activities were obtained. All these activities were still membrane-bound.
It is of interest that those fractions which exhibited lysolecithin acyltransferase activity contained receptors with high affinity for concanavalin A (Tab. 7).

Further evidence for the existence of membrane areas of distinct structure and function was obtained from the analysis of vesicles shed from intact thymocytes. These vesicles are shed to a small extent spontaneously from thymocytes. Interestingly, sublytic concentrations of a lysolecithin analog (1-dodecyl-propanediol-3-phosphorylcholine, ET-12-H) which do not impair membrane-bound enzymes increase the yield of shedded vesicles. Thus 15-25% of the total activity of the plasma membrane enzymes: alkaline phosphatase (EC 3.1.3.1.), nucleotide pyrophosphatase (EC 3.1.4.1.) and γ-glutamyl transferase (EC 2.3.2.2.) could be removed without any increased liberation of lactate dehydrogenase. None of these activities of plasma membrane enzymes were solubilized but were still tightly bound to the membrane. Moreover, both spontaneously shed and detergent-induced vesicles exhibited all typical characteristics of plasma membranes. Thus, the cholesterol and phospholipid content and the distribution of individual phospholipids was almost identical in shed membranes (spontaneously shed

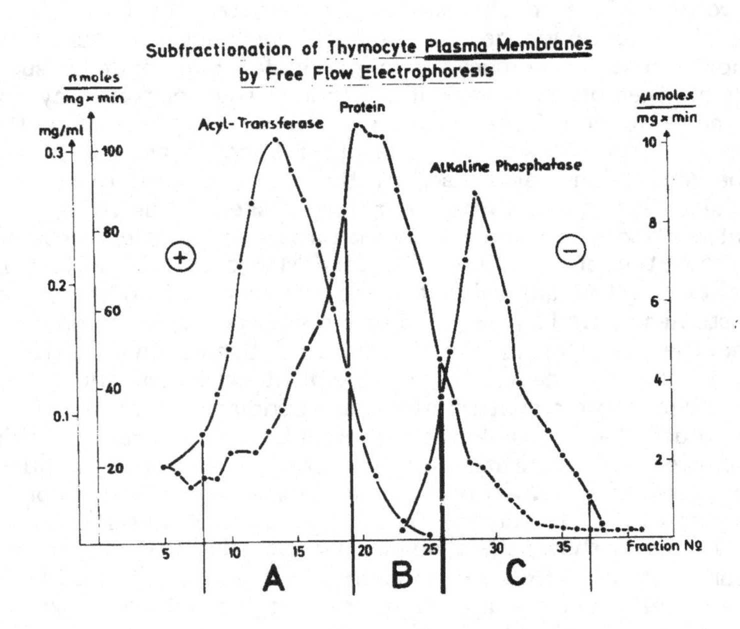

Fig. 4 Calf thymocyte plasma membranes (6-8 mg protein/ml) were separated by free-flow electrophoresis (Elphor Va P-5, Bender and Hobein, Munich, West-Germany). Protein is given as mg/ml. The specific activities of the acyl-CoA: lysolecithin acyltransferase are given as nmol . mg^{-1} . min^{-1} and of the alkaline phosphatase as µ moles . mg^{-1} . min^{-1}. For details of the method see Brunner et al. (1977).

Tab 7: Subfractionation of Plasma Membranes by Free-Flow Electrophoresis

Pooled fractions	A	B	C
Alkaline phosphatase	203	491	5120
Lysolecithin acyltransferase	71	9.5	4.9
Cholesterol (nmol/mg protein)	740	940	1110
Phospholipid (nmol/mg protein)	1320	1080	1220
Cholesterol/phospholipid (molar ratio)	0.55	0.86	0.88
Con A -binding sites (molecules/mg protein)	$2.0x10^{14}$	$4.2x10^{14}$	$8.2x10^{14}$
Con A -affinity (l/mol)	$82x10^{6}$	$37x10^{6}$	$22x10^{6}$

After electrophoretic separation of purified plasma membranes of calf thymocytes fractions were pooled (A,B,C) as indicated in Fig. 4. Specific enzyme activities are given as nmol x mg^{-1} x min^{-1}.

and detergent-induced vesicles) and conventional purified plasma membranes prepared after nitrogen cavitation (Tab. 8). Although these results show that the shed vesicles contain all constituents of the plasma membrane it should be emphasized that shed vesicles do not represent a statistical average of the entire plasma membrane. This is documented by the fact that the composition of spontaneously released vesicles and ET-12-H-induced vesicles differ. Of particular interest are the high specific activities of γ-glutamyl transferase in ET-12-H-induced vesicles (Fig. 5). With increasing ET-12-H concentrations this enzyme was enriched up to a maximum of 7-fold over spontaneously shed membranes. Another difference appeared in the composition of the phospholipid fatty acids (Tab. 9). The lowest amounts of unsaturated fatty acids were found in spontaneously shed vesicles while increasing amounts were

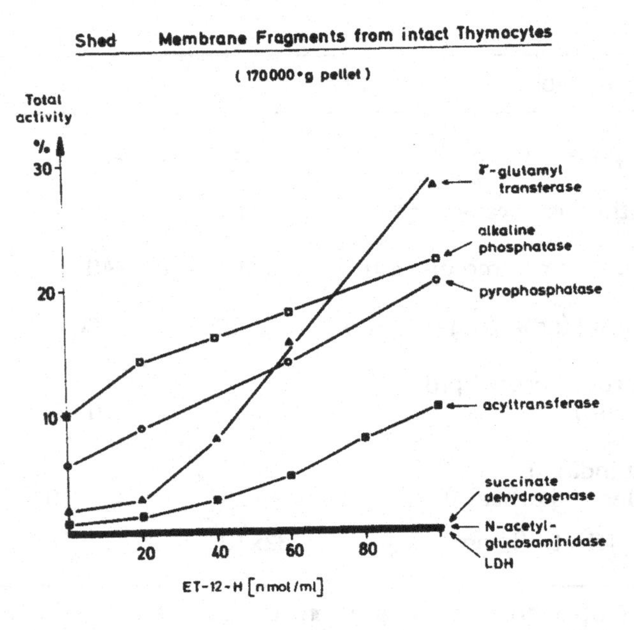

Fig. 5 Suspensions of calf thymocytes (1.5 x 10^8 cells/ml) in phosphate-buffered saline were incubated without and with increasing concentrations of 1-dodecyl-propanediol-3-phosphorylcholine (ET-12-H) 10 min at 37°C. After incubation intact cells were removed at 250 x g_{av}, 5 min, and two thirds of the supernatant were centrifuged at 9000 x g_{av}, 2 min. This centrifugation step removed broken cells and large fragments. Finally the supernatant was centrifuged at 170 000 x g_{av}, 60 min and the membraneous pellet was resuspended and analyzed for enzyme activities, lipids and proteins.

Table 8:
Comparison of the Composition of Membrane Vesicles Shed from Intact Thymocytes with Purified Plasma Membranes (Nitrogen Cavitation)

	Shedded membranes	Purified plasma membranes
Total lipids (nmol/mg protein)		
Cholesterol	320.1 ± 18.2	580.2 ± 25.3
Phospholipid	574.1 ± 28.6	980.2 ± 42.3
Molar ratio	0.557 ± 0.02	0.592 ± 0.03
Phospholipid distribution (%)		
Lysophosphatidyl choline	5.0	---
Sphingomyelin	12.0	12.1
Phosphatidyl choline	49.1	49.2
Phosphatidyl serine + phosphatidyl inositol	12.9	15.0
Phosphatidyl ethanolamine	21.0	23.7
Enzymes (nmol x mg^{-1} x min^{-1})		
Alkaline phosphatase	1250.1 ± 78.1	1150.2 ± 62.1
Nucleotide pyrophosphatase	420.1 ± 22.1	510.3 ± 28.3
γ-Glutamyl transferase	30.2 ± 2.0	31.5 ± 1.5
Mg^{++}-ATPase	18.0 ± 1.0	42.1 ± 2.0
Acyl-CoA:lysolecithin acyltransferase	6.0 ± 0.5	30.2 ± 1.9

Values are means ± standard deviation from 3 preparations with 3 determinantions each.

---

present in detergent-prepared vesicles and normal purified plasma membranes. Interestingly these differences were not restricted to distinct phospholipids but were detected in the total lipids and the major phospholipids-phosphatidyl choline and phosphatidyl ethanolamine-as well. The most likely explanation for these findings is that the plasma membrane of the thymocyte consists of areas of different structure and function which are anchored at different strengths within the membrane. These areas consist of different enzymes which are surrounded by lipids of distinct fatty acid composition. Obviously, those parts of the membrane which contain

Table 9:

Fatty Acid Composition of Shed Membranes and of Purified Plasma Membranes

Shed membranes induced without addition (control) and with addition of ET-12-H. Purified plasma membranes (PM) were prepared by the nitrogen cavitation method. Values are expressed as mol %.

	16:0	16:1	18:0	18:1	18:2	20:1 +18:3	20:4	22:5	22:6	Polyenoic/ saturated
Total lipids										
Control	44.9	1.7	18.9	25.4	7.2	---	1.7	---	---	0.139
ET-12-H 20 nmol/ml	40.3	1.8	18.3	30.8	6.8	---	2.0	---	---	0.150
ET-12-H 40 nmol/ml	38.2	2.3	18.6	29.2	6.3	1.1	3.2	1.0	---	0.204
PM	29.4	2.6	18.3	32.3	9.3	0.9	5.2	1.7	0.2	0.363
Phosphatidyl choline										
Control	52.6	3.0	11.6	26.3	5.6	0.9	---	---	---	0.101
ET-12-H 20 nmol/ml	56.9	3.5	7.5	27.5	3.6	0.9	---	---	---	0.070
ET-12-H 40 nmol/ml	54.4	3.6	9.1	26.1	5.8	0.9	---	---	---	0.106
PM	41.1	2.6	9.5	33.2	11.1	0.9	1.6	---	---	0.269
Phosphatidyl ethanolamine										
Control	31.7	---	19.4	30.1	16.4	---	2.6	---	---	0.370
ET-12-H 20 nmol/ml	23.7	---	22.4	33.2	13.9	1.1	5.7	---	---	0.451
ET-12-H 40 nmol/ml	24.7	---	21.9	31.5	11.4	---	10.5	---	---	0.470
PM	13.0	3.6	21.7	33.3	11.8	1.3	15.2	---	---	0.779

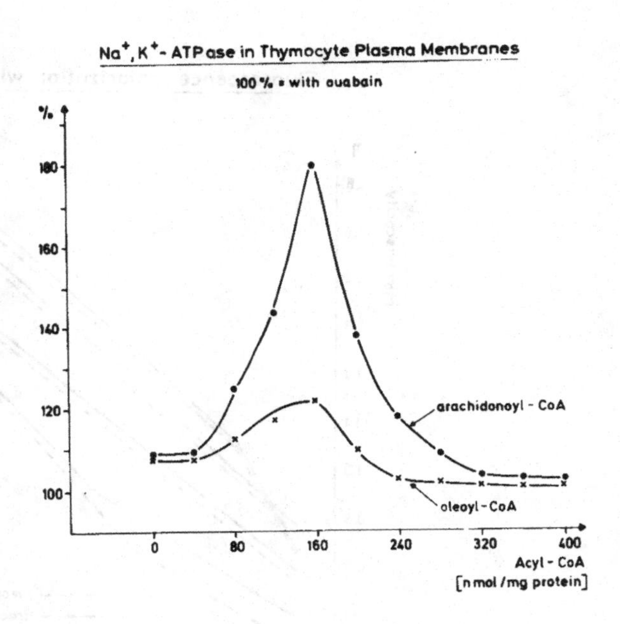

Fig. 6 Purified plasma membranes were incubated without and with increasing concentrations of oleoyl-CoA or arachidonoyl-CoA as indicated. Experiments with labelled acyl-CoA's showed that up to a concentration of 240 nmol/mg protein about 98% of the acyl-CoA was utilized and predominantly incorporated into phosphatidyl ethanolamine. After incubation the plasma membranes were washed and the ATPase activity was determined.

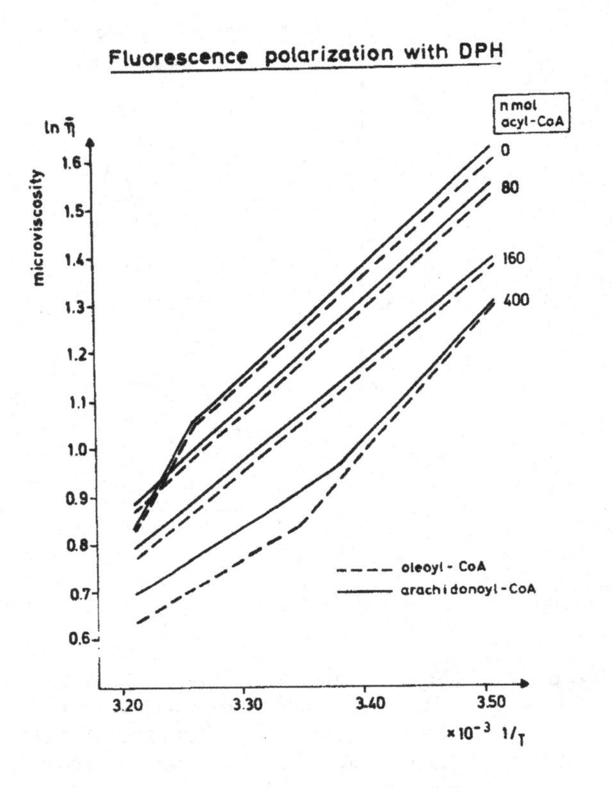

Fig. 7 Determination of the microviscosity of plasma membranes after incorporation of acyl-CoA's into phospholipids (see Fig. 6).
The microviscosity was determined according to Shinitzky and Barenholz (1978).

relatively saturated lipids and thus are more rigid are preferentially shed. These findings support the notion that lipid-protein interactions cannot be sufficiently described by general fluidity parameters (Kimelberg, 1975; Dornand et al., 1974) but that more specific interactions occur. Therefore, we have attempted to answer the question of whether changes in phospholipid fatty acids caused by the acyl-CoA: lysolecithin acyltransferase could alter the activity of membrane-bound enzymes. Using two acyl-CoA's of different unsaturation we changed the fatty acid composition of membrane phospholipids via the acylation-reaction of lysolecithin and then measured the Na^+, K^+-ATPase activity (Fig. 6). The fluidity of the lipid phase was determined using the fluorescence polarization of 1,6-diphenyl-1,3,5-hexatriene (DPH) (Fig. 7). While the incorporation of both acyl-CoA's (oleoyl- and arachidonoyl-CoA) resulted in a continuous but almost identical decrease in the microviscosity (about 30%) the Na^+, K^+-ATPase activity was elevated only when arachidonic acid was incorporated.

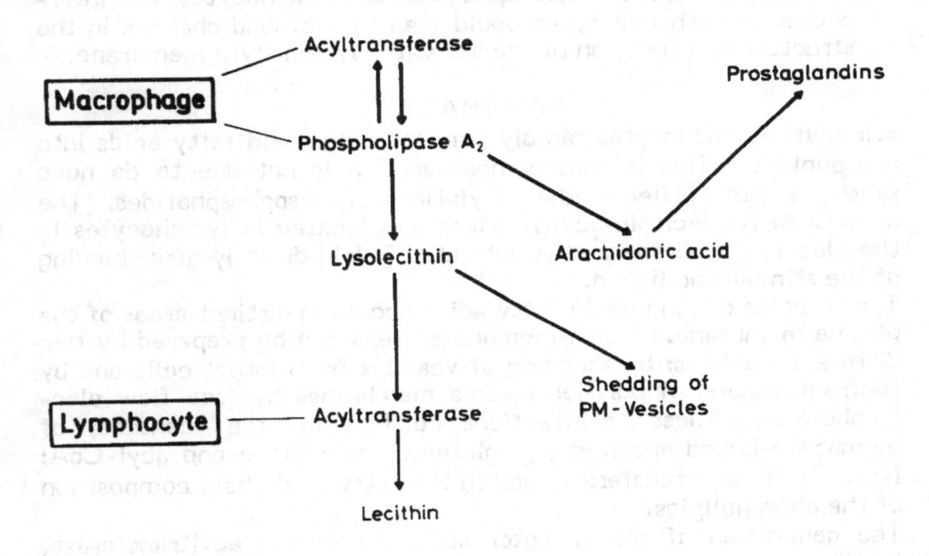

Fig. 8

CONCLUSIONS

The dynamics of membrane lipids during stimulation of lymphocytes may be rationalized as follows:

1. The re-acylation reactions are initiated in lymphocytes directly

after binding of the stimulating ligand and lead to membrane phospholipids of higher unsaturation. Although these changes reduce the overall microviscosity of the lipid phase we have evidence that more specific alterations of membrane-bound enzymes are achieved. For example Na^+, K^+-ATPase is selectively activated when an optimal amount of arachidonic acid is transferred to phospholipids.

2. Only macrophages - not lymphocytes - exhibit phospholipase A activity. However, even in stimulated macrophages the availability of lysophosphatides and free fatty acids is controlled by the lysolecithin acyltransferase. Only when the acyltransferase is inhibited the activated phospholipase A causes a progressive degradation of phospholipids (Fig. 8).
3. Liberated lysocompounds can modulate the composition of the lymphocyte plasma membrane by the induction of shedding of membrane vesicles.

   Although it is not yet established whether these compounds are transferred from the macrophage to the lymphocytes, the interaction of both cell types could lead to profound changes in the structure and function of the activated lymphocyte membrane.

ABSTRACT

Stimulated lymphocytes rapidly transfer long chain fatty acids into phospholipids. This increased incorporation is not due to de novo synthesis but reflects the acylation of lysophosphatides. The acyl-CoA: lysolecithin acyltransferase is located in lymphocytes in the plasma membrane and is activated 3-fold directly after binding of the stimulating ligand.

The transfer of long chain fatty acids occurs in distinct areas of the plasma membrane. Plasma membrane areas can be prepared by two different methods: by shedding of vesicles from intact cells and by subfractionation of purified plasma membranes by free-flow electrophoresis. These subfractions differ in the content of membrane-bound enzymes (γ-glutamyl transferase and acyl-CoA: lysolecithin acyltransferase) and in the fatty acyl chain composition of the phospholipids.

The generation of the acceptor substrate of the acyltransferase, lysolecithin, does not occur in lymphocytes but in macrophages, which interact during the stimulation with lymphocytes.

REFERENCES

Andersson, J., G.M. Edelman, G. Möller, and O. Sjöberg (1972) Eur. J. Immunol. 2, 233

Betel, I., and K.J. van den Berg (1972) Eur. J. Biochem. 30, 571

Bloj, B., R.D. Morero, R.N. Farias, and R.E. Trucco (1973) Biochim. Biophys. Acta 311, 67

Brunner, G., E. Ferber, and K. Resch (1976) Differentiation 5, 161

Brunner, G., H.-G. Heidrich, J.R. Golecki, H.C. Bauer, D. Suter, P. Plückhahn, and E. Ferber (1977) Biochim. Biophys. Acta 471, 195

de Kruyff, B., W.J. de Greef, R.V.W. van Eyk, R.A. Demel, and L.L.M. van Deenen (1973) Biochim. Biophys. Acta 298, 479

Dornand, J., J.C. Mani, M. Mousseron-Canet, and B. Pau (1974) Biochimie 56, 1425

Ferber, E. (1973) In D. Chapman and D.F.H. Wallach (eds.) Biological Membranes Vol. 2, Academic Press, New York and London, p. 221

Ferber, E., and K. Resch (1973) Biochim. Biophys. Acta 296, 335

Ferber, E., G.G. de Pasquale, and K. Resch (1975) Biochim. Biophys. Acta 398, 364

Ferber, E., C.E. Reilly, and K. Resch (1976) Biochim. Biophys. Acta 448, 143

Goldemberg, A.L., R.N. Farias, and R.E. Trucco (1973) Biochim. Biophys. Acta 291, 489

Greaves, M.F., and S. Bauminger (1972) Nature New Biology 235, 67

Habu, S., and M.C. Raff (1977) Eur. J. Immunol. 7, 451

Kimelberg, H.K. (1975) Biochim. Biophys. Acta 413, 143

Matthes, M.L., W. Ax, and H. Fischer (1971) Z. ges. exp. Med. 154, 253

McElhaney, R.N., J. de Gier, and E.C.M. van der Neut-kok (1973) Biochim. Biophys. Acta 298, 500

Meerpohl, H.-G., M.L. Lohmann-Matthes, and H. Fischer (1976) Eur. J. Immunol. 6, 213

Mendelsohn, J., S.A. Skinner, and S. Kornfeld (1971) J. Clin. Invest. 50, 818

Munder, P.G., M. Modolell, E. Ferber, and H. Fischer (1966) Biochem. 344, 310

Munder, P.G., E. Ferber, M. Modolell, and H. Fischer (1969) Int. Arch. Allergy 36, 117

Patton, S. (1970) J. Theoret. Biol. 29, 489

Peters, J.H., and P. Hausen (1971a) Eur. J. Biochem. 19, 502

Peters, J.H., and P. Hausen (1971b) Eur. J. Biochem. 19, 509

Pickett, W.C., R.L. Jesse, and P. Cohen (1977) Biochim. Biophys. Acta 486, 209

Quastel, M.R., and J.G. Kaplan (1971) Exp. Cell Res. 63, 230

Ray, T.K., V.P. Skipski, M. Barclay, E. Essner, and F.M. Archibald (1969) J. Biol. Chem. 244, 5528

Resch, K., and E. Ferber (1972) Eur. J. Biochem. 27, 153

Resch, K., and E. Ferber (1975) In A.S. Rosenthal (ed.) Immune Recognition, Academic Press, New York and London p. 281

Resch, K., E.W. Gelfand, K. Hansen, and E. Ferber (1972) Eur. J. Immunol. 2, 598

Schairer, H.U., and P. Overath (1969) J. Mol. Biol. 44, 209

Shinitzky, M., and Y. Barenholz (1978) Biochim. Biophys. Acta 515, 367

van den Berg, K.J., and I. Betel (1973) Exp. Cell. Res. 76, 63

Vogt, W. (1978) In C. Galli, G. Galli and G. Porcellati (eds.) Advances in Prostaglandin and Thromboxane Research Vol 3, Raven Press, New York p. 89

Whitney, R.B., and R.M. Sutherland (1972) Cellular Immunol. 5, 137

Wilson, G., and C.F. Fox (1971) J. Mol. Biol. 55, 49

This investigation was supported by a grant from the Stiftung Volkswagenwerk.

PHOSPHATIDYLINOSITOL AS A SOURCE OF DIACYLGLYCEROL TO PROMOTE MEMBRANE FUSION IN EXOCYTOSIS

J. N. Hawthorne and M. Pickard

Department of Biochemistry, University Hospital

and Medical School, Nottingham NG7 2UH, U.K.

INTRODUCTION

Twenty years ago an enzyme hydrolysing phosphatidylinositol by the phospholipase C route was first reported (Kemp *et al.*, 1959) and shown to be activated by calcium ions. The products of the reaction are diacylglycerol and inositol phosphate. Subsequent work has shown that the enzyme is widely distributed in mammalian tissues and that it is the key enzyme in what has been called the 'phosphatidylinositol effect' (Hokin & Hokin, 1953). This is the increased turnover of phosphatidylinositol, often measured by ^{32}P incorporation, in response to activation of various types of plasma membrane receptors. For recent reviews, see Michell (1975) and Hawthorne and Pickard (1979). The physiological significance of the effect remains uncertain, though Michell and others believe that it is associated with calcium gating (Michell *et al.*, 1977; Berridge & Fain, 1979). A conference volume presents other points of view (Wells & Eisenberg, 1978). Our purpose here is not to discuss why the effect is there. Instead we shall consider what part the phospholipase C might play, since it is reasonable to assume that the phosphatidylinositol effect always begins with this enzyme. Systems in which activation of cell-surface receptors leads to hydrolysis of phosphatidylinositol are summarised below.

PHOSPHATIDYLINOSITOL HYDROLYSIS FOLLOWING RECEPTOR ACTIVATION

In several tissues a decrease in phosphatidylinositol concentration following stimulation has been measured chemically. Hokin-Neaverson (1974) showed this in mouse pancreas when muscarinic receptors were activated by acetylcholine, as did Jones and Michell (1974) when similar receptors in rat parotid gland were stimulated. Jafferji and Michell (1976) obtained similar results in smooth muscle of guinea-pig ileum, though in this case the ratio of phosphatidylinositol phosphate to phosphate in a spot on a chromatogram containing most of the tissue lipids was presented. Activation of α-adrenergic receptors in rabbit iris muscle by noradrenaline also reduced phosphatidylinositol concentration (Abdel-Latif *et al.*, 1976). As in mouse pancreas there was a concomitant increase in the concentration of phosphatidic acid. After 60 min incubation with 0.1 mM acetylcholine/0.1 mM eserine there was a 20% loss of phosphatidylinositol (expressed per mg protein) from rat brain synaptosomes (Warfield & Segal, 1978). In this work the lipid was estimated by deacylation and gas-liquid chromatography of the trimethylsilyl derivative of glycerophosphorylinositol.

In a number of other systems loss of phosphatidylinositol has been measured by pre-labelling with $^{32}P_i$ or $[^3H]$-inositol and determining loss of radioactivity from the appropriate spot on a chromatogram or t.l.c. plate. Examples are rat cerebral cortex *in vivo*, where the loss was greatest in a synaptic vesicle fraction (Lunt & Pickard, 1975), electrically stimulated guinea-pig brain synaptosomes (Pickard & Hawthorne, 1978) and blowfly salivary gland stimulated by 5-hydroxytryptamine (Berridge & Fain, 1979).

SIGNIFICANCE OF THE PHOSPHOLIPASE C REACTION

The products of the reaction were first considered to be diacylglycerol and inositol 1-phosphate (Kemp *et al.*, 1961) but Dawson *et al.* (1971) showed that a cyclic phosphate was produced:

phosphatidylinositol → diacylglycerol + D-inositol-1,2-cyclic phosphate

The cyclic phosphate is a major product at pH 5 but as the

pH approaches 7 it makes up only half of the water-soluble phosphate, the remainder being inositol monophosphate. A diesterase hydrolysing the cyclic phosphate to D-inositol-1-phosphate is present in mammalian tissues (Dawson & Clarke, 1972). Though it is a distinct enzyme, it resembles cyclic-AMP diesterase in requiring Mg^{2+} and having an alkaline pH optimum.

The significance of phosphatidylinositol breakdown in response to activation of receptors could therefore be in (a) provision of cyclic inositol phosphate, (b) provision of diacylglycerol, (c) removal of phosphatidylinositol itself or (d) some combination of these alternatives. Cyclic inositol phosphate had obvious attractions as another 'second messenger' (Michell & Lapetina, 1972) but has not survived long in this role. The problems are reviewed dispassionately by Michell (1975). One is the difficulty of estimating this labile compound and the methods of Koch-Kallnbach and Diringer (1977) should improve matters.

The removal of phosphatidylinositol and the provision of diacylglycerol are best considered together in the context of this meeting on control of membrane fluidity.

PRODUCTION OF DIACYLGLYCEROL IN STIMULATED TISSUES

In one or two tissues where a phosphatidylinositol effect occurs there is now evidence that the stimulus leads to diacylglycerol production. Under physiological conditions it is unlikely that very much of this lipid will accumulate since the normal turnover of phosphatidylinositol will involve resynthesis, in which diacylglycerol kinase provides the first step.

Synaptosomes, or nerve-ending particles, provide a model system for the study of transmitter release in brain tissue. When they are depolarized electrically or by high KCl concentrations transmitter is released. Most workers consider that the release involves exocytosis - fusion of transmitter vesicle membrane with the plasma membrane of the nerve ending followed by release of the vesicle contents and subsequent retrieval of the vesicle membrane. Tracer studies *in vivo* (Lunt & Pickard, 1975; Pickard & Hawthorne, 1978) showed that the phosphatidylinositol of the vesicle membrane had a much more rapid turnover than any other brain phosphatidylinositol. Electrical stimulation of synaptosomes which had been labelled *in vivo* with ^{32}P caused loss of this vesicle phosphatidylinositol. At the same time labelled phosphatidate was lost from a 'microsomal' membrane fraction quite different from the vesicle fraction and diacylglycerol concentration increased in two fractions rich in synaptosomal outer membrane. In the fraction from which labelled phosphatidate was lost on stimulation there was also a loss of diacylglycerol and this fraction was rich in the diacylglycerol kinase which forms phosphatidate from diacylglycerol.

The microsomal fraction may well contain fragments of plasma membrane as well as Golgi type membranes but electron microscopy indicated hardly any contamination of the synaptic vesicle fraction. It was suggested (Pickard & Hawthorne, 1978) that the phospholipid changes were associated with transmitter release by exocytosis. Conversion of phosphatidylinositol to diacylglycerol in the vesicle membrane would produce a more fluid outer monolayer and contribute to the fusion of this membrane with the plasma membrane of the nerve ending. The increased fluidity of the membrane would also allow proteins to move away from the region of the vesicle in contact with the plasma membrane. The phosphatidate changes were considered to be part of the vesicle membrane retrieval process.

MOLECULAR MECHANISMS IN EXOCYTOSIS

The suggestion that diacylglycerol derived from phosphatidylinositol (or possibly from other phospholipids) might contribute to membrane fusion originated from the work of Allan and Michell (1975) on the erythrocyte. It is discussed further by Michell *et al.* (1976). Allan *et al.* (1975) showed that a phospholipase C from *Cl. perfringens* attacked the phosphatidylcholine and sphingomyelin of human erythrocytes without significant cell lysis. However, the cells became spherical and internal membrane vesicles were seen, changes attributed to the accumulation of ceramide and diacylglycerol in the membrane. Subsequently (Allan *et al.*, 1976) treatment of human red cells with the calcium ionophore A23187 was also shown to increase the diacylglycerol content of the membrane. Again there was a change in morphology, but in this case microvesicles budded off projections from the cell surface. It seemed that production of diacylglycerol on the inner face of the membrane led to outward vesiculation, while the same lipid produced on the outer face caused inward vesiculation. In a more recent study Allan *et al.* (1978) provide evidence for the transbilayer diffusion of diacylglycerol.

While these membrane fusion processes are clearly not physiological, they may provide models for the role of diacylglycerol in exocytosis. It seems quite likely that in the synaptosome model entry of calcium ions triggers the hydrolysis of phosphatidylinositol to diacylglycerol and that this diacylglycerol plays a part in membrane fusion. This is unlikely to be the whole story. Production of a more fluid membrane will not in itself move membrane proteins away from the sites of contact between vesicle and plasma membranes, though it would make such movement easier. "Cytoskeletal" elements containing contractile proteins and possibly microtubules may be important in exocytosis (Orci & Perrelet, 1978) and the interaction of calcium ions with membrane phospholipids should not be ignored (Portis *et al.*, 1979).

Finally it should be made clear that the production of diacylglycerol from phosphatidylinositol may not be the central feature of the phospholipid effect seen so widely in response to cell membrane activation. We do not yet know whether phosphatidylinositol always makes the same contribution to the physiological response. It is interesting that in the platelet, diacylglycerol is generated within five seconds of the response to thrombin (Rittenhouse-Simmons, 1979) and may be involved in release of serotonin from secretory granules.

ABSTRACT

Conversion of phosphatidylinositol to diacylglycerol probably increases membrane fluidity and contributes to the fusion of transmitter vesicle membrane and nerve ending plasma membrane for release of transmitter. After injection of $^{32}P_i$ *in vivo* guinea-pig brain synaptosomes containing labelled phosphatidic acid and phosphatidylinositol could be prepared. Under the conditions used, other phospholipids incorporated little ^{32}P. The most highly labelled phosphatidylinositol was in the vesicle membrane and the most radioactive phosphatidate was in a microsomal fraction rich in diacylglycerol kinase. Electrical stimulation of the isolated synaptosomes caused transmitter release and loss of both these pools of labelled phospholipid. Simultaneously the diacylglycerol content of the synaptosomal plasma membrane increased. The possible role of diacylglycerol in exocytosis is discussed.

REFERENCES

Abdel-Latif, A.A., M. P. Owen, and J. L. Matheny (1976). *Biochem. Pharmacol.* 25, 461.
Allan, D., and R. H. Michell (1975). *Nature, Lond.* 258, 348.
Allan, D., M. G. Low, J. B. Finean, and R. H. Michell (1975). *Biochim. Biophys. Acta.* 413, 309.
Allan, D., M. M. Billah, J. B. Finean, and R. H. Michell (1976). *Nature, Lond.* 261, 58.
Allan, D., P. Thomas, and R. H. Michell (1978). *Nature, Lond.* 276, 289.
Berridge, M.J., and J. N. Fain (1979). *Biochem. J.* 178, 59.
Dawson, R. M. C., and N. Clarke (1972). *Biochem. J.* 127, 113.
Dawson, R. M. C., N. Freinkel, F. B. Jungalwala, and N. Clarke (1971). *Biochem. J.* 122, 605.
Hawthorne, J. N., and M. R. Pickard (1979). *J. Neurochem.* 32, 5.
Hokin, M. R., and L. E. Hokin (1953). *J. Biol. Chem.* 203, 967.
Hokin-Neaverson, M. (1974). *Biochem. Biophys. Res. Commun.* 58, 763.
Jafferji, S.S., and R. H. Michell (1976). *Biochem. J.* 154, 653.
Jones, L. M., and R. H. Michell (1974). *Biochem. J.* 142, 583.
Kemp, P., G. Hübscher, and J. N. Hawthorne (1959). *Biochim. Biophys. Acta.* 31, 585.
Kemp, P., G. Hübscher, and J. N. Hawthorne (1961). *Biochem. J.* 79, 193.
Koch-Kallnbach, M. E., and H. Diringer (1977). *Hoppe-Seyler's Z. Physiol. Chem.* 358, 367.
Lunt, G. G., and M. R. Pickard (1975). *J. Neurochem.* 24, 1203.
Michell, R. H. (1975). *Biochim. Biophys. Acta.* 415, 81.
Michell, R. H., and E. G. Lapetina (1972). *Nature (Lond) New Biol.* 240, 258.
Michell, R. H., D. Allan, and J. B. Finean (1976). *Adv. Exp. Med. Biol.* 72, 3.
Michell, R. H., S. S. Jafferji, and L. M. Jones (1977). *Adv. Exp. Med. Biol.* 83, 447.
Orci, L., and A. Perrelet (1978). *In* G. Poste and G. L. Nicolson (eds.), *Membrane Fusion.* Elsevier/North Holland, Amsterdam.
Pickard, M. R., and J. N. Hawthorne (1978). *J. Neurochem.* 30, 145.
Portis, A., C. Newton, W. Pangborn, and D. Papahadjopoulos (1979). *Biochemistry.* 18, 780.
Rittenhouse-Simmons, S. (1979). *J. Clin. Invest.* 63, 580.

Warfield, A. S., and S. Segal (1978). *Proc. Natl. Acad. Sci. U.S.A.* 75, 4568.
Wells, W. W., and F. Eisenberg Jr. (1978). *Cyclitols and Phosphoinositides*. Academic Press, New York.

GLYCOSPHINGOLIPID DOMAIN FORMATION AND LYMPHOID CELL ACTIVATION.

C. Curtain, F.D. Looney and J.A. Smelstorius.

CSIRO Division of Chemical Technology,

69 Yarra Bank Road, Melbourne, Vic. Australia.

INTRODUCTION.

Peripheral blood lymphocytes are stimulated to mitosis following interaction of plasma membrane receptors with a wide range of ligands, including the seed-derived lectins, phytohaemagglutinin (PHA) and concanavalin-A (Con-A). A series of biochemical events follow ligand binding, including an influx of Ca^{++} into the cell (Allwood *et al* 1971), an increased uptake of metabolites (Greaves and Janossy, 1972), increased synthesis of cyclic nucleotides (Edelman, 1974), increased turnover of phosphatidyl inositol and increased RNA and DNA and protein synthesis, culminating in mitosis (Diamenstein and Ulmer, 1975; Whitney and Sutherland, 1972; Alford, 1970). Ligand-induced receptor clustering ("patching and capping") is one of the earliest observable phenomena following ligand binding. Spin label and immunofluorescence studies have shown that patching is accompanied by clustering of glycosphingolipids (Curtain *et al* 1978; Curtain, 1979). The immunofluorescence studies were done by making fluorescein-labelled antibodies to cerebroside or gangliosides and using these to counterstain B-lymphocytes which had been patched or capped with rhodamine-labelled anti-immunoglobulin antiserum. Both the labelled anti-glycosphingolipid and the labelled anti-immunoglobulins showed identical localization. Sphingolipid clustering was also inferred from spin-label studies in which it was shown that mitogenic ligands induced spin-spin interaction in methyl ester spin-label stearic acid probes inserted into lymphocyte membranes. It was shown in model systems that the methyl ester probes preferentially localized in glycolipid-rich regions. For spin-spin interaction to occur the spin probes must lie < 15 x 10^{-8}cm apart (Devaux and McConnell, 1972). Hence the spin-spin interaction induced by mitogenic ligands must reflect marked condensation of membrane regions probed by the methyl ester probes.

STUDIES WITH SPIN-LABELLED SPHINGOLIPIDS.

Recently we have synthesized spin labelled sphingolipids using standard methods (Sharom and Grant, 1975) and have been able to demonstrate this clustering effect with these (Fig.1).

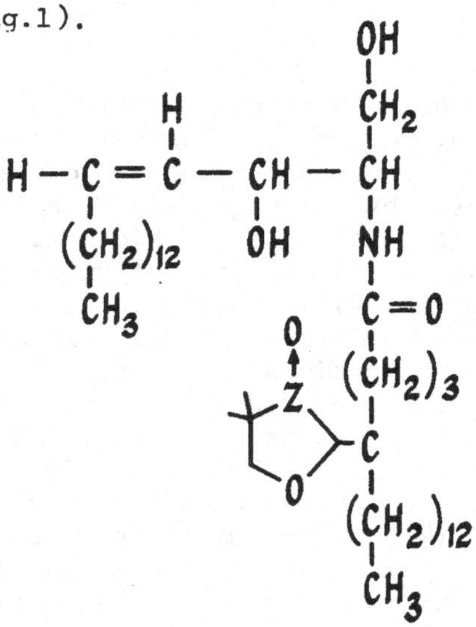

FIG. 1. Labelled ceramide prepared by condensing 5 nitroxide-stearic acid and sphingosine. Labelled cerebroside can be similarly prepared using nitroxide fatty acids and psychosine.

Marked spin-spin interaction is seen in the spectra of ceramide or cerebroside spin labels when ligands such as Con A or PHA are added to lymphocyte membranes which contain the probes (Fig.2).

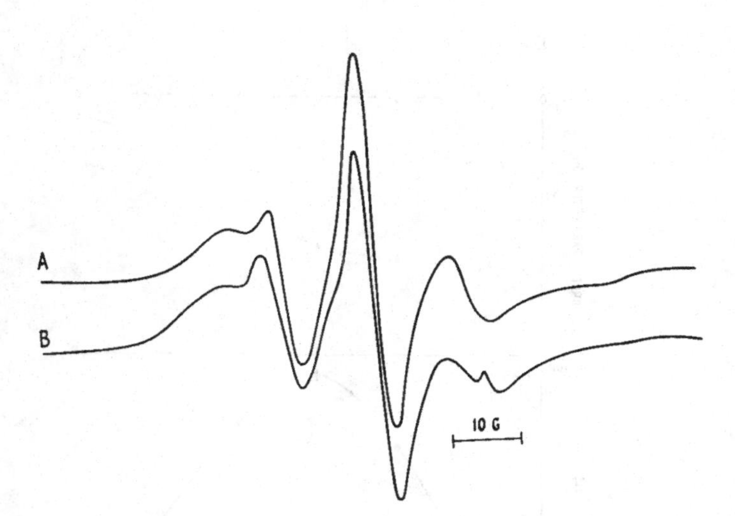

FIG.2. ESR spectrum of 5 nitroxide ceramide in lymphocyte membrane, A before and B after adding 2.5 μg/ml of PHA to the cells.

Ligand binding also increases the hydrocarbon chain motion in regions probed by spin labelled phospholipids. The magnitudes of both this effect and the sphingolipid clustering closely parallel patch and cap formation and disappearance as followed by monitoring with fluorescein-labelled ligands (Fig.3). Both the increase in phospholipid motion and sphingolipid clustering do not occur if the cells are treated with microtubule-disrupting agents such as colchicine.

SPHINGOLIPID CLUSTERING IN SENSITIZED MAST CELLS

The sphingolipid clustering effect is also observed when sensitized mast cells are challenged with allergen. The system used was density gradient-purified rat peritoneal mast cells, sensitized with reagin from the serum of rats which had been injected with ovalbuminin and pertussis vaccine. The addition of reagin

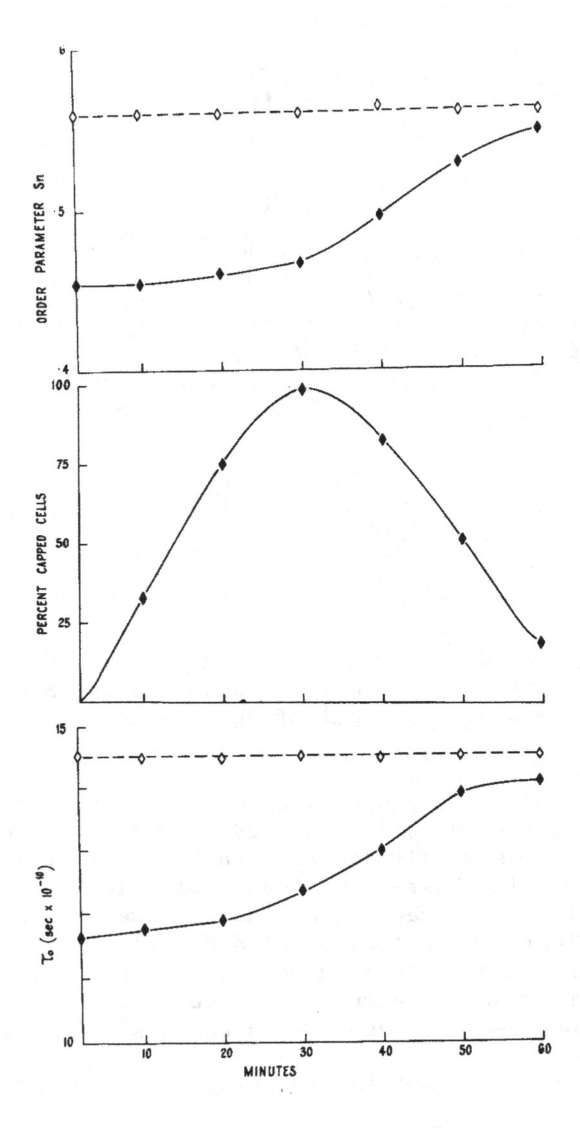

FIG. 3. Rotational correlation time (τ_o), percentage capped cells and order parameter (Sn) at various times after the addition of 2.5 µg/ml of Con-A.

to the cells had no effect upon the spectrum of 5-nitroxide ceramide present in the membrane. When ovalbumin was added spin-spin interaction immediately occurred, similar to that seen in Fig.2. The spin-spin interaction could be inhibited by the addition of certain antiallergic drugs which appear to exert their therapeutic effect by preventing mast cell degranulation and histamine release, possibly by blocking Ca++ influx into the challenged cell.

The two drugs which were used were disodium cromoglycate (Intal, Fisons) and 7' oxo-7-thiomethoxy-xanthone-2-carboxylic acid (Syntex). When added to normal mast cells both drugs caused a decrease in the rate of motion in the region probed by cerebroside or ceramide spin labels. From the shape of the spectra (Fig.4) there were signs of a very slow component, possibly due to binding of the glycosphingolipid to the relatively rigid drug molecule. No drug induced effect was observed in membrane regions probed by phospholipid spin labels.

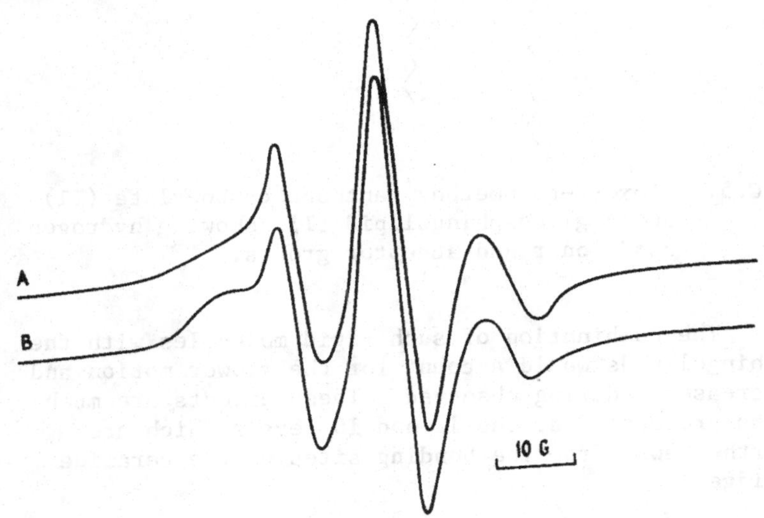

FIG.4. Spectra of 5 nitroxide-ceramide in membranes of A, mast cells and B, mast cells suspended in 12 μg/ml disodium cromoglycate.

Both cromoglycate and the xanthone carboxylic acid are rigid ring structures containing hydrogen bond-accepting oxygen atoms (Fig.5). They are also amphiphilic and would be capable of inserting into the outer leaflet of the membrane bilayer. Here they could accept hydrogen bonds from the secondary amide and the hydroxyl groups of the ceramide bridge of the sphingolipids.

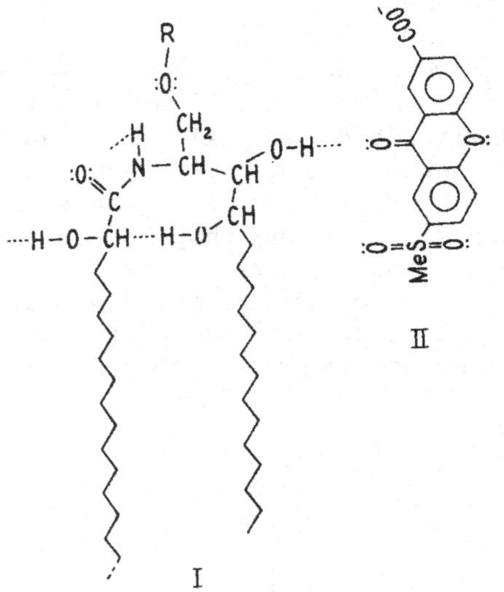

FIG.5. 7'oxo-7-thiomethoxyxanthone carboxylate (II) and a glycosphingolipid (I) showing hydrogen bond donor and acceptor groups.

The combination of such rigid molecules with the sphingolipids would account for the slower motion and increased ordering observed. These effects are much less pronounced at the 12 and 16 levels which are further away from the bonding sites on the ceramide bridge.

THE ROLE OF SPHINGOLIPID CLUSTERING IN CELL ACTIVATION

There has been a great deal of evidence and speculation about the prime role of Ca^{++} influx in both initiating the chain of events leading to lymphocyte mitosis (Greaves et al, 1972; Malm, 1978; Freedman and Raff 1975), and in the triggering of mast cell degranulation (Foreman and Gomperts, 1975). No clear cut mechanism has emerged however for this calcium "gating" effect of a wide range of agents. Could this be based on glycosphingolipid clustering? If it were, how does clustering occur in response to diverse mitogenic agents in the peripheral lymphocyte and in the challenged, sensitized mast cell? One explanation may be that because of their variety and complexity the glycosphingolipids may combine with a wide range of agents and that the clustering may be a fortuitous cross reaction with the ligand causing the particular glycosphingolipid to be incorporated into the ligand-receptor patch. We feel that we disproved this possibility in the case of the fluorescent antibody study of anti-immunoglobulin-induced clustering because the antiglobulin used was shown by a quite sensitive method to have no antiglycosphingolipid activity (Curtain 1979). Alternatively the glycosphingolipids may be associated with a wide range of receptor molecules, possibly via cytoskeletal elements. There are two pieces of evidence pointing to the involvement of the latter. The first is that the clustering effect does not occur in the colchicine-treated cells. The second is that Sela et al (1978) showed that anti-G_{m1} ganglioside-induced capping in rat thymocytes was inhibited by cytochalasin B. Of interest, but not directly relevant, is the observation that cytochalasin B inhibits the lectin-induced decrease in ordering of phospholipids in duck erythrocytes. Presumably the decrease in ordering is similar to that observed in the phospholipids of the lectin treated lymphocyte and which we explained by the removal of the more rigid sphingolipids into the clusters (Curtain et al, 1978). Human erythrocytes which do not have microfilaments show no changes in either phospholipids or glycosphingolipids after lectin treatment.

Inhibition of sphingolipid clustering by the anti-allergic drugs may be explained by the fact that these drugs combine with the sphingolipids in the mast cell membrane, possibly dissociating them from reagin receptors. We have some evidence from immuno-electron microscopy that allergen-induced micropatch formation still occurs on the membranes of challenged, sensitized, drug-treated cells, although Ca^{++} influx and degranulation does not follow challenge. This is further evidence of the importance of sphingolipids in Ca^{++} "gating". A clue to possible molecular mechanisms may be found in the fact that gangliosides bind Ca^{++} (Van Heyningen, 1963). Also, because of packing considerations concentration of large head group sphingolipids, such as gangliosides, may lead to considerable local modification of membrane curvature and perhaps even to pore formation as has been suggested for lysolecithin (Israelachvili, 1978). Like lysolecithin gangliosides form micelles because of their large hydrophilic headgroup to hydrophobic chain ratio (Formisano _et al_, 1979). Membrane curvature modification could also be encouraged by the asymmetry of the hydrocarbon chains of sphingolipids, where the sphingosine base chain extends only 14 carbon atoms into the bilayer, whilst fatty acid chain is C17 or longer. In the lymphocyte more than 50% of glycosphingolipid fatty acid chains contain 22 or more carbon atoms (Stein and Marcus, 1977). To sum up, a mechanism for Ca^{++} influx, common to both mitogen treated lymphocytes and allergen challenged, sentitized mast cells, may be the clustering of glycosphingolipids which could be linked to the relevant receptors by cytoskeletal elements. Pore formation would follow the clustering of large head group glycosphingolipids such as gangliosides and these pores would be lined with negatively charged N-acetyl neuraminic acid groups. Conceivably these charged pores could form Ca^{++} selective channels.

ABSTRACT

Studies carried out with spin labelled lipid probes have shown that glycosphingolipids form clusters when lymphoid cells react with a range of ligands. This effect has been shown with peripheral blood lymphocytes and mitogenic plant lectins and sensitized mast cells after challenge with antigen. The sphingolipid clustering in the mast cell was prevented by antiallergic drugs of the cromone series. It is suggested that these act by blocking hydrogen bond donor sites on the sphingolipids, thus either preventing self association or displacing them from reagin receptors. Since the cromone drugs block the Ca^{++} influx that occurs on mast cell activation it seems likely that the sphingolipid clusters may be part of the Ca^{++} "gate". Since certain classes of di and trisialogangliosides bind calcium they could be excellent candidated for the role. Gangliosides are "wedge shaped" molecules which readily form micelles and it is possible that their clustering in the membrane could lead to the formation of pores lined with Ca^{++} binding groups.

REFERENCES

Alford, R.H. (1970). J. Immunol., 104, 698

Allwood, G., Asherson, G.L., Davey, M.J. and Goodford, P. (1971) Immunology, 21, 509.

Curtain, C.C., Looney, F.D., Marchalonis, J.J., and Raison, J.K. (1978). J. Membrane Biol. 44, 211.

Curtain, C.C. (1979). Immunology 36, 805.

Devaux, P. and McConnell, H.M., (1972). J. Am. Chem. Soc. 94, 4475.

Diamanstein T. and Ulmer, A. (1975). Immunology 28, 121.

Edelman, G.M. (1974) in Control of proliferation in Animal Cells. Cold spring Harbor Symposium.

Freedman and Raff, M. (1975). Nature 255, 378.

Foreman, J.C. and Gomperts, D.B. (1975). Int. Archs. Allergy Appl. Immunol. 49, 179.

Formisano, S., Johnson, M.L., Lee, G., Aloj, S.M. and Edelhoch, H. (1979) Biochemistry, 18, 1119.

Greaves, M.F., and Janossy, G. (1972). Transplant Rev. 11, 87.

Greaves, M.F., Bauminger, S., Janossy, G. (1972) Clin. Exp Immunol. 10, 537.

Israelachvili, J.N. (1978) Light Transducing Membranes (Ed. D. Deamer) (Academic Press, New York) 91.

Malm, T.M. (1978). Speculations in Science and Technology. 1, 15.

Sela, B-A., Raz. A. and Geiger, B. (1978). Eur. J. Immunol. 8, 268.

Sharom F.J. and Grant, C.W.M. (1975). Biochem. Biophys Res. Comm.67, 1501.

Stein, K.F. and Marcus, D.M. (1977). Biochemistry 16, 5285.

Van Heyningen, W.E. (1963). J. Gen. Microbiol, 31, 275.

Whitney, R.B. and Sutherland, R.M. (1972). J. Cell. Physiology, 80, 329.

PHOSPHOLIPID METABOLISM OF RAT GASTRIC MUCOSA

Momtaz K. Wassef, Y. N. Lin and M. I. Horowitz

Department of Biochemistry
New York Medical College
Valhalla, NY 10595

ABSTRACT

Phospholipids of rat gastric mucosa comprised about 30% of the total extractable lipids; phosphatidylcholine (PC) constitutes about half of the phospholipids. Disaturated (dipalmitoyl)-PC made up about 31% of the PC species, a concentration comparable to that reported for lung tissue. Gastric mucosa exhibited three distinguishable phospholipid deacylating enzyme activities: lysophospholipase, phospholipase A_1 and phospholipase A_2 and an acylase activity owing to a lysophospholipase-transacylase enzyme complex. The lysophospholipase hydrolyzed 1-palmitoyl lyso-PC to free fatty acid and glycerophosphorylcholine (GPC). This enzyme had an optimum pH of 8.0, was heat labile, did not require Ca^{2+} for maximum activity and was not inhibited by bile salts or buffers of high ionic strength. Phospholipase A_2 and phospholipase A_1 deacylated dipalmitoyl-PC to the corresponding lyso compound and free fatty acid. The specific activity of phospholipase A_2 was 2-4-fold higher than that of phospholipase A_1 under all the conditions tested. Both activities were enhanced 4-7.5-fold in the presence of bile salts at alkaline pH and 11-18-fold at acidic pH. Phospholipase A_2 hydrolyzed diarachidonyl-PC>PC-18:2>>PC-18:1>PC-16:0. No phospholipase C or D activities were detected with dipalmitoyl-PC as a substrate. The lysophospholipase-transacylase enzyme complex displayed both a lysophospholipase and a transacylase activity. The transacylase converted, stoichiometrically, 2 moles of 1-palmitoyl lyso-PC to one mole each of dipalmitoyl-PC and GPC in the absence of ATP, CoA, fatty acids, bile salts or detergents. The transacylase-lysophospholipase activities were both located in the cytosol fraction and were associated at a constant ratio throughout several purification steps, including the isoelectrofocusing procedure.

The high levels of dipalmitoyl-PC and the presence of deacylating enzymes and of a lysophospholipase-transacylase enzyme complex in gastric mucosa and the lung suggest similarities between gastric and lung phospholipid metabolism. Dipalmitoyl-PC is generally accepted as the major active component of pulmonary surfactant, but the exact role of dipalmitoyl-PC in gastric mucosa awaits further documentation.

INTRODUCTION

Gastric mucosa membranes may be exposed to noxious agents which disrupt lipid constituents. For example, regurgitation of bile acids and lysolecithin from the small intestine was implicated in the pathogenesis of gastric erosions (Janowitz, 1969; Hamza and DenBesten, 1972; Johnson and McDermott, 1974; Kivilaakso et al., 1976) and of back-diffusion of H^+ from the lumen into the gastric mucosa (Davenport, 1970; Ivey et al., 1970; Ivey, 1973; Eastwood, 1975). Despite the importance of gastric phospholipids, little is known about their composition and metabolism (Scheithaur et al., 1971; Slomiany et al., 1975; Wassef et al., 1978, 1979; Lin et al., 1979). We report here on gastric mucosal phospholipid content, composition and its metabolizing enzymes.

MATERIALS AND METHODS

Labeled ^{14}C-phosphatidylcholine substrates were obtained from Applied Science Labs., Inc., Pa. and from New England Nuclear, Boston, Mass. Purified lipolytic enzymes were from Boehringer Manheim, Indianapolis, Ind. Other supplies and chemicals were from Sigma Chemical Co., St. Louis, Mo. and from other retailers.

Extraction and fractionation of lipids. Stomachs of female Sprague-Dawley rats (120-150 g), starved for 24 hr, were removed, the rumen was discarded and the stomachs were inverted and washed with ice-cold EGTA. The fundus and pylorus were lightly scraped and lipids of the mucosa scraping were immediately extracted and fractionated as described (Wassef et al., 1979).

Quantitative analyses of the phospholipids were made by two-dimensional thin-layer chromatography (Rouser et al. 1976). Specific spray reagents were used, according to Kates (1972). Lipid amino nitrogen determination (Lea and Rhodes, 1955) was performed on PE and LPE separated by TLC and eluted from the silica gel by the Bligh and Dyer procedure (1959). Individual phospholipid components were aspirated from the plates and the fatty acids were transmethylated and fractionated by gas liquid chromatography (GLC). Identification of unsaturated fatty acids was further confirmed by catalytic hydrogenation (Kates, 1972) and re-examination of the resulting saturated acids by GLC.

Pure phosphatidylcholines were isolated and the ester bond and phosphorus contents were determined (Wassef et al. 1979). Disaturated PC was separated and quantitated as described by Mason et al. (1972). Fatty acid positional distribution was determined by hydrolysis of total purified PC with phospholipase A_2 (EC 3.1.1.4) (Wells and Hanahan, 1969) and chromatography of the products on TLC (Wassef et al., 1978). Hydrolysis of PC with phospholipase C (3.1.4.3) was performed as described by Ottolenghi (1969). The resulting 1,2-diglycerides were fractionated by silver nitrate-impregnated TLC (Van Golde and Van Deenen, 1966), visualized by spraying with 0.01% Rhodamine 6G and the individual bands were eluted (Van Golde et al., 1967). The fatty acid positional distribution of the diglycerides, reflecting those of the original PC, was determined by hydrolysis with a lipase (EC 3.1.1.3) from Rhizopus arhizus (Van Golde et al., 1967). The reaction products were separated by TLC (Wassef et al., 1978), transmethylated and the fatty acid methyl esters were analyzed by GLC (Wassef et al., 1979).

Enzyme Preparation and Subcellular Fractionation. Gastric mucosa scrapings were obtained from fasted rats as described above. For activities of lysophospholipase-transacylase enzyme complex and oleoyl-CoA:LPC acyltransferase, pooled scrapings (4 to 20 rats) were fractionated and analyzed as described by Lin et al. (1979).

For lysophospholipase and phospholipases A_1 and A_2 activities, pooled scrapings were immediately suspended in 3 ml of ice-cold 0.05 M glycine/NaOH buffer, pH 9.5 and sonicated for one min, at 4°C. Enzymatic activities were determined as described by Wassef et al. (1978).

RESULTS

Lipid composition of rat gastric mucosa. Total lipids extracted from gastric mucosa scrapings comprised 19.1% of the tissue dry weight. Fractionation of the total lipids according to their polarity using silicic acid column chromatography, is shown in Table I. Phospholipids constituted about one-third of the total extractable lipids of the gastric mucosa, whereas the glycolipid polarity group was only 15%. The neutral lipid fraction constituted more than half of the total lipids. Composition of the neutral lipid and glycolipid fractions are not fully characterized. The complex mixture of phospholipids obtained in fraction 3 of the silicic acid column, Table I, was resolved into individual components by chromatography on two-dimensional TLC as shown in Fig. 1; quantitative composition and tentative identification are given in Table II. None of the components separated by TLC gave positive α-naphthol reaction, strongly suggesting no glycolipid contamination in the phospholipid fraction of the silicic acid column. Phosphatidylcholine (PC) was the predominant phospholipid and comprised 41% of the total phospholipid content. Phosphatidylethanolamine (PE) was ca. 25%, two-thirds of which was plasmalogen. Sphingomyelin was about 12%, phosphatidylinositol (PI), phosphatidylserine and cardiolipin were about 4% each of the total phospholipids. Only traces of phosphatidic acid and phosphatidylglycerol could be detected. The ratios of PC/LPC was 10, whereas PE/LPE was 4. The rather high content of LPE and its ratio relative to PE was further confirmed by determining the lipid amino-nitrogen content of both components eluted from the TLC plates.

Table I

Separation of Lipids of Rat Stomach Mucosa Scrapings into Polarity Groups by Silicic Acid Column Chromatography

Fraction and eluting solvent [a]	% of Total Lipids	% of Tissue [b]
1. Chloroform (Neutral Lipids)	55.7	10.5
2. Acetone (Lipids of Intermediate Polarity)	14.8	2.8
3. Methanol (Polar Lipids)	29.5	5.8

[a] See Rouser et al. (1976) for fraction composition;
[b] Oven dry-weight basis.

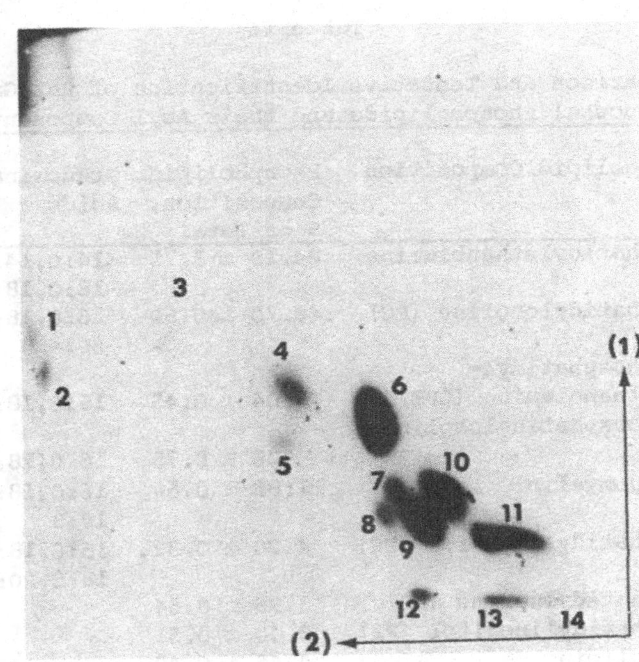

Fig. 1.

Two-dimensional TLC chromatogram of phospholipids of rat gastric mucosa. Plates were developed in (1) CH_2Cl_3/MeOH/28% NH_4OH (65:25:5 v/v), and in (2) $CHCl_3$/acetone/MeOH/HAc/H_2O (4:3:1:1:0.5 v/v). Components were detected by charring. Spots #1,2,7 and 13 are unknowns; 14, origin; 3, cardiolipin; 4, PE; 5, PG; 8, PI; 9, LPE; 10, sphingomyelin; 11, LPC; 12, PS.

The fatty acid composition of the individual phospholipid components separated by TLC is shown in Table II. Palmitic acid comprised about half of the total fatty acids of the choline containing phospholipids and was the major fatty acid in all the components, except phospholipids containing amino groups, where the 18 carbon fatty acid class was predominant (Table II). Fatty acid residues of the lysophospholipids were almost completely saturated (95%-100%); lysophosphatidylcholine contained 52% of the fatty acids as 16:0 and 32% as 18:0, whereas fatty acids of lysophosphatidylethanolamine were 50%, as 18:0 and 35%, as 16:0 (Table II). The ratio of saturated fatty acids to unsaturated acids in the diacylphospholipids was between 3 and 4 to 1. The major unsaturated fatty acid was 18:1, followed by 18:2 and 18:3.

Table II

Comparison and Tentative Identification of Rat Gastric Mucosal Phospholipids and their Acyl Components

Phospholipid Composition	Phospholipid Composition, % of Total	Predominant Fatty Acids
Phosphatidylethanolamine (PE)	24.19 ± 3.75	<14:0,14:0,16:0, 18:0,18:1,20:4
Phosphatidylcholine (PC)	40.70 ± 0.60	16:0,18:0,18:1, 20:4
Lysophosphatidyl-ethanolamine (LPE)	6.04 ± 0.45	16:0,18:0
Lysophosphatidylcholine (LPC)	4.06 ± 0.75	16:0,18:0,22:0
Sphingomyelin	11.86 ± 0.54	16:0,18:0,18:2, 18:3
Phosphatidylglycerol (PG)	4.20 ± 0.32	16:0,18:0,18:1, 18:3,20:4
Phosphatidylserine (PS)	3.96 ± 0.54	
Phosphatidylinositol (PI)	3.40 ± 0.51	
Cardiolipin	4.45 ± 0.49	

The overall fatty acid composition of phosphatidylcholine from rat gastric mucosa and the positional distribution of the fatty acids among both ester positions were further investigated. Saturated fatty acids seemed to be largely esterified to C-1, whereas unsaturated fatty acids were predominantly present at C-2. However, although palmitic acid made up 69% of the fatty acids at C-1, it was found also esterified to C-2 at a rather unusually high concentration (44%), (Wassef et al., 1979), suggesting the presence of elevated levels of dipalmitoyl phosphatidylcholine. Disaturated phosphatidylcholine separated from unsaturated species by various procedures constituted 30% of the total phosphatidylcholine (Wassef et al., 1979).

Table III
Phospholipase A_1 and A_2 Activities Towards Various Phosphatidylcholines in Presence and Absence of 1.25 mM Sodium Deoxycholate (DC)

Phosphatidylcholine Species	Activity, % PC Hydrolysis				Fold Activation	
	Phospholipase A_1		Phospholipase A_2			
	NO DC	1.25 mM DC	NO DC	1.25 mM DC	Phosp A_1	Phosp A_2
16:0-PC	0.3	4.9	0.6	6.8	16.3	11.3
18:1-PC	1.4	4.1	1.1	7.5	2.3	6.8
18:2-PC	--	--	11.0	33.5	--	3.1
20:4-PC	1.1	0.6	11.0	37.0	0.5	3.4

Assay system contained 100 mM glycine/NaOH buffer, pH 9.5, 2 mM $CaCl_2$, 0.2 M NaCl, 0.1 mM *[^{14}C]* phosphatidylcholine, 1.25 mM sodium deoxycholate (DC) and 20 µg sonicate protein of rat gastric mucosa scraping in a final volume of 200 µl. Control experiments were treated exactly the same, except that the sonicate protein was replaced by 20 µl buffer. Incubations were carried out in a water bath at 37°C, with shaking.

<u>Phospholipase A_1 and phospholipase A_2.</u> The gastric mucosa sonicate displayed phospholipase A_2 and A_1 activi-

ties that deacylated, in presence of Ca^{2+}, various molecular species of phosphatidylcholines to the corresponding lyso compound and free fatty acids. Both activities displayed multiple pH optima and were enriched 4-7.5-fold in the presence of bile salts at alkaline pH and 11-18-fold at acidic pH (Wassef et al., 1978).

The fatty acid specificity of the phospholipase A_2 and A_1 activities of rat gastric mucosa was tested using dipalmitoyl-, dioleoyl-, diarachidonyl- and 1-acyl-2-linoleoyl phosphatidylcholines as substrates. These phosphatidylcholines are probably the physiological substrates for phospholipase A *in vivo*, since palmitic, oleic and linoleic acids are among the major fatty acid components of rat gastric mucosa. Sodium deoxycholate was essential for optimum activities of phospholipase A_1 and A_2. Table III shows that both phospholipase A_2 and A_1 activities toward deacylation of dipalmitoyl phosphatidylcholine were significantly enhanced in presence of deoxycholate. The extent of enhancement of either phospholipase activity was less pronounced as the degree of fatty acid unsaturation in phosphatidylcholines increased. Phospholipase A_1 activity was slightly inhibited towards deacylation of diarachidonyl phosphatidylcholine in presence of 1.25 mM sodium deoxycholate (Table III). The efficiency of hydrolysis of various phosphatidylcholines by phospholipase A_2 and A_1 can be calculated from Table III. In presence of bile salt, diarachidonyl phosphatidylcholine (20:4-PC) was the most preferred substrate for phospholipase A_2 (100%)>18:2-PC (91%) >>18:1-PC (20%)$\geq$16:0-PC (18%). Phospholipase A_1, on the other hand, showed the reverse preference for substrate hydrolysis in the presence of sodium deoxycholate: 16:0-PC (100%) $\geq$18:1-PC (84%)>>20:4-PC (12%). In the absence of bile salts, unsaturated phosphatidylcholines were hydrolyzed more rapidly than the saturated ones by both phospholipase A_1 and A_2.

Phospholipase C and phospholipase D. Phospholipase C and D activities were determined under conditions described previously (Ottolenghi, 1969; Kates, 1972) and under optimum conditions for phospholipase A_1 and A_2 activities from gastric mucosa (Wassef et al., 1978). Practically no significant phospholipase C or D activities were detected in rat gastric mucosa when dipalmitoyl-PC was used as a substrate. The same substrate was completely hydrolyzed when activities of phospholipase C from Bacillus cereus or phospholipase D from cabbage were examined under the same conditions used for testing the gastric mucosa enzymes.

Lysophospholipase. Incubation of mucosal sonicate with ^{14}C-lysophosphatidylcholine produced labeled free fatty acid at two optimal pH values. A large sharp peak at pH 8.0 and a smaller peak at pH 6.5-7.0 contained 27-30% and 7-9%, respectively, of total radioactivity as free fatty acid (Wassef et al., 1978). This lysophospholipase activity did not require Ca^{2+}, was heat labile and was inhibited (50%) by 5 mM bile salts (Wassef et al., 1978).

Lysophospholipase-LPC:LPC-transacylase enzyme complex. When 1-palmitoyl lysophosphatidylcholine (LPC) was incubated with mucosal sonicate in the absence of ATP, CoA, bile salts or detergents, dipalmitoyl-PC was produced together with free fatty acid and glycerophosphorylcholine (GPC). These reaction products strongly suggest the presence of a transacylase and a lysophospholipase enzymatic activities. Both of these activities were located in the cytosol fraction and remained associated at a constant ratio of about 10 (lysophospholipase/transacylase) throughout four purification steps, including the isoelectrofocusing procedure. The 100,000 x g and the 35-65% $(NH_4)_2SO_4$ precipitate were devoid of oleoyl-CoA:LPC acyltransferase activity (Lin et al., 1979). Both enzyme activities responded similarly when treated with metal ions, sodium taurocholate, triton X-100 or heated. They also had similar apparent K_m values of 0.25 mM for 1-palmitoyl-LPC (Lin et al., 1979). These observations strongly suggest that both the lysophospholipase and transacylase activities may reside in the same enzyme.

DISCUSSION

Our results indicate that sodium deoxycholate is essential for phospholipase A_1 and A_2 optimal activities to deacylate phosphatidylcholine, in keeping with the earlier observations of Magee et al. (1962) and Van Deenen et al. (1963). Studies on the effect of fatty acid unsaturation of phosphatidylcholines on the rate of hydrolysis by phospholipase A_2 revealed that a high degree of fatty acid unsaturation resulted in an increased rate of hydrolysis. Furthermore, the dependency on deoxycholate for a high rate of hydrolysis significantly decreased as the degree of unsaturation increased. These results suggest that the apolar acyl chain plays an important part in the mechanism of the enzymatic reaction. Perhaps the acyl chain affects the attachment of the substrate to the enzyme by an interaction with a hydrophobic region of the active center of the enzyme (Van Deenen et al., 1963; Magee et al., 1962). With regard to the activating effect of deoxycholate on phosphatidylcholine hydrolysis, micellar structure and accessibility of water and enzyme to the substrate are important factors as was discussed by Brockerhoff and Jensen (1974).

Our studies on the effect of different fatty acids at the 1-position of phosphatidylcholines on the rate of hydrolysis by phospholipase A_1 agreed with those reported by Woelk et al. (1974) and by Gatt (1968). In the presence of sodium deoxycholate, the rates of hydrolysis of unsaturated phosphatidylcholine by phospholipase A_1 were remarkably less than those of fully saturated substrates. It is worth noting that gastric phospholipases exhibit marked differences in activities according to the fatty acid composition of phosphatidylcholines presented as substrates. Phospholipase A_2 is more active against PC species containing an unsaturated acyl moiety at the *sn*-2-position than with disaturated PC. Since phospholipase A_2 activity far exceeds that of phospholipase A_1 activity against unsaturated substrate, this may facilitate the reconstruction of gastric membrane phosphatidylcholines with saturated rather than unsaturated fatty acids.

Gastric lysophospholipase, phospholipase A_1 and A_2 activities have been demonstrated (Wassef et al., 1978). However, the presence of an enzyme system that forms phosphatidylcholine via intermolecular transesterification of 2 moles of lysophosphatidylcholine is intriguing (Lin et al., 1979). The gastric transacylase preparation also exhibited a powerful lysophospholipase activity. Both activities were located in the cytosol fraction and were associated at a constant ratio throughout several purification steps, including the isoelectrofocusing procedure (Lin et al., 1979). Evidence obtained strongly suggests that the transacylating and deacylating activities may both be functions of the same enzyme just as they are suggested to be for the rat lung enzyme (Brumley and Van den Bosch, 1977; Vianen and Van den Bosch, 1978).

Analysis of the total lipids of rat gastric mucosa revealed that the phospholipid fraction constituted about 30% of the total extractable lipids. Phosphatidylcholine was the predominant phospholipid, followed by phosphatidylethanolamine and sphingomyelin. The unusual feature of gastric mucosal phospholipids is the presence of a high content of lysophospholipids [compared to lipids of other tissues (McMurray and Magee, 1972)] particularly lysophosphatidylethanolamine. Although precautions were exercised in preventing lipid degradation during extraction and fractionation, this high level of lyso PE could be due to hydrolysis of PE by selectively active phospholipase A_2 during the post-mortem time lag or during chemical processing of the lipids. Lysophospholipids, however, were reported to comprise a significant percentage of the total phospholipids of some subcellular organelles, ranging from 4.7% in the rough endoplasmic reciculum, 12.2% in the Golgi membranes (Keenan and Morré, 1970), to 24% in the chromaffin granules (Blaschko et al., 1968). Lysolipids (particularly LPC) are implicated in the induction of membrane fusion (Poole et al., 1970) and in the enhancement of glycosyltransferase activity (O'Doherty, 1978). The exact mechanism of this process, however, remains to be elucidated.

Of interest are the high levels of saturated fatty acids as compared to the unsaturated fatty acids esterified to the individual phospholipids. Choline-containing compounds had palmitic acid as the major fatty acid with lesser amounts of stearic and oleic acids. Dipalmitoyl PC constituted 30% of the total PC (Wassef et al., 1979). This strikingly high level of dipalmitoyl-PC was reported only for certain tissues, particularly the lung (Montfoort et al., 1971; Mason, 1973; Okano and Akino, 1978). Dipalmitoyl phosphatidylcholine, however, is generally accepted as the major active component of pulmonary surfactant (for review, see Van Golde, 1976), but the exact role of this lipid in the gastric mucosa and its mode of biosynthesis await further investigation.

The high levels of dipalmitoyl phosphatidylcholine and the presence of a lysophospholipase-transacylase enzyme complex in gastric mucosa (Lin et al., 1979) and the lung (Vianen and Van den Bosch, 1978) suggest similarities between gastric and lung phospholipid metabolism. These similarities are intriguing when one considers the embryonic relationship of these two tissues. Owing to the common entodermal origin of stomach and lung (Emery, 1969) it will be of interest to examine whether these two tissues exhibit other similarities in their lipid biochemistry.

This research was supported by grant AM 15565-08 from the National Institute of Arthritis, Metabolism and Digestive Diseases.

REFERENCES

Blaschko, H., D. W. Jerome, A. H. T. Robb-Smith, A. D. Smith and H. Winkler (1968). Clin. Sci. 34, 453.

Bligh, E. G. and W. J. Dyer (1959). Can. J. Biochem. Physiol. 37, 911.

Brockerhoff, H. and R. J. Jensen (1974). In H. Brockerhoff and R. J. Jensen (eds.), Lipolytic Enzymes. Academic Press, Inc., N.Y.

Brumley, G. and H. Van den Bosch (1977). J. Lipid Res. 18, 523.

Davenport, H. W. (1970). Gastroenterol. 59, 505.

Eastwood, G. L. (1975). Gastroenterol. 68, 1456.

Emery, J. (1969). In J. Emery (ed.), The Anatomy of the Developing Lung. Lavenham Press, Ltd., Lavenham.

Gatt, S. (1968). Biochim. Biophys. Acta 159, 304.

Hamza, K. N. and L. DenBesten (1972). Surgery 71, 161.

Ivey, K. J. (1973). Acta-Hepato-Gastroenterol. 20, 524.

Ivey, K. J., L. DenBesten and J. A. Clifton (197). Gastroenterol. 59, 683.

Janowitz, H. D. (1969). Gastroenterol. 57, 356.

Johnson, A. G. and S. J. McDermott (1974). Gut 15, 710.

Kates, M. (1972). In T. S. Work and E. Work (eds.), Techniques of Lipidology. American Elsevier Co., Inc., N.Y.

Keenan, T. W. and D. J. Morré (1970). Biochemistry 9, 19.

Kivilaakso, E., C. Ehnholm, T. V. Kalima and M. Lempinen (1976). Surgery 79, 65.

Lea, C. H. and D. N. Rhodes (1955). Biochim. Biophys. Acta 17, 416.

Lin, Y. N., M. K. Wassef and M. I. Horowitz (1979). Archiv. Biochem. Biophys. 193, 213.

Magee, W. L., J. Gallia-Hatchard, H. Sanders and R. H. S. Thompson (1962). Biochem. J. 83, 17.

Mason, R. J. (1973). Ann. Rev. Respir. Dis. 107, 678.

Mason, R. J., G. Huber and M. Vaughan (1972). J. Clin. Invest. 65, 68.

McMurray, W. C. and W. L. Magee (1972). Ann. Rev. Biochem. 41, 129.

Montfoort, A., L. M. G. Van Golde and L. L. M. Van Deenan (1971). Biochim. Biophys. Acta 125, 496.

O'Doherty, P. J. A. (1978). Lipids 13, 297.

Okano, G. and T. Akino (1978). Biochim. Biophys. Acta 528, 373.

Ottolenghi, A. C. (1969). In J. L. Lowenstein (ed.) Methods in Enzymology. Academic Press, N.Y., Vol. 14, 188.

Poole, A. R., J. I. Howell and J. A. Lucy (1970). Nature 227, 810.

Rouser, G., G. Kritchevsky and A. Yamamoto (1970). In G. V. Marinetti (ed.), Lipid Chromatographic Analysis. Marcell Dekker, Inc., N. Y., Vol. 3, 713.

Scheithaur, E. M., Waldron-Edward and S. C. Skoryna (1971). Gastroenterol. 60, 713.

Slomiany, A., B. L. Slomiany and M. I. Horowitz (1975). Clinica Chimica Acta 59, 215.

Van Deenen, L. L. M., G. H. De Haas and C. H. Th. Heemskerk (1963). Biochim. Biophys. Acta 67, 295.

Van Golde, L. M. G. (1976). Annu. Rev. Respir. Dis. 114, 977.

Van Golde, L. M. G., V. Tomasi and L. L. M. Van Deenen (1967). Chem. Phys. Lipids 1, 282.

Van Golde, L. M. G. and L. L. M. Van Deenen (1966). Biochim. Biophys. Acta 125, 496.

Vianen, G. M. and H. Van Den Bosch (1978). Arch. Biochem. Biophys. 190, 373.

Wassef, M. K., Y. N. Lin and M. I. Horowitz (1978). Biochim. Biophys. Acta 528, 318.

Wassef, M. K., Y. N. Lin and M. I. Horowitz (1979). Biochim. Biophys. Acta 573, 222.

Wells, M. A. and D. J. Hanahan (1969). In J. L. Lowenstein (ed.), Methods in Enzymology. Academic Press, New York. Vol. 14, 178.

Woelk, H., K. Peiler-Ichikawa, L. Binaglia, G. Goracci and G. Porcellati (1974). Hoppe-Seyler's Z. Physiol. Chem. 355, 1535.

EFFECT OF TEMPERATURE ON THE BIOSYNTHESIS OF 3-SN-PHOSPHATIDYLCHOLINE BY *FUSARIUM OXYSPORUM* F. SP. LYCOPERSICI

Alan C. Wilson and Leslie R. Barran

Research Branch, Agriculture Canada

Ottawa, Ontario K1A OC6, Canada

ABSTRACT

Phosphatidylcholine (PC) and phosphatidylethanolamine (PE) are the two major phospholipids present in hyphae of *Fusarium oxysporum* accounting for 70-75% of the total phospholipid fraction. The ratio of PC to PE varied as a function of growth temperature and age of the cells. Both the methylation and choline pathways for PC synthesis were shown to be present in whole cells of *F. oxysporum* and the latter was the major pathway for PC synthesis. Both pathways for PC synthesis were sensitive to growth temperature shifts as the fungus readjusted its phospholipid composition at the specific temperature to which the shift was made.

INTRODUCTION

Microorganisms respond to changes in growth temperature by changing their cellular lipid composition. Generally, lowered growth temperature is associated with increased fatty acid unsaturation and this change has been interpreted as a response by the organism to maintain the requisite membrane fluidity at the lower temperature (Cronan and Vagelos, 1972). Other effects of temperature on cellular lipid composition are also important in maintaining the integrity of cellular membrane function. These include changes in fatty acid chain length and branching, as well as alterations in the phospholipid head group composition and sterol content of cell membranes (Fulco and Fujii, this volume; Nozawa, this volume; Miller and de la Roche, 1976).

Temperature-induced changes in the membrane lipid composition of Fusarium oxysporum f. sp. lycopersici (Barran, et al, 1976) allowed the fungus to maintain the fluidity of its membrane lipids within close tolerances (Miller and de la Roche, 1976; Miller, this volume). A marked effect on the ratio of the two major phospholipids, phosphatidylcholine (PC) and phosphatidylethanolamine (PE) (which constituted 70-75% of the total phospholipid) was observed when the organism was grown in the temperature range of 15-37°C. Since the changes in PC and PE appeared to be reciprocal, it was thought that PC may be formed by methylation of PE, and the presence of a methylation system for PC synthesis was in fact demonstrated (Barran et al, 1976).

PC synthesis has been shown to occur via the choline pathway (Kennedy and Weiss, 1956), and via the methylation of PE (Bremer and Greenberg, 1961). In order to understand the effect of growth temperature on cellular PC levels in F. oxysporum the biosynthesis of PC was studied in whole cells.

PC Synthesis. PC synthesis in whole cells of F. oxysporum was measured by following the incorporation of [Me-^{3}H]-choline (choline pathway) and [Me-^{3}H]-methionine (methylation pathway) into PC. The schemes for PC synthesis via these pathways are shown below:

Methylation pathway:

$$\text{L-methionine} + \text{ATP} \xrightarrow[\text{adenosyl-transferase}]{\text{methionine}} \text{Ado-Met} + \text{PP} + \text{Pi}$$

$$\text{PE} + 3\ \text{Ado-Met} \xrightarrow[\text{system}]{\text{PE methyltransferase}} \text{PC} + 3\text{S-adenosyl-homocysteine}$$

Choline pathway:

$$\text{Choline} + \text{ATP} \xrightarrow[\text{kinase}]{\text{choline}} \text{phosphocholine} + \text{ADP}$$

$$\text{phosphocholine} + \text{CTP} \xrightarrow[\text{cytidyl-transferase}]{\text{phosphocholine}} \text{CDP-choline} + \text{PPi}$$

$$\text{CDP-choline} + \text{diacylglycerol} \xrightarrow[\text{transferase}]{\text{phosphocholine}} \text{PC} + \text{CMP}$$

METHODS

Growth of cells. Mycelia of *F. oxysporum* were grown on a modified Fries medium (Miller, 1971) as previously described (Barran *et al*, 1976).

PC synthesis by whole cells. Cells grown for 18 h at 25 or 37°C in the presence of [^{33}P]-orthophosphate to label the phospholipids (Barran *et al*, 1976), were equilibrated at the appropriate incorporation temperature (15,25 or 37°C) for 20 min. The cells were then incubated with L-[Me-^{3}H]-methionine (150 µCi, 10 µmol per flask) or [Me-^{3}H]-choline (62.5 µCi, 10 µmol per flask). Aliquots were removed and centrifuged at 800 x g for 2 min and the labelled lipids were extracted from the cells with boiling isopropanol (Kates, 1972). The lipid extract in chloroform was washed three times with 2.5 ml of methanol-water (10:9 by volume). Radioactivity was determined with a Beckman LS800 Scintillation Counter with Aquasol as the counting fluid.

Chromatography of labelled lipids. Phospholipids were separated by TLC on silica gel G in chloroform-methanol-28% ammonia (65:25:5 by volume) and chloroform-acetone-methanol-acetic acid-water (30:40:10:10:5). Neutral lipids were separated with petroleum ether-diethylether-acetic acid (80:20:1). Radioactive spots were located with a Panax Radio -TLC Scanner and scraped from the chromatogram for counting. Lipids were identified with group specific sprays (Kates, 1972) and by comparison of their R_f values with those of authentic standards.

RESULTS AND DISCUSSION

Levels of PC and PE during growth. At 15 and 25°C the changes in the levels of PC and PE were similar (Table I). The PC to PE ratio varied from about 0.7 at mid-log to about 1 at late-log, and to 2 at the late-stationary phase. The level of PC in 37°C grown cells were higher than that of PE throughout the growth cycle. The ratio of PC to PE varied from about 1 at mid-log to 2 in the late-log phase and to about 3 in the late-stationary phase. Since the changes in the levels of PC and PE were similar in 15 and 25°C grown cells, further studies were carried out with 25

Table I. Levels of PC and PE during growth of *F. oxysporum* at 15, 25 and 37°C.

Growth temperature (°C)	Phospholipid	% of Total Phospholipid		
		mid-log	late-log	late-stationary
15	PC	27	35	50
	PE	40	40	22
25	PC	27	37	48
	PE	43	37	25
37	PC	36	44	48
	PE	30	24	18

Table II. Incorporation of radioactivity from L-(Me-^{3}H)-methionine into lipid by whole cells of *F. oxysporum* grown at 25 and 37°C

Growth temp. (°C)	Incubation temp. (°C)	time (min)	^{3}H-methyl incorporation, nmole/g dry weight			
			Neutral Lipid	MePE	DiMePE	PC
25	25	0	8	2.6	3.7	4.7
		5	26	7.7	9.1	12.7
		10	39	17.6	22.4	25.8
		20	89	28.8	43.3	55.8
		40	153	53.1	77.0	162.5
25	37	0	7	2.5	3.4	5.0
		5	34	9.2	11.6	18.6
		10	59	21.0	25.2	39.7
		20	111	33.9	56.6	96.8
		40	113	49.5	92.7	266.5
37	25	0	24	0.5	2.1	3.1
		5	29	9.2	10.2	13.2
		10	77	5.0	13.9	11.2
		20	136	7.5	28.1	15.6
		40	200	22.9	53.2	35.2
37	37	0	13	3.0	4.9	6.3
		5	30	9.3	16.4	20.1
		10	57	13.9	21.7	39.6
		20	104	16.3	39.0	83.4
		40	136	19.7	59.4	193.8

and 37°C grown cells.

Methylation pathway. The synthesis of PC via the methylation pathway is shown in Table II. F. oxysporum rapidly incorporated the label into phospholipid and a neutral lipid fraction (NL). A major component of NL was a sterol (probably ergosterol). The only phospholipids labelled were PC, monomethylphosphatidylcholine (MePE) and dimethylphosphatidylcholine (DiMePE); lyso PC was also present as a minor component.

The results can be explained if it is noted that hyphal cells have a higher sterol content as the growth temperature is lowered (Miller and de la Roche, 1976), and that 37°C grown late-log phase cells have a higher PC content than cells grown at 25°C (see Table I). Thus when 25°C grown cells are incubated at 25°C with labelled methionine, both PC and NL (of which sterol is a major component) continue to be synthesized. When incorporation was carried out at 37°C, NL synthesis was shut off (after 20 min) and PC synthesis continued as expected. When cells were grown at 37°C and incorporation was carried out at 25°C, NL synthesis is now required, therefore it continues, but PC synthesis which is not required (over the relatively short incubation period) is curtailed.

Choline pathway. PC synthesis via the choline pathway showed that when 25°C grown cells were incubated at 15, 25 and 37°C, PC synthesis continued (Table III). PC synthesis was expected to continue at the lower temperature (15°C) since the PC content of 15 and 25°C grown late log-phase cells are similar. The cells therefore do not have to adjust their PC content when the growth temperature was lowered. Cells grown to the late-log phase at 37°C continued to synthesize PC when incorporation was measured at 37°C. However, PC synthesis was shut off as expected when incorporation was measured at 15 and 25°C.

Significant changes in the levels of the two major phospholipids (PC and PE) were observed in hyphal cells of F. oxysporum during growth at three temperatures spanning the growth temperature range. Both the methylation and choline pathways for PC synthesis were shown to be present in F. oxysporum and the latter pathway was found to be the major pathway for PC synthesis, Both pathways were shown

Table III. Incorporation of [^{3}H-Me]-choline into PC by whole cells of *F. oxysporum* grown at 25 and 37°C

Growth temp. (^{o}C)	Incubation temp. (^{o}C)	Choline Incorporation into PC nmole/g dry wt. of cells — Incubation Time (min) 5	10	20	40
25	15	32	55	91	154
	25	60	102	175	270
	37	45	97	189	373
37	15	51	125	272	285
	25	195	292	548	545
	37	156	309	309	1053

to be sensitive to growth temperature shifts as the organism sought to readjust its phospholipid and presumably sterol composition at the specific temperature to which the shift was made. The mechanism whereby the cells are able to modulate PC synthesis as a function of temperature is unknown. Control may be exerted at the levels of the enzymes in the respective pathways, or, at the level of the specific substrates (e.g. availability of CTP in the choline pathway). The temperature shifts could, for example, affect the levels of LPC and LPE in the membranes. These lipids have been shown to modulate the activity of phosphocholine cytidyl-transferase and hence affect PC synthesis (Choy and Vance, 1978).

Changes in the levels of PC and PE as a response to growth temperature is a part of an overall change in *F. oxysporum* membrane lipid composition. All of these changes represent a homeostatic response by the organism when grown at different temperatures (Miller and de la Roche, 1976; Miller, this volume). The activity of the fatty acid desaturases in *F. oxysporum* were studied as a function of growth temperature (Miller and Wilson, 1978; Miller, this volume). Further studies on the activities of the enzymes involved in PC biosynthesis as a function of temperature are in progress, and should allow for a better understanding of the mechanism whereby *F. oxysporum* modulates its cell membrane composition.

REFERENCES

Barran, L.R. and I.A. de la Roche (1979). *Trans. Br. Mycol. Soc.* 73, 12.

Barran, L.R., Miller, R.W. and I.A. de la Roche (1976). *Can. J. Microbiol.* 22, 557.

Bremer, J. and D.M. Greenberg (1961). *Biochim. Biophys. Acta* 46, 205.

Choy, P.C. and D.E. Vance (1978). *J. Biol. Chem.* 253, 5163.

Cronan, Jr. J.E. and P.R. Vagelos (1972). *Biochim. Biophys. Acta* 265, 25.

Kates, M. (1972). In T.S. Work and E. Word (eds.), Laboratory Techniques in Biochemistry and Molecular Biology. North Holland/American Elsevier, New York, 275 pp.

Kennedy, E.P. and S.B. Weiss (1956). J. Biol. Chem. 222, 193.

Miller, R.W. (1971). Arch. Biochem. Biophys. 146, 256.

Miller, R.W. and I.A. de la Roche (1976). Biochim. Biophys. Acta 443, 64.

Miller, R.W. and A.I. Wilson (1978). Can. J. Biochem. 56, 1109.

IN VIVO MODIFICATION OF PHOSPHOLIPID POLAR HEAD GROUPS AND ALTERATIONS IN MEMBRANE ACTIVITIES AND CELLULAR EVENTS

F. Snyder,* T-c. Lee, M. L. Blank, and C. Moore

Medical and Health Sciences Division, Oak Ridge Assoc. Univ., Oak Ridge, TN 37830 (USA)

ABSTRACT

Assessment of the functional role of the polar head group portion of membrane phospholipids in vivo is a difficult task. Our laboratory has approached this problem by using lipid analog precursors to alter the composition of the polar head groups. We have shown that N-isopropylethanolamine (IPE), a competitive inhibitor of choline, can be incorporated into 1,2-diacyl-sn-glycero-3-phosphoisopropylethanolamine (phosphatidyl-IPE) of membranes from cultured L-M cells and rat liver. After IPE treatment, we observed profound inhibition of certain enzymatic activities (e.g., Δ9 acyl-desaturase and 5'-nucleotidase), but not in others (e.g., NADH-cytochrome c reductase, NADPH-cytochrome c reductase, and succinic dehydrogenase). The changes in enzyme activities do not appear to be directly related to membrane fluidity or the formation of phosphatidyl-IPE.

Experiments initiated to determine the time-dependent changes in metabolism caused by IPE in order to gather a more complete assessment of how IPE might be affecting the enzyme activities revealed that the following changes occurred in the IPE-treated cells: (1) immediate inhibition of both the cellular uptake of [^{3}H]choline and its incorporation into phosphatidylcholine, (2) a decrease in the incorporation of [^{3}H]thymidine into DNA as early as 2 h

*To whom correspondence should be addressed

after initiating the choline block, (3) inhibition of the cellular uptake of [^{3}H]uridine and incorporation into RNA 16 to 24 h after addition of the IPE, and (4) stimulation of the cellular uptake of [^{3}H]leucine and an inhibition of its incorporation into protein, which reached a maximum (68% of controls) 8 h after IPE treatment.

Collectively, our data indicate that the effect of the choline analog, IPE, on selected enzyme activities associated with the membranes is not caused by the presence of an unnatural polar head group in the membrane where the enzyme resides, but instead is apparently manifested via a derangement in the metabolism of phosphatidylcholine or choline, and this serves an important signal in the modulation of the expression of specific membrane-bound enzyme activities.

INTRODUCTION

The influence that choline and ethanolamine head groups in phospholipids exert on the properties of cell membranes is still poorly understood. Although numerous investigators have reported that unnatural amino-alcohols can be incorporated into phospholipids of a variety of animal systems (Ansell and Chojnacki, 1966; Ansell et al., 1965; Bell and Strength, 1968; Bridges and Ricketts, 1965, 1967; Chojnacki, 1964; Chojnacki and Ansell, 1976; Cornatzer et al., 1949; Geer and Vovis, 1965; Hodgson and Dauterman, 1964; Hodgson et al., 1969; Longmore and Mulford, 1960; McArthur et al., 1947; Mehendale et al., 1970; Willetts, 1974) and thus alter their head group composition, these studies did not determine the membrane distribution of the phospholipid analogs or whether the analogs altered any biochemical properties of the membranes.

Our experiments were designed to answer the following questions: Can unnatural analogs of polar head groups be incorporated into membrane phospholipids in significant quantities? Do the amino alcohol analogs modify the composition of phospholipid classes and their acyl moieties? What effect does polar head group modification have on cellular metabolism and membrane properties? What mechanisms are involved? We selected *N*-isopropylethanolamine (IPE) as the phospholipid modifier for cultured L-M cells and rat liver because it was structurally similar to choline. Others (Åkesson, 1977; Engelhard et al., 1976; Ferguson et al., 1975; Glaser et al., 1974; Schroeder and

Vagelos, 1976; Schroeder et al., 1976) have used N,N-dimethylethanolamine, N-monomethylethanolamine, and ethanolamine in cultured L-M cells and hepatocytes in related studies.

This paper describes highlights of our results on how the IPE analog alters the composition of polar head groups in membrane phospholipids (Blank et al., 1975, 1976; Lee et al., 1975, 1978; Moore et al., 1978a,b) and how such modifications affect membrane enzymatic properties and cellular metabolism. The data indicate that the inhibitory effect of IPE on selected membrane-bound enzyme activities is not caused by a polar head group modification in the membrane where the enzyme resides, but instead is apparently manifested via a derangement in the metabolism of phosphatidylcholine or choline before any significant amount of phospholipid analog is synthesized.

METHODS

Details for the experimental design and procedures used in most of the experiments reported in this paper have been described earlier (Blank et al., 1975, 1976; Lee et al., 1975, 1978; Moore et al., 1978a,b). Head group modifications in membrane phospholipids were accomplished in two ways: (a) L-M fibroblasts were incubated in serum-free medium 199 in the absence or presence of 5 mM or 10 mM IPE (Blank et al., 1975) or (b) fasting rats were injected daily for 4 days with IPE (20 mg/kg) and then killed 1 h after the last injection to remove and process the liver (Lee et al., 1975); all other pertinent information and appropriate references are provided in the Results and Discussion.

RESULTS AND DISCUSSION

Effect of IPE on Phospholipid Classes of L-M Cells and Rat Liver

IPE has been shown to be incorporated into phosphatidyl-IPE in L-M cells (Blank et al., 1975, 1976; Lee et al., 1978) and rat liver (Lee et al., 1975; Moore et al., 1978b). The level of this phospholipid analog reaches about 8% of the total lipid phosphorus when L-M cells are cultured for 24 to 48 h in media containing 10 mM IPE. In liver, phosphatidyl-IPE levels reached 9% of the total lipid phosphorus after fasted rats were given 4 daily intraperitoneal injections of IPE; under these conditions we also found about 3 to 4% phosphatidyl-N-methyl-IPE (Lee et al., 1975; Moore et al., 1978b). These data show that phosphatidyl-IPE participates in the methylation pathway found in the liver, but which is absent in L-M cells. Moreover, we found that when N-methyl-IPE was injected intraperitoneally into rats, approximately 19% of the microsomal phospholipid-P was found as phosphatidyl-N-methyl-IPE (Moore et al., 1978b); label from [$^{14}CH_3$]methionine was also incorporated into the methylated phospholipid analog.

Distribution of Phosphatidyl-IPE and Phosphatidyl-N-methyl-IPE in Membrane Fractions

IPE markedly affects the composition of membrane phospholipid classes by the fact that it forms phosphatidyl-IPE and its methylated derivative at the expense of the choline, and to some extent, the ethanolamine phosphoglycerides (Blank et al., 1975; Lee et al., 1975). After 4 daily injections of IPE, the deposition of phosphatidyl-IPE is highest (8 to 9% of total lipid-P) in microsomes and mitochondria from rat liver (Lee et al., 1975), whereas the plasma membranes contain only half as much (3.6%). The overall distribution of phospholipid classes in microsomes and mitochondria from livers of IPE-treated rats is similar to that found in the total homogenates (Table I). In a time study with monolayers of L-M cells incubated with 10 mM IPE, microsomal phosphatidyl-IPE reached a maximum of 8 mol % phospholipid-P by 24 h, whereas a concomitant decrease (15 mol %) in microsomal phosphatidylcholine occurred during the same period (Lee et al., 1978).

Labeling of phosphatidyl-IPE in subcellular fractions of rat liver after an intraperitoneal injection of [1,2-^{14}C]-IPE indicated that, as with other phospholipids, a rapid equilibration occurs between the microsomal and mitochondrial fractions (Lee et al., 1975). However, as can be seen from the 24-h values in Table II, the phosphatidyl-IPE is ultimately transferred to the plasma membrane, presumably by intracellular transfer (exchange) proteins. Results from the [1,2-^{14}C]IPE experiments also demonstrated that a portion of the isopropyl moiety of IPE was removed after injection, since approximately 22% of the label was found in phosphatidylethanolamine within 1 h after administering the [1,2-^{14}C]IPE.

TABLE I

EFFECT OF IPE ON PHOSPHOLIPID CLASS COMPOSITION IN RAT LIVER

Phospholipid Class	Total homogenates		Subcellular fractions from IPE-treated		
	Control	IPE-treated	Microsomes	Mitochondria	Plasma Membranes
Sphingomyelin	4.9	8.2	8.6	5.7	16
Phosphatidylcholine	46	37	44	35	33
Phosphatidylinositol + Phosphatidylserine	15	14	16	16	16
Phosphatidylethanolamine	29	24	19	24	23
Phosphatidyl-IPE	0	9.0	8.4	9.1	3.6
Phosphatidyl-*N*-methyl-IPE	0	3.2	4.6	3.4	3.2
Cardiolipin	5.5	4.5	<0.5	6.8	1.6

The controls for the total homogenates contain 0.85 mg phospholipid phosphorus/g tissue and IPE-treated samples contain 0.83 mg phospholipid phosphorus/g tissue. Values are expressed as the percentage of total phospholipid phosphorus. [Data from Lee et al. (1975).]

TABLE II

LABELING OF PHOSPHATIDYL-IPE IN SUBCELLULAR FRACTIONS OF RAT LIVER AFTER ADMINISTRATION OF [1,2-^{14}C]IPE

Membrane fraction	dpm/μg Phospholipid-P	
	1 h	24 h
Mitochondria	25	246
Microsomes	22	300
Plasma membranes	14	376

Rats were injected intraperitoneally with [1,2-^{14}C]IPE (12.8 μCi, 0.66 mg in 1 ml saline). [Data from Lee et al. (1975).]

Mechanism for the Formation of Phosphatidyl-IPE

The phosphatidyl-IPE could be formed either by the de novo biosynthetic pathway or by a base-exchange reaction. To differentiate these two pathways, we incubated one group of monolayers for 24 h in normal media containing 100 μCi ^{32}P and 100 μCi [^{3}H]glycerol; the radioisotopes were then removed and pre-labeled monolayers then incubated for 4 h in normal media in the presence or absence of IPE. Another batch of cells was incubated for 4 h in the presence of both radioisotopes in media with or without IPE. The amount of label incorporated into each phospholipid class is shown in Table III. When IPE was present during the second 4-h incubation in fresh media containing no label, very little of the ^{3}H or ^{32}P was in phosphatidyl-IPE. In contrast, when IPE was present during the initial 4-h pulse interval, 13.9% of the ^{32}P and 5.0% of the ^{3}H of the total radioactivity incorporated into phospholipids were in phosphatidyl-IPE. These results document that phosphatidyl-IPE is primarily synthesized by the de novo pathway.

Effect of IPE on Acyl Composition of Liver Phospholipids

IPE did not cause any major changes in the acyl composition (Lee et al., 1975) of phosphatidylcholine or phosphatidylethanolamine in rat liver (Table IV). However, the

TABLE III

BIOSYNTHETIC PATHWAYS FOR THE FORMATION OF PHOSPHATIDYL-IPE: DE NOVO VERSUS BASE EXCHANGE

Condition		Phospholipid classes (dpm · 10^{-5}/monolayer)			
		PC	PS + PI	PE	P-IPE
24-h labeling, then 4 h in normal media	^{32}P	27.3	6.11	17.4	—
	^{3}H	31.3	7.44	13.6	—
24-h labeling, then 4 h in IPE media	^{32}P	24.7	7.06	17.7	0.37 (0.7%)
	^{3}H	26.2	8.11	12.8	0.27 (0.6%)
4-h labeling in normal media	^{32}P	3.38	1.54	1.05	—
	^{3}H	11.3	2.93	3.38	—
4-h labeling in IPE media	^{32}P	1.30	1.94	0.97	0.79 (13.9%)
	^{3}H	11.2	4.01	6.12	1.25 (5.0%)

The values are averages of duplicate determinations on two separate monolayers. The values in parentheses represent the percentage of the total labeled lipids that are present as phosphatidyl-IPE. Abbreviations: PC, phosphatidylcholine; PS, phosphatidylserine; PI, phosphatidylinositol; PE, phosphatidylethanolamine; P-IPE, phosphatidyl-IPE. [Data from Lee et al. (1978).]

acyl composition of the phosphatidyl-IPE had considerably less 18:0 and 20:4 acids and more 18:1 and 18:2 acids than either phosphatidylcholine or phosphatidylethanolamine. In contrast, phosphatidyl-N-methyl-IPE had an acyl composition almost identical to phosphatidylcholine.

TABLE IV

EFFECT OF IPE ON THE FATTY ACID COMPOSITION OF PHOSPHOLIPIDS FROM RAT LIVER HOMOGENATES

Acyl chain	PC	PE	P-IPE	P-CH_3-IPE
16:0	20 (19)	12 (15)	13	20
18:0	20 (22)	26 (24)	18	20
18:1	9 (8)	6 (6)	12	9
18:2	18 (14)	9 (8)	35	23
20:4	24 (28)	29 (31)	15	23
22:6	4 (6)	12 (13)	5	2

Values are expressed as weight % of total methyl esters. Numbers in parentheses are control values. Samples used for analysis are the same as those described in Table I. Abbreviations are: PC, phosphatidylcholine; PE, phosphatidylethanolamine; P-IPE, phosphatidyl-IPE; P-CH_3-IPE, phosphatidyl-N-methyl-IPE. [Data from Lee et al. (1975).]

Time-Dependent Effects of IPE on Cellular Metabolism in L-M Cells

Our results (Lee et al., 1978) have shown a time-dependent sequence of events in the metabolism of choline, thymidine, uridine, and leucine by L-M fibroblasts grown in monolayer cultures (Table V). The first and most striking alteration is the immediate block in choline uptake by IPE, which is also reflected by the decreased incorporation of choline into phosphatidylcholine of the treated cells. Subsequently (∿ 2 h), but before the appearance of phosphatidyl-IPE, there is a decrease in the labeling of DNA with $[^3H]$thymidine. Later (> 8 h), and after the accumulation of phosphatidyl-IPE could be detected, there is a decreased incorporation of uridine into RNA and leucine into protein.

TABLE V

TIME-DEPENDENT EFFECTS OF IPE ON L-M CELL METABOLISM

Choline cellular uptake	Immediate
Choline into phosphatidylcholine	Immediate
Thymidine into DNA	2 h
Appearance of phosphatidyl-IPE	4-8 h
Uridine into RNA and leucine into protein	> 8 h

See Lee et al. (1978) for details.

Another major metabolic alteration that IPE causes in L-M cells is the significant deposition of neutral lipids, which consists of both triacylglycerols and alkyldiacylglycerols. The appearance of alkyldiacylglycerols in cells that normally do not contain ether-linked neutral type lipids indicates that IPE either affects the redistribution of ether lipids from phospholipids via the diacylglycerol branch point enzyme or alkyldihydroxyacetone-P synthase activity. Schroeder and Vagelos (1976) and Esko et al. (1977) have also observed an increase in alkyldiacylglycerols in L-M cells fed N,N-dimethylethanolamine, N-monomethylethanolamine, or ethanolamine as substitutes for choline.

One other profound metabolic effect of IPE was the decrease in desaturation of stearic acid in L-M cells. L-M cell monolayers incubated with 10 mM IPE were able to desaturate only 40% as much $[1-^{14}C]$stearic acid as those cells not exposed to IPE (Blank et al., 1976). Also, the activity of stearoyl-CoA desaturase assayed in cell-free homogenates of L-M cells treated with IPE exhibited a decrease to the same extent as found for the desaturation of stearic acid in the intact cells when compared to control cultures. The earliest significant decrease in the desaturation of saturated fatty acids occurred at 4 to 8 h after treating L-M cells with IPE. At this time, only 0.2 to 1.2% of microsomal phospholipids consisted of phosphatidyl-IPE.

Altered Enzymatic Properties of Membranes from IPE-Treated Cells.

A number of experiments were conducted to determine whether membrane modification(s) or some other IPE-related event caused specific alterations in membrane properties. We selected the stearoyl-CoA desaturase as a test system (Blank et al., 1976), since our earlier metabolic work had demonstrated that IPE markedly suppressed fatty acid desaturation. Moreover, the desaturase system contains three protein components (cytochrome $\underline{b}_5$, cytochrome $\underline{b}_5$ reductase, and the terminal desaturase protein) that are coordinated in microsomal electron transport. Thus, the desaturase system made it possible to evaluate three separate protein components that are required to carry out the function of desaturating fatty acids. Results in Table VI demonstrated that only the terminal desaturase activity (cyanide-sensitive protein) was decreased in the L-M cells grown in the presence of IPE. The levels of the other proteins, NADH- or NADPH-dependent cytochrome $\underline{c}$ reductase and cytochrome $\underline{b}_5$, were unaffected. These data show that the IPE treatment can cause a highly specific alteration in the membrane protein activities of an integrated enzyme complex.

In view of the desaturase findings, we conducted an experiment to evaluate other membrane-bound enzymes in the liver of rats injected with IPE (Moore et al., 1978a). Moreover, since the desaturase system had a lipid requirement, we wanted to examine an enzyme activity (5'-nucleotidase) that is not lipid-dependent. The data in Table VII clearly show that the two microsomal enzymes from rat liver (NADH- and NADPH-cytochrome $\underline{c}$ reductase), as was found in the L-M cells, were unaffected by the IPE treatment; succinic acid dehydrogenase activity was also unaltered by IPE. However, 5'-nucleotidase in the plasma membranes of rat liver was decreased by 20 to 50% after the administration of IPE. Arrhenius plots (Fig. 1) of the 5'-nucleotidase activity from the control and IPE-treated rats paralleled each other and had break points in the curves at identical temperatures (35, 37, 45, and 50°C). These results suggest that the membrane phospholipid modification caused by IPE did not affect a conformational change in the enzyme.

TABLE VI

EFFECT OF IPE ON COMPONENTS OF THE STEAROYL-CoA DESATURASE SYSTEM IN L-M CELL MICROSOMES

Component	Control	IPE-treated
NADH-cytochrome *c* reductase (μmol/min/mg protein)	0.27	0.27
NADPH-cytochrome *c* reductase (nmol/min/mg protein)	9.2	9.7
Cytochrome b_5 (nmol/mg protein)	0.023	0.024
Stearoyl-CoA desaturase (nmol/min/mg protein)	0.27	0.14

IPE (10 mM) was incubated with L-M cells for 24 h. All values are based on measurements made at two or three protein concentrations. [Data from Blank et al. (1976).]

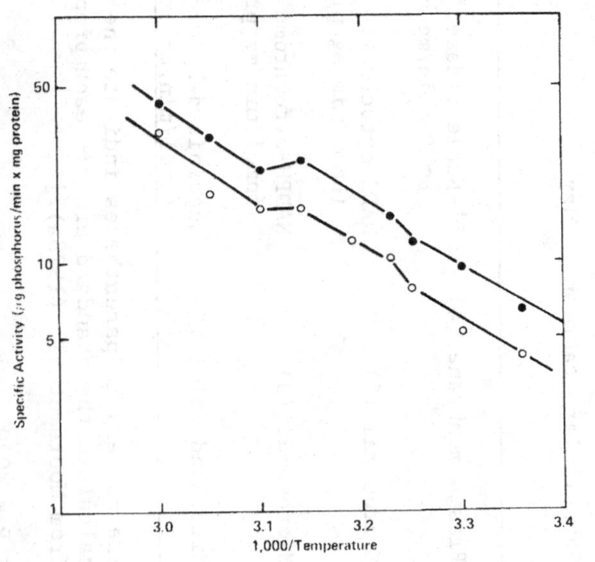

Fig. 1. Arrhenius plot of 5'-nucleotidase activity in plasma membranes from control (•) and IPE-injected (o) rats. [Data from Moore et al. (1978a).]

TABLE VII

EFFECT OF IPE ON ENZYME ACTIVITIES IN DIFFERENT SUBCELLULAR FRACTIONS OF RAT LIVER

Subcellular fraction	Enzyme	Specific activities ± S.E.	
		Control	IPE-treated
Plasma membrane (4)	5'-Nucleotidase (μg P_i/min/mg protein)	20.8 ± 0.64*	12.4 ± 1.87*
Microsomal (3)	NADH-cytochrome c reductase (μmol/min/mg protein)	0.4 ± 0.03	0.4 ± 0.04
Microsomal (3)	NADPH-cytochrome c reductase (nmol/min/mg protein)	71.7 ± 0.54	71.2 ± 0.22
Mitochondrial (3)	Succinic dehydrogenase (nmol/min/mg protein)	10.0 ± 0.17	10.6 ± 0.03

The numbers in parentheses indicate the number of subcellular preparations used to calculate the standard error; each preparation was assayed in duplicate. [Data from Moore et al. (1978a).]

* P = <0.01.

A Lineweaver-Burk plot of the 5'-nucleotidase activity in control and IPE-injected rats demonstrated that the affinity of the enzyme for its substrate was unaffected by IPE treatment (Fig. 2); the K_m was 0.3 mM for both the IPE and control preparations (Moore et al., 1978a). However, the maximum velocity was about 50% lower than controls in the IPE preparations. The curves also reflect that the inhibition was noncompetitive and indicated that the lower activity in the IPE preparations was due to either a decreased amount of enzyme or its inactivation through an irreversible interaction of the enzyme with the phospholipid analog in the membrane.

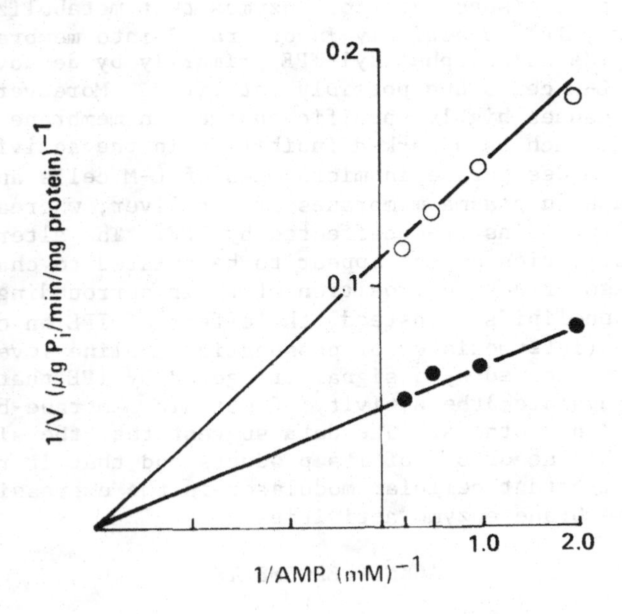

Fig. 2. Lineweaver-Burk plot of 5'-nucleotidase activity in control (●) and IPE-injected (o) rats. [Data from Moore et al. (1978a).]

We tested whether the phosphatidyl-IPE in the plasma membrane was acting as a noncompetitive inhibitor of 5'-nucleotidase activity by removing the base groups of the glycerophosphatides by phospholipase C treatment (Moore et al., 1978a). After the complete removal of phosphatidyl-IPE and 80% of the other phospholipids present in the membrane (sphingomyelin made up the remaining 20%), the same degree of inhibition of 5'-nucleotidase activity was found in the IPE preparations as previously observed in the membranes possessing the full complement of phospholipids. These results rule out any interaction between phosphatidyl-IPE and 5'-nucleotidase as an explanation of the inhibition. Also, free IPE (1, 5, and 10 mM) by itself when added to control preparations had no effect on the 5'-nucleotidase activity (Moore et al., 1978a).

CONCLUSIONS

IPE interferes with choline metabolism by participating as a competitive substrate for enzymes that metabolize choline. Thus, IPE is actively incorporated into membrane phospholipids as phosphatidyl-IPE primarily by de novo synthesis in L-M cells and possibly rat liver. Moreover, IPE treatment causes highly specific changes in membrane protein properties, such as a marked inhibition in the activity of stearoyl-CoA desaturase in microsomes of L-M cells and 5'-nucleotidase in plasma membranes of rat liver, whereas other functional proteins are unaffected by IPE. The altered protein activities do not appear to be related to changes in the class or acyl composition of their surrounding membrane phospholipids. Instead, the effect of IPE on choline metabolism (intermediates or phosphatidylcholine levels) appear to be caused by a signal triggered by IPE that affects (regulates)the activity of certain membrane-bound enzymes and not others. The data suggest that the signal involves choline or one of its products and that it represents an important cellular modulator in the expression of selected membrane enzyme activities.

ACKNOWLEDGEMENTS

This work was supported by the U.S. Department of Energy (Contract No. DE-AC05-76OR00033), American Cancer Society (Grant BC-70J), National Cancer Institute (Grant CA-11949-10), and National Cancer Institute (Training Grant CA-09104-05).

REFERENCES

Åkesson, B. (1977). *Biochem. J.* 168, 401.

Ansell, G. B. and Chojnacki, T. (1966). *Biochem. J.* 98, 303.

Ansell, G. B., Chojnacki, T., and Metcalfe, R. F. (1965). *J. Neurochem.* 12, 649.

Bell, O. E., Jr. and Strength, D. R. (1968). *Arch. Biochem. Biophys.* 123, 462.

Blank, M. L., Piantadosi, C., Ishaq, K. S., and Snyder, F. (1975). *Biochem. Biophys. Res. Commun.* 62, 983.

Blank, M. L., Lee, T-c., Piantadosi, C., Ishaq, K. S., and Snyder, F. (1976). *Arch. Biochem. Biophys.* 177, 317.

Bridges, R. G. and Ricketts, J. (1965). *Biochem. J.* 95, 41P.

Bridges, R. G. and Ricketts, J. (1967). *J. Insect. Physiol.* 13, 835.

Chojnacki, T. (1964). *Acta Biochim. Pol.* 11, 11.

Chojnacki, T. and Ansell, G. B. (1967). *J. Neurochem.* 14, 413.

Cornatzer, W. E., Artom, C., and Crowder, M. (1949). *Fed. Proc.* 8, 192.

Engelhard, V. H., Esko, J. D., Storm, D. R., and Glaser, M. (1976). *Proc. Natl. Acad. Sci. USA* 73, 4482.

Esko, J. D., Gilmore, J. R., and Glaser, M. (1977). *Biochemistry* 16, 1881.

Ferguson, K. A., Glaser, M., Bayer, W. H., and Vagelos, P. R. (1975). *Biochemistry* 14, 146.

Geer, B. W. and Vovis, G. F. (1965). *J. Exp. Zool.* 158, 223.

Glaser, M., Ferguson, K. A., and Vagelos, P. R. (1974). *Proc. Natl. Acad. Sci. USA* 71, 4072.

Hodgson, E. and Dauterman, W. C. (1964). *J. Insect Physiol.* 10, 1005.

Hodgson, E., Dauterman, W. C., Mehendale, H. M., Smith, E., and Khan, M. A. O. (1969). *Comp. Biochem. Physiol.* 29, 343.

Lee, T-c., Blank, M. L., and Snyder, F. (1978). *Biochim. Biophys. Acta* 529, 351.

Lee, T-c., Blank, M. L., Piantadosi, C., Ishaq, K. S., and Snyder, F. (1975). *Biochim. Biophys. Acta* 409, 218.

Longmore, W. J. and Mulford, D. J. (1960). *Biochem. Biophys. Res. Commun.* 3, 566.

McArthur, C. S., Lucas, C. C., and Best, C. H. (1947) *Biochem. J.* 41, 612.

Mehendale, H. M., Dauterman, W. C., and Hodgson, E. (1970). *Int. J. Biochem.* 1, 429.

Moore, C., Lee, T-c., Stephens, N., and Snyder, F. (1978a) *Biochim. Biophys. Acta* 531, 125.

Moore, C., Blank, M. L., Lee, T-c., Benjamin, B., Piantadosi, C., and Snyder, F. (1978b). *Chem. Phys. Lipids* 21, 175.

Schroeder, F. and Vagelos, P. R. (1976). *Biochim. Biophys. Acta* 441, 239.

Schroeder, F., Perlmutter, J. F., Glaser, M., and Vagelos, P. R. (1976). *J. Biol. Chem.* 251, 5015.

Willetts, A. (1974). *Biochim. Biophys. Acta* 362, 448.

PART V

HOMEOSTATIC REGULATION OF MEMBRANE FLUIDITY

HOMEOSTATIC CONTROL OF MEMBRANE LIPID FLUIDITY IN *FUSARIUM*

R.W. Miller

Research Branch, Agriculture Canada

Ottawa, Ontario K1A 0C6

In considering the physical properties of membrane lipids in eukaryotic fungi, several problems arise which limit interpretation of analytical, spectroscopic and electron microscopic data. These limitations are imposed both by the complex membrane architecture of the cells and by our present technical inability to definitively separate and isolate specific cellular membranes free of all other membrane components. Figure 1 illustrates the complex distribution of the cellular membranes and storage lipid droplets. In this freeze-fracture replica we can see that the inner fracture face (IFF) of the plasma membrane contains many 10-15 nm particles presumed to be membrane proteins. Fixation of the cells with 2% glutaraldehyde in the temperature range, 5-38°, does not alter the random distribution of these 10-15 nm particles.

The presence of a tough chitin containing cell wall makes it necessary to either physically disrupt the cell wall of vegetative cells or spores or to enzymatically digest the wall of hyphal cells in order to obtain membrane preparations. In the case of the physical methods, by which all data reported herein have been obtained, membrane fragments may be separated on the basis of density by differential centrifugation. Table 1 shows the ranges of the major components of such preparations. Neutral lipids, including sterols and triacylglycerol are present in all membrane fragments purified in this way. Although some artifacts of cell extraction may be introduced by these

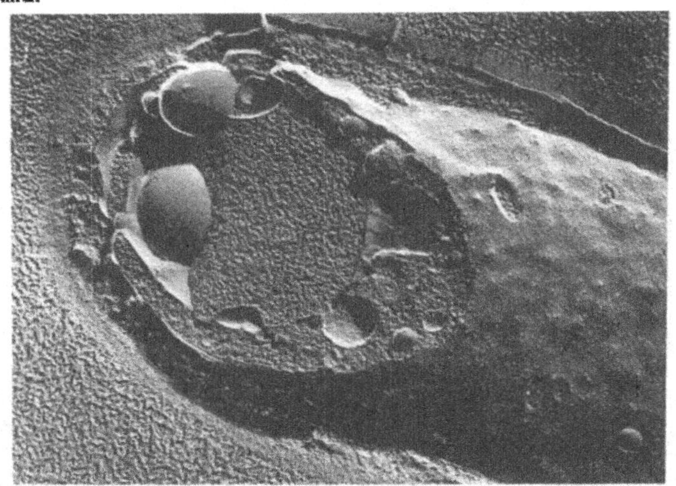

Figure 1: Longitudinally fractured Fusarium cell showing distribution of particles of cytoplasmic membrane. Some infolding of the plasma membrane is present. Smooth, particle-free areas suggest exposure of small lipid bodies or vesicles. Fractured tip shows smooth surface of large lipid bodies and a small area of the inner fracture face of endoplasmic reticulum which is in contact with lipid bodies.

procedures, a major portion of the cell triglycerides are separated from the membranes by centrifugation. These low density lipids are assumed to arise from the lipid storage droplets shown in Figure 1. From the data of Table 1, it is apparent that phospholipids, while assumed to be a major structural component of all fungal membranes, combine with various proportions of membrane neutral lipids and proteins depending on growth stage and conditions.

Plasma membrane ghosts isolated from protoplasts of other fungi by gentle osmotic shock also contain large amounts of sterols and triglycerides (Rank et al. 1978). Hence centrifugally isolated membranes, while not representative of any specific membrane, may be regarded as having compositions and physical properties which are averages of the intracellular membranes from which they have been derived.

Ergosterol is the major sterol of Fusarium oxysporum (Madhosingh et al. 1972, Miller and de la Roche, 1976). In other fungi ergosterol appears to be an essential structural component of membranes. For example, anaerobically

Table 1. Fungal Membrane Composition

Major component	Amount (% dry weight)	Function
Proteins	55 - 60	Metabolite active transport, electron transport, energy conservation, etc.
Neutral Lipid	24 - 30	?
Total Sterol[1]	7 - 20	? Condensing effect, reduction in transition enthalpy - passive permeability
Triacylglycerol[1]	12 - 19	? Disordering effects, fluidity effects
Phospholipids	12 - 19	Bilayer structure, disperse other lipids, fluidity, effectors for membrane enzymes, substrate for desaturases, etc.

[1] Present in all membrane fractions and isolated fungal plasma membranes.

grown yeasts have a nutritional requirement for ergosterol which cannot be completely replaced by cholesterol (Nes et al. 1978).

All analytical and spectroscopic data included in this article apply to log phase cells of Fusarium oxysporum unless otherwise specified. Additional compositional and spectroscopic changes in macroconidia and chlamydospores have been observed previously (Miller, 1977). Macroconidia and chlamydospores are asexually produced spores which form under specific environmental conditions. The lipid composition of the spores reflect both the age, temperature and nutritional conditions of formation and thus can be quite different from the vegetative hyphae and the associated single or double celled microconidia.

Growth of F. oxysporum to log phase at 3 test temperatures in the range 15°-37°C causes large, reproducible variations in the proportions of specific membrane lipids (Barran et al. 1975, Miller and de la Roche, 1976). Tables 2 and 3 show the wide temperature-induced variations in lipid composition of the total membrane fraction from pressure cell disrupted mycelia grown at 3 temperatures. The membranes were pelleted by centrifugation at 105,000 x g after first removing denser cellular debris, walls, etc., by centrifugation at 400 x g. Significant amounts of neutral lipids remain associated with the membranes even after washing with hypotonic media and resedimentation. It is believed that these components are present in intact fungal membranes although not necessarily in the same proportions observed in the isolated membranes.

The molar ratio of sterols and phospholipids was highly dependent on growth temperature (Miller and de la Roche, 1976). This ratio varied from 2.9 to 0.36, almost a 10-fold range, in log phase cells (Table 2). Low temperatures favored sterol production relative to phospholipid.

Total fatty acid polyunsaturation varied inversely with the growth temperature of log phase cells as reflected in the calculated number of double bonds per residue (Δ/mol). Table 3 shows that phospholipid fatty acids were, in general, more unsaturated than those of the triacylglycerol component in log phase cells. This component accounted for nearly 50% of isolated total membrane fatty

Table 2. Phospholipid and sterol content of isolated membranes of *F. oxysporum* mycelia (18 hrs)

Growth Temperature °C	Growth Stage	Phospholipid	Sterol	Sterol Ester	Total Sterol / phospholipid (mol ratio)
		μmoles			
15	early log	0.64	1.6	0.26	2.9
25	late log	5.1	2.8	0.30	0.61
37	late log	3.2	1.0	0.17	0.36

Table 3. Fatty acid composition of phospholipids and triglycerides of total mycelial membrane preparation

Growth Temp.	Lipid Class	Fatty Acid Composition (%)					Δ/mol
°C		16:0	18:0	18:1	18:2	18:3	
15	PL	19.9	1.8	7.8	52.9	17.5	1.66
25	PL	18.5	1.9	7.1	66.1	6.3	1.58
37	PL	19.5	0.6	41.7	36.6	1.6	1.19
15	TG	13.8	7.9	27.9	38.6	11.8	1.41
25	TG	13.1	7.1	29.9	43.0	6.9	1.36
37	TG	16.1	7.9	44.7	28.8	2.5	1.10
15	Total	17.1	4.3	16.2	46.8	15.7	1.56
25	Total	16.5	3.7	16.1	57.1	6.5	1.50
37	Total	17.5	4.1	43.2	32.4	2.1	1.14

Less than 1% digalactosyl diglycerides present

acids. The most marked changes in unsaturation occurred in 37° grown cells as compared to 15° or 25° grown cells.

The major phospholipids of Fusarium membranes were phosphatidylcholine (PC) and phosphatidylethanolamine (PE) under all growth conditions (Wilson and Barran, 1979). These accounted for 75% of all phospholipids. Table 4 illustrates differences in unsaturation of the fatty acids of the 2 major phospholipids. PE had only 1.45 double bonds per mol fatty acid as compared with 1.79 for PC in 25° grown cells.

Table 5 summarizes temperature-induced lipid changes in log phase vegetative hyphae of F. oxysporum and lists some expected effects of these changes on membrane fluidity. Both sterols and membrane proteins would be expected to have condensing effects on regions of the membrane with which they interact. Sterols specifically reduce phase transition enthalpies and in the amounts present in Fusarium membranes would be expected to prevent observation of any abrupt thermotropic change in membrane fluidity. This effect may be countered by fatty acid unsaturation in both the phospholipids and the triacylglycerols although the function of the latter in regulating membrane fluidity is, at this point, speculative.

Biosynthetic pathways for phospholipids, sterols and fatty acid unsaturation appear to respond in different ways to growth temperature. Growth at low temperature favors polyunsaturated fatty acid chains which may, by virtue of the reduced hydrophobic bonding energy and ordering of these chains, accommodate larger amounts of the sterol. The counter effects of fatty acid unsaturation and sterol suggest that membrane fluidity may be controlled primarily by the relative amounts of these 2 components. The effects of variation in the ratio of PE to PC on the fluidity of the cellular membranes cannot be so readily predicted since the distribution of these phospholipids in the lipid bilayers is not known with certainty, and cannot be assumed to be random or homogeneous. Hence, little can be concluded from the analytical data at a single growth temperature as to the effect of temperature-induced alterations in the relative amounts of these components.

Biosynthetic pathways important in the control of membrane fluidity - Although little is known of the effects of

Table 4. Fatty acid composition of PE and PC isolated from log phase mycelia (25°)

Fatty acid	Phosphatidyl-ethanolamine	Phosphatidyl-choline	Membrane
		mol %	
16:0	24.0	8.8	18.5
18:0 18:1	11.4	9.9	1.9 7.1
18:2	61.0	75.0	66.2
18:3	4.6	7.0	6.3
Δ/mol	1.45	1.79	

Table 5. Summary of temperature-induced changes in membrane lipid composition

Membrane component biosynthetic pathway	Product	Observed growth temperature for maximum product *in vivo*	Expected effect on membrane lipids
Fatty acid desaturases			
a. stearoyl-	18:1	37°	diminished fluidity
b. oleyl-	18:2	25°	increased fluidity
c. linoleyl-	18:3	15°	
Phospholipid biosynthesis			
a. phosphatidyl-ethanolamine	PE	15°	
b. methylation	PC	37°	variable
c. *de nova* phos-phatidylcholine	PC	37°	
Sterol biosynthesis	ergosterol	15°	a. fluidity decrease b. condensing effect

growth temperature on enzymes of sterol biosynthesis during early log phase growth, it has been established that growth at low temperature favors accumulation of sterols in late log or stationary cultures (Madhosingh, 1977). Moreover, the activity of a key enzyme of sterol biosynthesis, hydroxmethylglutaryl-CoA reductase, is maximal in 15° grown cells (Madhosingh, 1977). Thus, it would appear that sterol accumulation under these conditions is a deliberate strategy in Fusarium, possibly connected with enhanced multiple unsaturation of phospholipids in low temperature grown cells.

Two pathways for biosynthesis of phosphatidylcholine, i.e., transmethylation of PE and from CDP-choline and diacyllycerides are reported in this volume (Wilson and Barran, 1979). It would appear that both of these pathways are temperature sensitive.

Studies were carried out to determine the effect of growth temperature on the enzymatic desaturation of fatty acids. The microsomal desaturases which were identified are responsible for the production of oleic acid and linoleic acid, both major fatty acids of Fusarium phospholipids. The fatty acid composition of the microsomal fraction can be seen to vary (Table 6) with growth temperature in a manner similar to that of the total membrane fraction (Wilson et al., 1979). From this composition we might expect maximal activity for desaturation of stearic acid in 37° grown cells and a lower activity for conversion of oleyl substrates to linoleyl containing products. By similar reasoning maximal accumulation of linoleyl phospholipids in 15 and 25° grown cells requires high activity of the 18:1 desaturase and implies a lower 18:2 desaturase activity. The latter enzyme has not yet been identified in Fusarium microsomes. Table 7 shows that these expectations are realized for 25 and 37° grown cells but both microsomal desaturase specific activities in 15° grown cells are anomalously low. The total phospholipid content of the latter cells in log phase is quite low, however, and this may partially explain the unexpected lack of desaturase activities in these cells. Assaying the enzymes at 15° did not result in increased activities.

The fungal desaturases preferentially utilize NADPH as an electron donor, are most active at pH 8.5, and are inhibited by both cyanide and thenoyltrifluoracetone. The

Table 6. Fatty acid composition of lipids from microsomal membranes of Fusarium oxysporum grown at 15, 25 and 37°C†

Growth temp. °C	Fatty acid composition, mol %						Δ/mol*
	<16:0	16:0	18:0	18:1	18:2	18:3	
15	6.6	14.2	2.6	15.4	54.0	7.1	1.48
25	2.8	14.9	4.4	17.5	51.9	8.6	1.48
37	2.6	15.6	5.2	58.3	17.1	1.4	0.98

*Unsaturation Δ/mol = (% monoene)/100 + 2 (% diene)/100 + 3 (% triene)/100.

†Reprinted with permission from Wilson, Adams and Miller, 1979.

latter reagent specifically inhibits certain non-heme iron electron carriers. Cytochrome b_5 does not appear to be involved in electron transport associated with the 18:0 and 18:1 desaturases of Fusarium (Wilson and Miller, 1978; Wilson et al., 1979). Also there is considerable evidence that the 2 fungal desaturases differ in their substrate specificity. The stearoyl desaturase acts directly on stearoyl-coenzyme A while the oleyl desaturase appears to convert an oleyl-phospholipid to the corresponding linoleyl phospholipid (Wilson et al., 1979). This difference in substrate specificity may account for the lower polyunsaturation of the membrane triacylglycerides at a given growth temperature. The oleyl desaturase appears to be strongly inhibited by free fatty acids and requires the addition of bovine serum albumin or soluble Fusarium proteins for full activity.

Although the metabolic control of membrane fluidity in Fusarium clearly involves several separate pathways operating in concert, all of these pathways respond to environmental, notably growth temperature, factors. The available evidence suggests that this control effect is essential and that a teleological interpretation of the lipid analytical results is justified. Spectroscopic probes of membrane average fluidity support this hypothesis.

Table 7. Microsomal stearoyl-CoA and oleoyl-CoA desaturase activities of Fusarium oxysporum grown at 15, 25 and 37°C†

Growth temp. °C	Desaturase activity, pmol/min/mg protein	
	18:0 → 18:1	18:1 → 18:2
15	17 ± 1	6 ± 2
25	248 ± 30	95 ± 11
37	440 ± 76	32 ± 8

Incubations(0.5 ml) contained 0.5 to 2.0 mg of microsomal protein, 40 μM [1-^{14}C]stearoyl-CoA or 80 μM [1-^{14}C]oleoyl-CoA, 1 mM NADPH and 50 mM Tris-HCl buffer, pH 8.5. Incubations with 18:1 CoA also contained 20 mM $MgCl_2$ and 3 mg BSA. Rates of oleate or linoleate formation were determined at 25°C (± S.D.).

†Reprinted with permission from Wilson, Adams and Miller, 1979.

Spectroscopic probes of membrane fluidity - Figure 2 illustrates some of the spin probes which have proven useful for the investigation of membrane lipid fluidity in Fusarium. In general, these compounds are non-toxic to the organism and may be applied to the study of live cells as well as isolated membrane preparations. All of the probes are reduced to non-paramagnetic forms to some extent by actively respiring cells. Hence, non-toxic oxidants must be added to live cell suspensions to reoxidize probe molecules to the free radical forms. The 3 stearate probes report on membrane fluidity at different depths in the hydrophobic, inner portion of the membrane leaflet while the cholestane probe provides information on the distribution of membrane sterols in the leaflet. The small, partitioning probes such as Tempone report the relative average volumes and viscosities of intracellular aqueous and lipid phases (Miller, 1978).

Liposome spin probe models - Liposomes were prepared with varying proportions of soybean lecithin, mixed commercial triacylglycerols and ergosterol or with mixtures of PE and

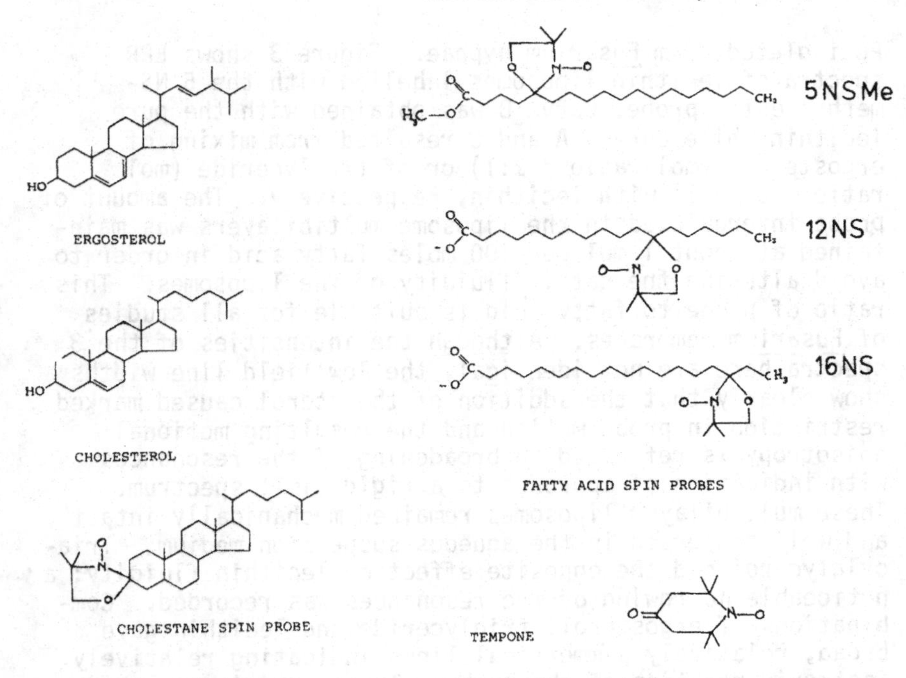

Figure 2: Spin probes useful in determination of membrane lipid fluidity in Fusarium.

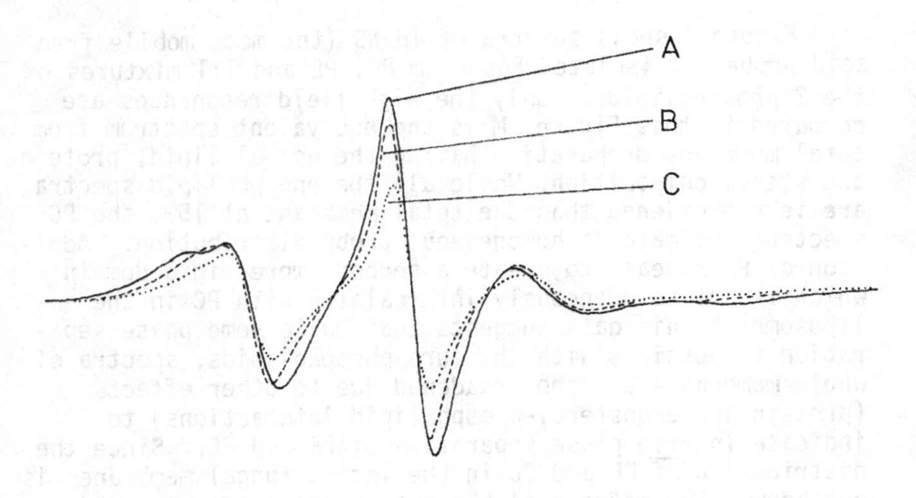

Figure 3: EPR spectra of 5 NS-methyl ester in multibilayer liposomes dispersed in 0.05 M sodium phosphate buffer, pH 6.9.

PC isolated from Fusarium hyphae. Figure 3 shows EPR spectra of lecithin liposomes labelled with the 5 NS-methyl ester probe. Curve B was obtained with the pure lecithin while Curves A and C resulted from mixing of ergosterol (mol ratio = 2:1) or of triglyceride (mol ratio = 0.85:1) with lecithin, respectively. The amount of probe intercalated in the liposome multibilayers was maintained at about 1 mol per 100 moles fatty acid in order to avoid altering the native fluidity of the liposomes. This ratio of probe to fatty acid is suitable for all studies of Fusarium membranes. Although the intensities of the 3 spectra here are not identical, the low field line widths show clearly that the addition of the sterol caused marked restriction in probe motion and the resulting motional anisotropy is reflected in broadening of the resonances with indication of approach to a rigid limit spectrum. These multibilayer liposomes remained mechanically intact and well dispersed in the aqueous suspension medium. Triacylglycerol had the opposite effect on lecithin fluidity; a noticeable narrowing of the resonances was recorded. Combinations of ergosterol, triglyceride and lecithin gave broad, relatively symmetrical lines indicating relatively isotropic rotation of the probe. These experiments show that both triglycerides and ergosterol can be accommodated in relatively large proportions in multibilayer liposomes.

Figure 4 shows spectra of 16 NS (the most mobile free acid probe) in isolated Fusarium PC, PE and 1:1 mixtures of the 2 phospholipids. Only the high field resonances are compared in this Figure; M is the equivalent spectrum from total membrane preparation having the normal lipid, protein and sterol composition. While all the phospholipid spectra are less broadened than the total membrane at 15°, the PC spectrum indicates a homogeneous probe distribution. Addition of PE appears to create a second, more rigid domain which is not homogeneously intercalated with PC in the liposomes. This data suggests that while some phase separation is possible with the pure phospholipids, spectra of whole membranes are too broadened due to other effects (protein and ergosterol-phospholipid interactions) to indicate in vivo phase separation of PE and PC. Since the distribution of PE and PC in the intact fungal membranes is not known, the effects of the net charge difference and differences in unsaturation between the 2 phospholipids on fluidity cannot be determined unambiguously.

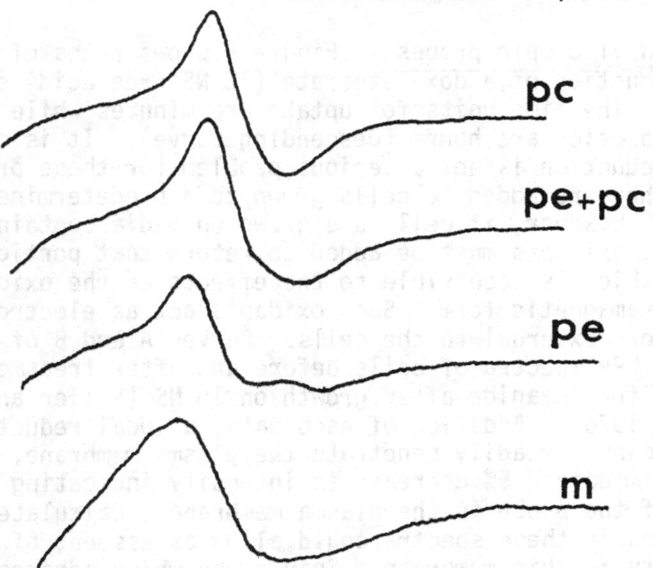

Figure 4: High field ESR signal from 16NS intercalated in *Fusarium* phospholipids. From Miller and de la Roche (1976)

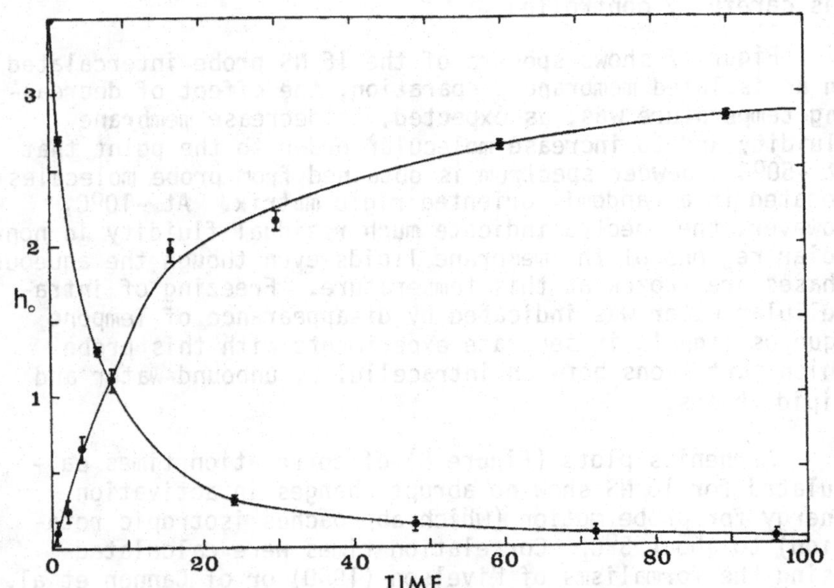

Figure 5: Uptake and reduction of 16NS by *Fusarium* cells. ESR signal height is plotted vs. time in min (uptake) and in hr (reduction). From Miller (1976).

In vivo spin probes - Figure 5 shows rates of uptake and reduction of a doxylstearate (16 NS free acid) spin probe. The time units for uptake are minutes while those for reduction are hours (descending curve). It is clear that reduction is not a serious problem for these probes when they are added to cells grown to a predetermined stage. However, if cells are grown on media containing the probes, oxidants must be added to return that portion of the probe which is accessible to the effects of the oxidant to the paramagnetic form. Such oxidants act as electron acceptors external to the cells. Curves A and B of Figure 6 show EPR spectra of cells before and after treatment with 200 μM ferricyanide after growth on 16 NS (Miller and de la Roche, 1976). Addition of ascorbate, a local reductant which cannot readily penetrate the plasma membrane, results in an immediate 5% decrease in intensity indicating reduction of the probe in the plasma membrane. Calculated differences in these spectra could allow assessment of probe fluidity in this membrane. Spin probe which adheres to dense cell wall fractions after cell disruption is totally reducible by ascorbate. Due to the pH dependence of the free acid probe-membrane interactions, the pH of all samples was carefully controlled.

Figure 7 shows spectra of the 16 NS probe intercalated in an isolated membrane preparation; the effect of decreasing temperature was, as expected, to decrease membrane fluidity and to increase molecular order to the point that at -50°C a powder spectrum is obtained from probe molecules located in a randomly oriented rigid matrix. At -10°C, however, the spectra indicate much residual fluidity in non-polar regions of the membrane lipids even though the aqueous phases are frozen at this temperature. Freezing of intracellular water was indicated by disappearance of Tempone aqueous signals in separate experiments with this probe which partitions between intracellular, unbound water and lipid phases.

Arrhenius plots (Figure 8) of correlation times calculated for 16 NS show no abrupt changes in activation energy for probe motion (which approaches isotropic rotation) to about 5°C. Correlation times were calculated using the formalisms of Kivelson (1960) or of Cannon *et al.* (1975). These methods cannot be used for the other stearate spin probes (12 NS, 5 NS) because of the increasing effects of rotational anisotropy as the reporter group

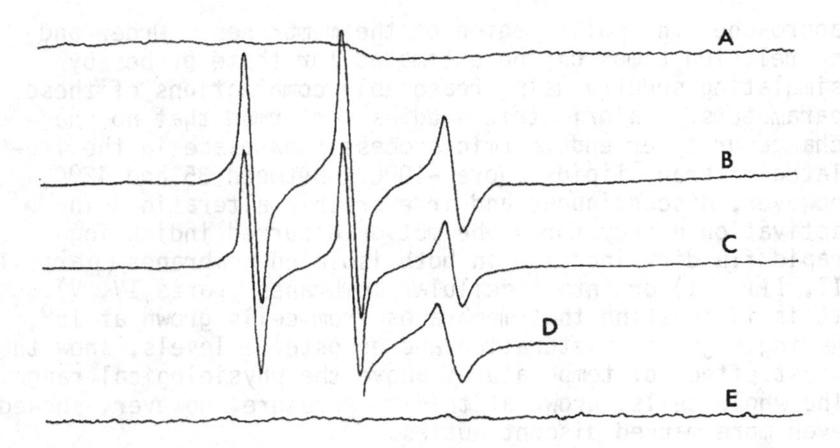

Figure 6: ESR spectra of 16-nitroxyl stearate. Spectrum A was obtained from whole cells before any treatment while Spectrum B was observed after treatment with 0.2 mM potassium ferricyanide. Treatment of cells with potassium ferricyanide and 1% sodium dodecyl sulfate did not remove significant label from whole conidia (Spectrum D) but the same treatment eliminated any signal from broken cell material (Spectrum E). From Miller and de la Roche (1976).

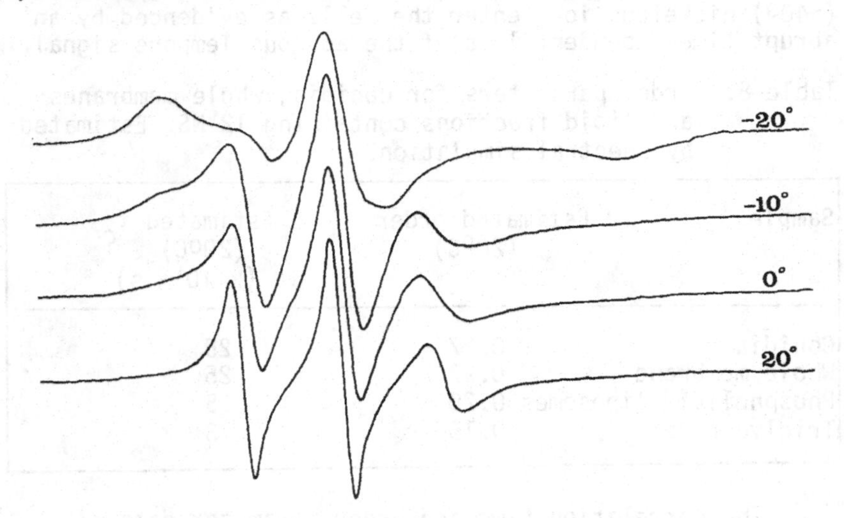

Figure 7: Effect of temperature on ESR spectra of 16-nitroxyl stearate in total membrane preparation. From Miller and de la Roche (1976).

approaches the polar region of the membranes. Order and correlation times can be estimated for these probes by simulating spectra using reasonable combinations of these parameters. Calorimetric studies confirmed that no phase change or other endothermic process took place in the isolated membrane lipids above -10°C. Between 35 and 42°C, however, discontinuous and irreversible alterations in activation energy for probe motion occurred indicating rapid fluidity increase in both isolated membranes (parts I, II, III, VI) or intact cellular membranes (parts IV, V). It is interesting that membranes from cells grown at 15°, having highest unsaturation and ergosterol levels, show the least effect of temperatures above the physiological range. The whole cells, grown at this temperature, however, showed even more marked discontinuties.

Figure 9 provides correlating data on the temperature-induced loss of plasma membrane impermeability to divalent cations. The probe used here was Tempone; the partitioning behavior of this small, isotropically tumbling probe is temperature dependent. Throughout the physiological temperature range the decrease in the ratio (R) indicates a gradual decrease in the signal arising from the intracellular aqueous phase while near the lethal temperature (>40°) nickelous ions enter the cells as evidenced by an abrupt time-dependent loss of the aqueous Tempone signal.

Table 8. Order parameters for conidia, whole membranes and lipid fractions containing 12 NS. Estimated by spectral simulation.

Sample	Estimated order (20°C)	Estimated τ_c (20°C) (x 10^{10} s)
Conidia	0.27	25
Whole membrane	0.27	25
Phospholipid liposomes	0.20	5
Triglyceride	0.15	3

The correlation time and order parameter data of Table 8 represents pairs of values calculated for the 12 NS probe located in whole cells, total membrane, mixed phospholipid liposomes and isolated storage lipids.

Average intact cellular membrane spectra and isolated total membrane spectra may be closely simulated with similar or identical parameters. Isolated lipids show the expected enhancement in fluidity due to the absence of sterol and protein. The triglycerides show practically no order because stable bilayer leaflets are not formed. In this case, correlation time was confirmed by application of a formalism appropriate for isotropic tumbling.

If we examine correlation times calculated for 16 NS in either isolated membranes or whole cells (Table 9), we see that they remain relatively constant within the physiological range both at the temperature of growth or at 25°, as taken from the Arrhenius plot. Any error in this presumed regulation of fluidity appears to be slightly on the side of increased rigidity in the 15° grown, sterol-rich membranes.

Figure 10 illustrates spectra obtained with the cholestane probe which is thought to associate specifically with endogenous sterol molecules. If a phase separation involving ergosterol were occurring at the lower temperatures islands of isolated sterol in a closely packed configuration, might be expected. There is no indication of spin-exchange broadening of probe signals in the spectra recorded at low temperature. Although spectra between 0 and 30° indicate anisotropic motion of the probe, at 40° the spectra indicate isotropic motion. It is in this region that irreversible, temperature-induced destruction of membrane properties occurs.

ABSTRACT

The combined spin probe data clearly indicate that the rates of motion of membrane lipids, and hence the average fluidity of membrane lipids, is controlled within close tolerances. Neither phase separations nor abrupt physical state changes are indicated by the available spectroscopic, calorimetric or electron microcopic data for fungal eukaryotes. The analytical data, together with the newly developing understanding of temperature dependent alterations in the metabolic pathways responsible for the biosynthesis of fungal membrane lipids, indicate that these metabolic responses lead to necessary changes in the ratios and distribution of the membrane lipids. The entire process of regulation of lipid biosynthesis and assembly would be

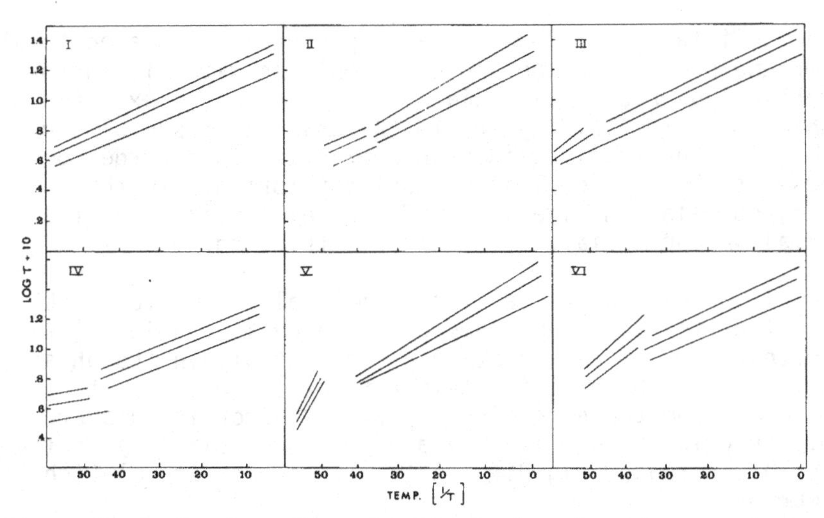

Figure 8: Arrhenius plots 16-NS. Isolated membranes from cells grown at: I, 15°; II, 20°; III, 25°; VI, 37°. Whole cells grown at: IV, 15°; V, 25°. Top and bottom lines, τ_c according to: Cannon et al. (1975); middle line; Kivelson (1960).

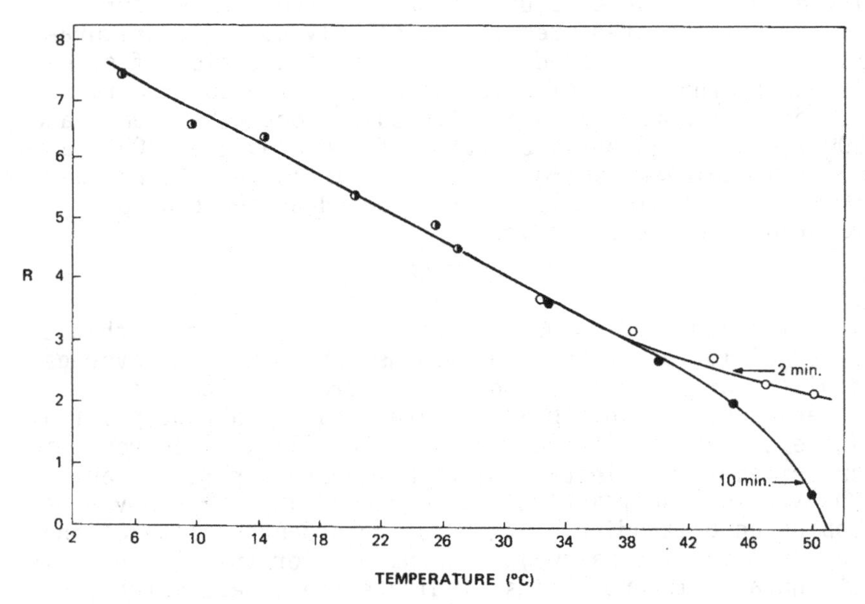

Figure 9: Temperature dependence of ratio (R) of intracellular tempone ESR signals from probe in aqueous and lipid phases. Open circles, R after 2 min. Closed circles, 10 min.

expected to be under genetic control. Unfortunately, this cannot be studied in Fusarium oxysporum due to the lack of a controlled sexual reproductive process for these eukaryotes. Other related fungi should be more amenable to the genetic approach.

Table 9. Rotational correlation times for 16 NS probe in isolated membranes and intact cells

Growth Temperature °C (Tg)	In vivo at 25°	In vivo at Tg	Membranes at 25°	Membranes at Tg
	$\tau_c \times 10^{10}$ sec			
15	11.7	12.8	10.0	12.0
25	10.2	10.2	10.7	10.7
37	-	-	11.7	10.9

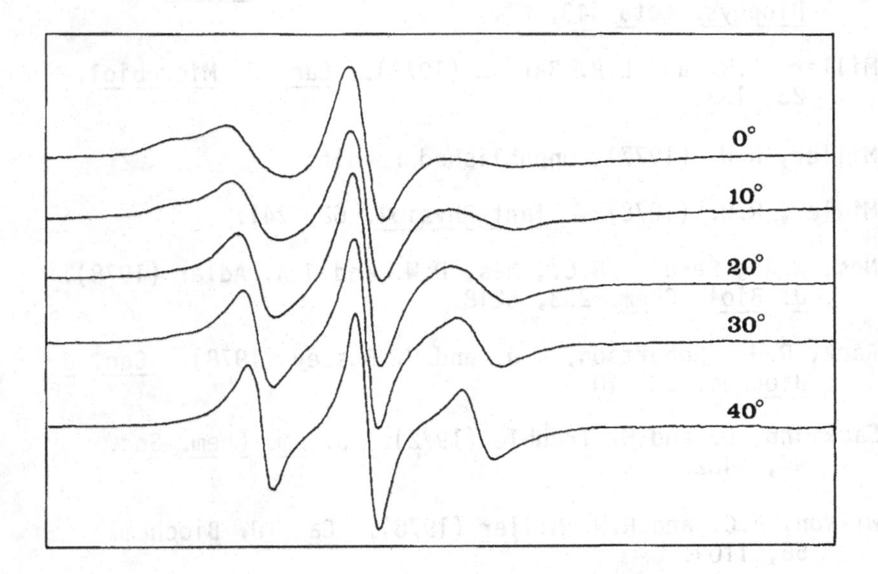

Figure 10: Effect of temperature on ESR spectra of cholestane probe. Spin labelled total membrane preparation from 25°C grown F. oxysporum hyphae was dispersed in 0.05 M Tris buffer, pH 7.5. ESR spectra were recorded after 4 min equilibration at the indicated temperatures. From Miller and de la Roche (1976).

REFERENCES

Barran, L.R., R.W. Miller and I.A. de la Roche (1975). Can. J. Microbiol. 22, 557.

Cannon, B., C.F. Polnaszek, K.W. Butler, L.E.G. Eriksson and I.C.P. Smith (1975). Archives Biochem. Biophys. 167, 505.

Kivelson, D. (1960). J. Chem. Phys. 33, 1094.

Madhosingh, C., Lepage, M. and Migicovsky, B.B. (1972). Can. J. Microbiol. 18, 1679.

Madhosingh, C. (1977). Agric. Biol. Chem. 41, 1233.

Miller, R.W. (1976) Spectroscopy Letters 9, 895.

Miller, R.W. and I.A. de la Roche (1976). Biochim. Biophys. Acta 443, 64.

Miller, R.W. and L.R. Barran (1977). Can. J. Microbiol. 23, 1373.

Miller, R.W. (1977), unpublished results.

Miller, R.W. (1978). Plant Physiol. 62, 741.

Nes, W.R., Sekula, B.C., Nes, D.W. and J.H. Adler (1978). J. Biol. Chem. 253, 6218.

Rank, G.H., Robertson, A.J. and H. Bussey (1978). Can. J. Biochem. 56, 1036.

Sackmann, E. and H. Traüble (1972). J. Am. Chem. Soc. 94, 4482.

Wilson, A.C. and R.W. Miller (1978). Can. J. Biochem. 56, 1109.

Wilson, A.C. and L.R. Barran (1979). This volume.

Wilson, A.C., Adams, W.C. and R.W. Miller (1979). Can. J. Biochem. in press.

HOMEOVISCOUS ADAPTATION IN PSYCHROPHILIC, MEOSPHILIC AND THERMOPHILIC YEASTS

Kenneth Watson

Department of Chemistry and Biochemistry
James Cook University of North Queensland
Townsville 4811 Australia

ABSTRACT

Thermophilic and psychrophilic yeasts were differentiated on the basis of their ability, in the case of thermophiles, and inability, in the case of psychrophiles, to (a) grow under strictly anaerobic conditions and (b) produce stable respiratory-deficient mutants either spontaneously or on treatment with acriflavine or ethidium bromide. A possible correlation between these properties and the polyunsaturated fatty acid composition of the membrane phospholipids was proposed. There were marked changes in the fatty acyl composition of membrane phospholipids isolated from the thermophilic yeast, *Torulopsis bovina*, when monitored at different stages of growth under aerobic and anaerobic conditions. During aerobic induction of anaerobically grown cells, the direct desaturation of palmitic acid residues on phosphatidylcholine to palmitoleic acid appeared to be a mechanism whereby cells adjusted their membrane fluidity.

INTRODUCTION

Yeasts may be classified into psychrophilic, mesophilic and thermophilic species on the basis of their temperature limits of growth (Watson et al., 1976; Arthur & Watson, 1976). Previous studies have established that a characteristic of psychrophilic yeasts is a high content of polyunsaturated fatty acids in the membrane phospholipids (Kates & Baxter, 1962; Watson et al., 1976; Arthur & Watson, 1976). Conversely, thermophilic yeasts, whilst rich in mono-unsaturated fatty acids (Arthur & Watson, 1976), have little or no polyunsaturated fatty acid. The temperature dependence of glucose transport into psychrophilic and thermophilic yeasts has recently been shown to be correlated with these differences in membrane lipid composition (Morton et al., 1978).

In this paper we summarize additional properties of psychrophilic and thermophilic yeasts which may be exploited to further differentiate between these two thermal categories. These include the ability or inability to grow anaerobically and to produce stable respiratory-deficient mutants. The availability of psychrophilic, mesophilic and thermophilic yeasts belonging to the same genus, namely *Torulopsis*, offered the opportunity to test the concept, introduced by Sinesky (1974) and termed homeoviscous adaptation, of microbial membrane lipids maintaining a near constant fluidity at the growth temperature. Alterations, with growth phase, of the fatty acyl composition of membrane phospholipids was monitored in yeasts grown aerobically and anaerobically. Preliminary results using cells from *T. bovina*, a thermophilic yeast, indicated that during aerobic induction of anaerobically grown cells, the direct desaturation of palmitic acid residues on phosphatidylcholine to palmitoleic acid was a mechanism whereby cells adjusted their membrane fluidity.

METHODS

Organisms

All organisms were obtained from the Centraalbureau voor Schimmelcultures, Delft, The Netherlands, and were identified by their CBS numbers.

(a) Aerobic growth. The composition of the growth media was as previously described (Arthur & Watson, 1976). Cells were grown in 2 liter flasks and shaken in orbital shakers operating at 180 rev./min under controlled temperature conditions. An inoculum of 3 mg dry weight cells per liter was used and cells were harvested at different stages of growth as indicated in the text.

(b) Anaerobic growth. The composition of the media was the same as for the aerobic cultures except for the addition of ergosterol (20 mg/l) and unsaturated fatty acid - oleic, linoleic or linolenic acids (30 mg/l). Purified nitrogen, essentially free of oxygen after passage through an Oxy-trap (Alltech Associates) at 60°C, was sprayed through the media throughout the growth period which was from 3 to 7 days for the psychrophiles (15°C) and 1 to 2 days for the mesophiles (30°C) and thermophiles (37°C). An inoculum of 250 mg dry weight of cells per liter was used for the psychrophiles and 10 mg dry weight of cells per liter for the mesophiles and thermophiles. The heavy inoculum used for the psychrophiles was a deliberate measure to obtain sufficient cells for subsequent lipid analysis. In oleic acid supplemented media, the presence of 40-60% polyunsaturated fatty acid clearly demonstrated that the psychrophilic cells had undergone a minimal number of divisions.

Cells for aerobic induction were harvested at mid-exponential phase, suspended in the standard medium containing 0.2% glucose at a concentration of 10-15 mg dry weight per ml and suspensions shaken in an orbital shaker at 180 rev./min. Samples for dry weight and lipid analysis were taken at zero time and at intervals up to 6 hr.

Isolation of Respiratory-Deficient Mutants

Cells were incubated with shaking at appropriate temperatures, as indicated above, in standard media with ethidium bromide or acriflavine at concentrations from 1 to 200 µg/ml. Appropriate dilutions of an exponentially growing culture were plated out onto solid media containing 2% glycerol and 2% glucose as carbon sources. Petite colonies were characterized by their size and were further identified by the tetrazolium overlay procedure (Ogur et al., 1957). Alternatively, cells were exposed to the highest concentration of mutagen compatible with growth by the drop method. A drop of mutagen (10 mg/ml) was placed onto solid medium containing 2% glycerol and 0.2% glucose as carbon sources, and after 24 hr a small loopful of yeast culture was streaked across the top. After incubation at the appropriate temperature, cells from the margin of growth were streaked away from the mutagen and single colonies were picked after a further period of incubation. Small petite colonies were clearly distinguished from the large respiratory-competent colonies on this medium. Spontaneous petite formation was determined by streaking cultures onto the solid medium in the absence of mutagen. Putative respiratory-deficient mutants were selected on the basis of their small size and lack of colouration after treatment with the tetrazolium overlay procedure.

Lipid Analysis

Lipids were extracted from cells by a modification of the Bligh & Dyer (1959) procedure. The chloroform:methanol (2:1, v/v) extracts were washed with one quarter volume of 0.88% KCl and the lower phase taken for lipid analysis.

(a) Thin-layer chromatography. Phospholipids were separated by two dimensional thin-layer chromatography, essentially as described by Rouser et al. (1970). Phospholipid spots were detected by brief exposure of the plates to iodine and fatty acid methyl esters prepared by direct methylation of the spots with 5% HCl in methanol.

(b) Gas chromatography. Methyl esters were separated by gas chromatography on a 3 m Silar 10C column (Applied Sciences) operating at 175°C with an injection temperature of 230° and a carrier gas (nitrogen) flow of 40 ml/min. Area percentage of individual methyl esters was determined by a Hewlett-Packard model HP3380A integrator.

RESULTS

Table 1 summarizes the results on anaerobic growth of psychrophilic and thermophilic yeasts. The obligate psychrophilic yeasts were unable to grow under strictly anaerobic conditions. In some experiments, the one to two generations apparently undergone by the psychrophilic yeasts may be attributed to the initial heavy inoculum (see Methods). By contrast, all four thermophilic yeasts grew well anaerobically. The table includes, for comparative purposes, data on three mesophilic yeasts. The two *Torulopsis* sp., *candida* and *torresii*, did not grow anaerobically in contrast to *S. cerevisiae*.

Yeast		Anaerobic growth (no. of generations)
Psychrophiles		
Leucosporidium frigidum	CBS5270	0-2
Leucosporidium gelidum	CBS5272	0-2
Leucosporidium nivalis	CBS5266	0-2
Torulopsis austromarina	CBS6179	0-2
Torulopsis psychrophila	CBS5956	0-2
Mesophiles		
Torulopsis candida	CBS940	0-2
Torulopsis torresii	CBS5152	0-2
Saccharomyces cerevisiae	CBS1171	7-9
Thermophiles		
Torulopsis bovina	CBS2760	7-9
Torulopsis pintolopesii	CBS1787	7-9
Saccharomyces telluris	CBS2685	7-9
Candida slooffii	CBS2419	7-9

Table 1. Anaerobic growth of psychrophilic, mesophilic and thermophilic yeasts.

None of the psychrophilic yeasts listed in Table 1 produced stable respiratory-deficient mutants either spontaneously or on treatment with acriflavine or ethidium bromide. Two of the thermophilic yeasts, *T. pintolopesii* and *C. slooffii* are naturally-occurring respiratory-deficient yeasts (Bulder, 1964) and we have recently reported that *C. slooffii* resembles the cytoplasmic petite mutants of *S. cerevisiae* (Arthur et al., 1978). Spontaneous mutants of *S. telluris* and *T. bovina* were isolated at a frequency of 0.2 and 0.01% respectively compared with a frequency of 1% with *S. cerevisiae*. However, on treatment with mutagens, >90% conversion to respiratory-deficient cells was observed. Stable respiratory-deficient mutants were not obtained from the two mesophilic *Torulopsis* sp., *candida* and *torresii*. Table 2 summarizes the differentiating properties of psychrophilic and thermophilic yeasts.

	Psychrophiles	Thermophiles
Growth temperature	Max. 20°C	Min. 20°C
Anaerobic growth	-	+
Respiratory-deficient mutants	-	+
Polyunsaturated fatty acids	+	-

Table 2. Differentiating properties of psychrophilic and thermophilic yeasts.

The characterization of psychrophilic, mesophilic and thermophilic *Torulopsis* yeasts, offered the opportunity to test the hypothesis that changes in fatty acyl composition of membrane phospholipids is designed to produce membranes whose lipids have a constant fluidity at the temperature of growth - a phenomenon termed homeoviscous adaptation (Sinensky, 1974). As an indicator of membrane fluidity the unsaturated fatty acid index was calculated for phospholipids isolated from yeasts grown at different temperatures (Table 3). Two distinct patterns emerged from the results. Firstly, the long established observation in microorganisms of an increase in fatty acid unsaturation with decreasing growth temperature (Kates, 1964) was well illustrated by the results on *Torulopsis*. Secondly, for the same growth temperature, the phospholipids isolated from the different thermal categories of yeasts showed a similar unsaturation index. At 15°C, the two psychrophiles, *T. psychrophila* and *T. austromarina*, have a similar unsaturation index to that for the mesophile, *T. candida*. The question was then raised as to whether homeoviscous adaptation would apply throughout the growth phase. The thermophile, *T. bovina*, was selected for these studies since it is a facultative anaerobe (Table 1) and we could thus test both aerobically and anaerobically grown cells.

Fig. 1 shows changes with growth phase in the major fatty acids of phosphatidylcholine isolated from *T. bovina* grown aerobically at 37°C. As the cells approached the end of exponential growth there was a decrease in the amount of the monounsaturated fatty acids, palmitoleic (C 16:1) and oleic (C 18:1), resulting in a decrease in the unsaturation index from 0.77 to 0.50. Concurrent with these changes,

Yeast	Growth Temperature (°C)	Percentage of total fatty acids								Degree of unsaturation[a]
		Saturated				Unsaturated				
		$C_{12:0}$	$C_{14:0}$	$C_{16:0}$	$C_{18:0}$	$C_{16:1}$	$C_{18:1}$	$C_{18:2}$	$C_{18:3}$	
Psychrophiles										
T. psychrophila	0	-	-	7	tr[b]	tr	45	8	39	1.78
	15	-	-	14	tr	tr	50	18	17	1.37
T. austromarina	0	-	-	7	tr	7	29	13	44	1.87
	15	-	-	6	tr	3	48	33	10	1.44
Mesophiles										
T. candida	8	-	-	13	2	7	26	23	29	1.66
	15	2	2	12	1	13	27	27	16	1.42
	25	1	1	14	1	12	38	29	4	1.20
S. cerevisiae	30	1	2	10	3	53	31	-	-	0.84
Thermophiles										
T. bovina	37	1	5	12	4	50	28	-	-	0.78
T. pintolopesii	37	16	12	7	3	30	32	-	-	0.62

Table 3. Fatty acid composition of the phospholipid fraction isolated from psychrophilic, mesophilic and thermophilic yeasts.

[a] Degree of unsaturation calculated from (% monoene + 2[% diene] + 3 %[triene])/ 100.

[b] tr, <1%.

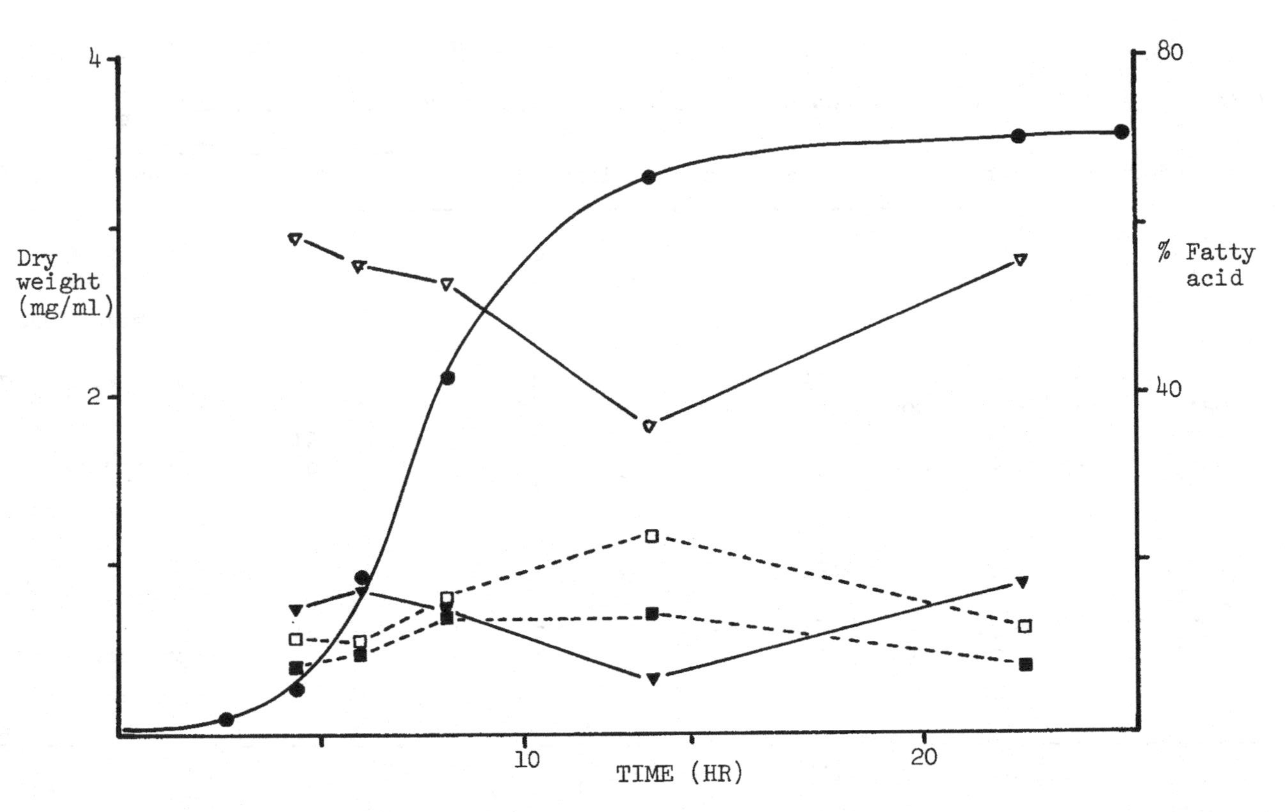

Fig. 1. Changes in fatty acid composition of phosphatidylcholine isolated from *T. bovina* grown aerobically. Growth curve ●. Fatty acid composition ▽ C 16:1, ▼ C 18:1, □ C 14:0, ■ C 16:0.

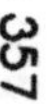

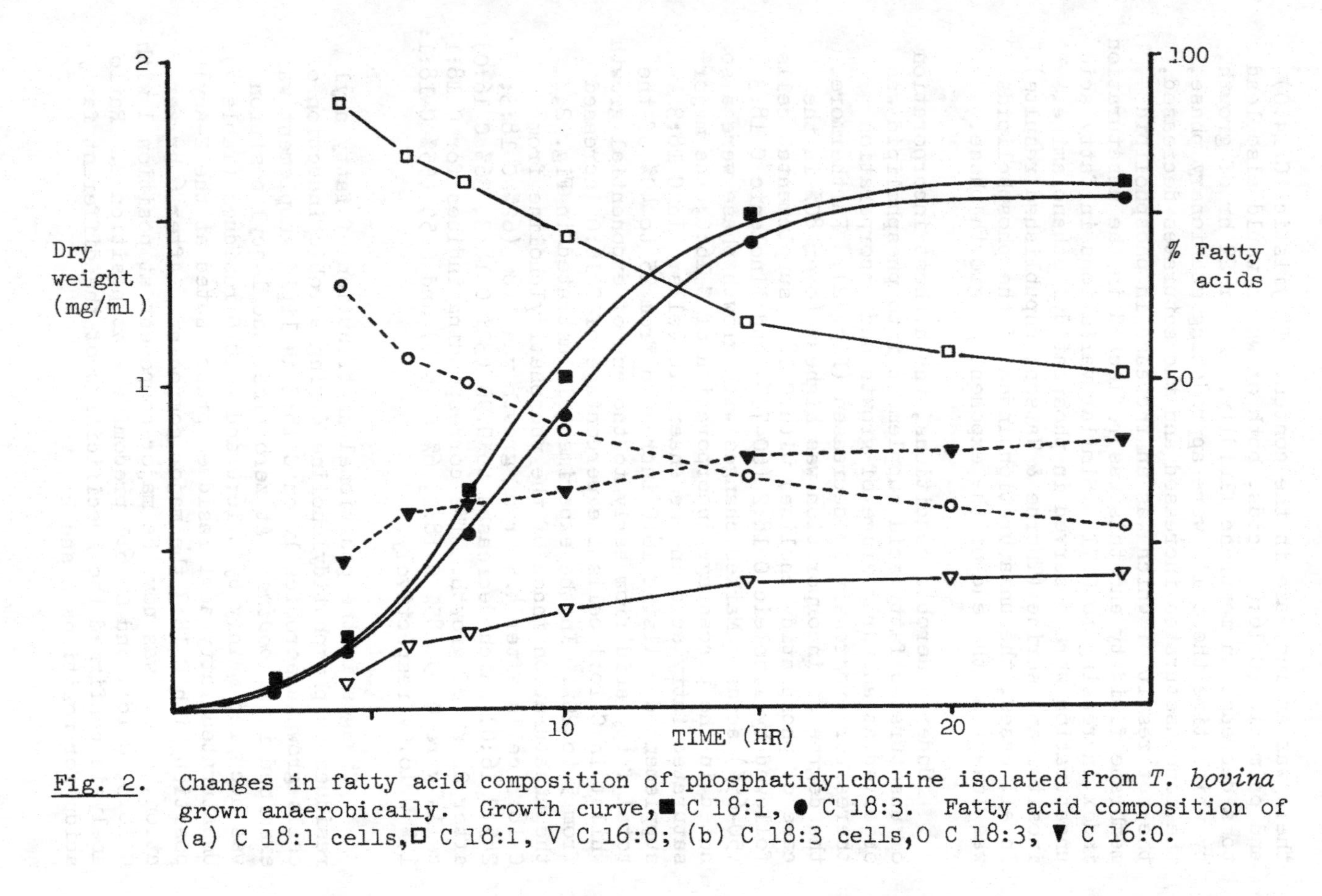

Fig. 2. Changes in fatty acid composition of phosphatidylcholine isolated from *T. bovina* grown anaerobically. Growth curve,■ C 18:1, ● C 18:3. Fatty acid composition of (a) C 18:1 cells,□ C 18:1, ▽ C 16:0; (b) C 18:3 cells,O C 18:3, ▼ C 16:0.

there was an increase in the content of myristic (C 14:0) and palmitic (C 16:0) acids, both of which would also lead to a decrease in membrane fluidity. After 22 hr of growth, by which time the cells were approaching stationary phase, the monounsaturates increased and the saturates decreased, the net result of which was an increase in phospholipid membrane fluidity and this was reflected in the unsaturation index increasing to 0.71. Similar variations in fatty acid unsaturation were observed in phosphatidyl ethanolamine, inositol and serine (Thorne & Watson, unpublished results). In all cases, the unsaturation index of the phospholipids was lowest at the end of the exponential growth phase.

Under anaerobic conditions, the highest incorporation of unsaturated fatty acid supplement into phospholipid was observed at an early stage of growth and incorporation thereafter progressively decreased (Fig. 2). Furthermore, the degree of incorporation was highest (over 80% in the case of phosphatidylcholine) with C 18:1 supplemented cells followed by linoleic C 18:2 (60-70%) and linolenic C 18:3 (50-65%) acids. Marked changes with growth phase were also noted in the percentage incorporation of C 16:0, the major saturated fatty acid in the anaerobic cells. In C 18:3 supplemented cells, C 16:0 increased from 25 to 40% of the total fatty acid from early to the end of exponential growth whilst in C 18:1 cells the percentage of C 16:0 increased from 5 to 20%. In the experiment illustrated in Fig. 2, the unsaturation index of the phosphatidylcholine from C 18:3 cells after 4.5 hr of growth was 1.86 (62% C 18:3; 24% C 16:0) which decreased to 0.90 (30% C 18:3; 36% C 16:0) after 20 hr of growth. The corresponding indices for C 18:1 cells were 0.89 (85% C 18:1; 6% C 16:0) and 0.55 (55% C 18:1; 17% C 16:0) respectively.

Changes in the positional distribution of fatty acyl residues on phosphatidylcholine during aerobic induction of cells grown anaerobically on C 18:3 as lipid supplement was examined in *T. bovina*. At zero time, the 1-acyl position was occupied mainly by C 16:0 and C 18:0 residues (Table 4). Unsaturated fatty acyl residues predominated at the 2-acyl position with C 18:3 the major component. After 6 hr aeration, C 16:1 was now the major component at position 1 with C 18:3, C 18:1 and C 16:1 predominant at position 2. Ratio analysis confirmed the specificities of the different fatty acids for positions 1 and 2.

Residue	Fatty acyl composition (percent of total)					
	Zero time			After 6 hr aeration		
	Position		Ratio	Position		Ratio
	1	2	2/1	1	2	2/1
14:0	7	6	0.86	10	4	2.5
16:0	54	7	0.13	15	4	0.27
16:1	<1	3	>3	38	18	0.47
18:0	23	2	0.09	9	2	0.22
18:1	<1	4	>4	7	14	2.0
18:3	8	74	9.3	10	54	5.4
others	7	4	-	11	4	-

Table 4. Positional distribution of fatty acyl residues on phosphatidylcholine at zero time and after 6 hr aeration.

DISCUSSION

Our results on anaerobic growth and induction of respiratory-deficient mutants in psychrophilic and thermophilic yeasts may be related to the experiments of Johnson & Brown (1972) and Moulin et al. (1975) which suggested a possible relationship between the fatty acid composition of yeast cells and their ability to produce respiratory-deficient mutants. These authors concluded that petite-negative yeasts (Bulder, 1964), in contrast to petite-positive species, contained polyunsaturated fatty acids. On the other hand, Bulder & Reinink (1974) concluded from their studies on a large number of petite-positive and petite-negative yeasts, predominantly of the genus *Saccharomyces*, that there was no direct causal relationship between fatty acid composition and ability to form respiratory-deficient mutants. Despite these discrepancies, we are led to the conclusion that there is a correlation between the presence of polyunsaturated fatty acids in the membrane phospholipids of obligate psychrophilic yeasts and their inability to grow anaerobically and produce stable respiratory-deficient mutants. The reverse situation holds for obligate thermophilic yeasts which have a low content of polyunsaturated fatty acids in their membrane phospholipids. While the picture appears to us to be clear-cut with respect to psychrophilic and thermophilic yeasts, the position requires clarification in the case of mesophiles. It is noteworthy in this respect that *T. candida* and *T. torresii*, which are rich in polyunsaturated fatty acids, do not grow anaerobically or give rise to stable respiratory-deficient mutants.

Notwithstanding the limitations of interpreting the fatty acid unsaturation index as an estimate of membrane fluidity, our results demonstrate that homeoviscous adaptation, as defined by Sinensky (1974), does not apply throughout the growth phase of aerobically and anaerobically grown yeast. Exceptionally high unsaturation values, approaching two, were found in phospholipids from cells grown anaerobically in media supplemented with C 18:3 fatty acid, and harvested early during exponential growth. Such high values for phospholipid unsaturation have generally only been recorded for psychrophilic yeasts (Kates & Baxter, 1962; Watson et al., 1976; Table 3) grown at low temperatures at which the high degree of lipid unsaturation would be advantageous, if not essential, to maintain the cellular membranes in a semi-fluid, functional state. The high unsaturation values were maintained only for a short period and, towards the end of exponential growth, the index had fallen to below one (Fig. 2), a value more compatible to that generally observed for cells at 37° (Arthur & Watson, 1976; Table 3).

It is now well established that many membrane-bound enzymes and proteins associated with transport processes require individual phospholipids (Sandermann, 1978) with, in some instances, specific fatty acyl residues for proper functioning (Finnerty, 1979). The changes in membrane phospholipid unsaturation occurring at different stages of growth in aerobic and anaerobic yeasts may thus reflect the requirements of newly synthesized membrane-bound enzymes and proteins for specific lipid environments. It was not surprising, therefore, to note that the most marked changes in phospholipid unsaturation occurred at the beginning and towards the end of exponential growth, the latter stage by which the biosynthetic activities of the cell would be decelerating. We conclude that the concept of cells having a near constant membrane fluidity at a given temperature can be discussed only with reference to specific segments of the growth phase in microbial systems.

Preliminary studies were conducted on changes in phospholipid membrane fluidity on aerobic induction of *T. bovina* cells grown anaerobically on C 18:3 as lipid supplement. After 6 h aeration, there was a net increase of 26% and 8% in C 16:1 and C 18:1 residues respectively (Table 4). During the same period, about 75% of the C 18:3 residues, which are not synthesized by *T. bovina* under aerobic conditions, were retained. It is evident, therefore, that the majority of the monounsaturated residues after the aeration period arose from phosphatidylcholine molecules originally present in the anaerobic cells. Moreover, the C 16:1 residues, which initially had a specificity for the 2-acyl position, now showed a specificity for the 1-acyl position (Table 4). Concurrent with these changes was a marked decrease in the percentage of C 16:0 residues at the 1-acyl position. The results indicated that direct desaturation of C 16:0 residues on the 1-acyl position of phosphatidylcholine to C 16:1 had occurred during the anaerobic to aerobic transition. A similar, though less strong, argument applies in the case of desaturation of C 18:0 to C 18:1 residues. Direct desaturation of C 16:0 residues on phospholipids has not previously been reported. There is evidence, however, from *in vitro* experiments, in yeasts (Talamo et al., 1973; Pugh & Kates, 1975) and other organisms (Gurr et al., 1969; Baker & Lynen, 1971; Slack et al., 1976) that such a desaturation can take place in C 18:1 residues on phospholipids.

Acknowledgements. This work was supported by grants from the Australian Research Grants Committee.

REFERENCES

Arthur, H., and K. Watson (1976). J. Bacteriol. 128, 56.

Arthur, H., K. Watson, C.R. McArthur, and G.D. Clark-Walker (1978). Nature 271, 750.

Baker, N., and F. Lynen (1971). Europ. J. Biochem. 19, 200.

Bligh, E.G., and W.J. Dyer (1959). Can. J. Biochem. Physiol. 37, 911.

Bulder, C.J.E.A. (1964). Antonie van Leeuwen. 30, 1.

Bulder, C.J.E.A., and M. Reinink (1974). Antonie van Leeuwen. 40, 445.

Gurr, M.I., M.P. Robinson, and A.T. James (1969). Europ. J. Biochem. 9, 70.

Finnerty, W.R. (1979). Adv. Microbial. Physiol. 18, 177.

Johnson, B., and C.M. Brown (1972). Antonie van Leeuwen. 38, 137.

Kates, M. (1964). Adv. Lipid Res. 2, 17.

Kates, M., and R.M. Baxter (1962). Can. J. Biochem. Physiol. 40, 1213.

Morton, H., K. Watson, and M. Streamer (1978). FEMS Microbiol. Lett. 4, 291.

Moulin, G., R. Ratomahenina, P. Galgy, and J. Bézard (1975). Folia. Microbiol. 20, 396.

Ogur, M., R. St. John, and S. Nagai (1957). Science, 125, 928.

Pugh, E.L., and M. Kates (1975). Biochim. Biophys. Acta 380, 442.

Rouser, G., S. Fleischer, and A. Yamamoto (1970). Lipids 5, 494.

Sandermann, H. (1978). Biochim. Biophys. Acta 515, 209.

Sinensky, M. (1974). Proc. Nat. Acad. Sci. U.S.A. 71, 522.

Slack, C.R., P.R. Roughan, and J. Terpstra (1976). Biochem. J. 155, 71.

Talamo, B., N. Chang, and K. Bloch (1973). J. Biol. Chem. 248, 2738.

Watson, K., H. Arthur, and W.A. Shipton (1976). J. Gen. Microbiol. 97, 11.

REGULATION OF MEMBRANE FLUIDITY IN ANAEROBIC BACTERIA

Howard Goldfine and Norah C. Johnston

Department of Microbiology, School of Medicine

University of Pennsylvania, Philadelphia, PA 19104

Since life on earth began in an anaerobic environment, some of the present day anaerobes presumably represent a closer link to the earliest epochs than organisms that evolved after the atmosphere became aerobic. A study of the membranes of anaerobes may reveal some of the most primitive ways in which living organisms have coped with environmental changes. A wealth of information has accumulated over the past 15 years showing the importance of membrane lipids in assuring continued membrane function over a wide range of temperatures. Many anaerobes belonging to such diverse groups as the gram positive clostridia, gram-negative bacteroides and cocci, and some spirochetes have been found to have substantial amounts of plasmalogens in addition to diacylphosphatides among their membrane lipid components (Goldfine and Hagen, 1972). Although these lipids which are characterized by a 1-O-alk-1'-enyl 2-acyl glycerophosphate structure (Fig. 1a), have been known for many years as constituents of mammalian tissues (Debuch and Seng, 1972), little is known about their precise functions.

In addition to the ethanolamine and choline plasmalogens characteristic of certain mammalian tissues such as heart, the central nervous system, and skeletal muscle, a wider variety of polar head groups have been found in bacteria. Among these are N-methylethanolamine (Baumann et al., 1965), glycerol (Hagen, 1974; Khuller and Goldfine, 1974; Clarke et al., 1976); and serine (van Golde et al., 1973; 1975). The acyl chains associated with

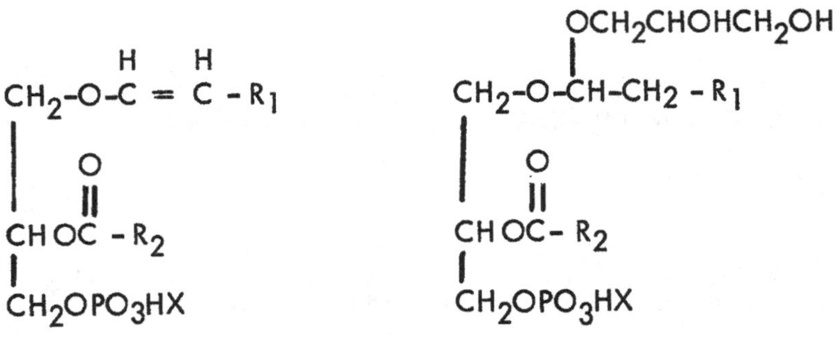

$$\begin{array}{l} \quad\quad\quad\; H \quad\; H \\ CH_2\text{-}O\text{-}C = C - R_1 \\ | \quad\quad O \\ | \quad\quad \| \\ CHOC - R_2 \\ | \\ CH_2OPO_3HX \end{array}$$

(a)

$$\begin{array}{l} \quad\quad\quad\quad OCH_2CHOHCH_2OH \\ \quad\quad\quad\quad | \\ CH_2\text{-}O\text{-}CH\text{-}CH_2 - R_1 \\ | \quad\quad O \\ | \quad\quad \| \\ CHOC - R_2 \\ | \\ CH_2OPO_3HX \end{array}$$

(b)

Fig. 1 a) Plasmalogen; b) Glycerol acetal of a plasmalogen.

bacterial plasmalogens are similar to those found in bacteria generally; straight chain saturated and mono-unsaturated, cyclopropane, and iso- and anteiso-branched chain fatty acids have been identified either in the plasmalogens or among the total fatty acids of organisms that contain plasmalogens (Goldfine and Hagen, 1972; Verkley et al., 1975). The alk-1-enyl chains are qualitatively similar to the acyl chains, but they often differ quantitatively. Up to now only a few species have been analyzed in detail (Goldfine, 1964; Kamio et al., 1970; Goldfine and Panos, 1971; Verkley et al., 1975). The questions we have asked are: 1) Are the membranes of these organisms fluid? 2) Is the fluidity regulated? 3) Since these anaerobes have both diacylphosphatides and plasmalogens, is the regulation of fluidity in their membranes a function of one or both types of lipids? Does it occur through regulation of the synthesis or incorporation of the acyl chains, the alk-1-enyl chains, the types and relative amounts of the polar head groups, or through other strategies of adaptation ?

THERMAL ADAPTATION IN CLOSTRIDIUM BUTYRICUM

Most of our work in this area has been done with Clostridium butyricum ATCC 6015, and since much of it has been published we will only briefly summarize the major findings. This strain grows over a limited range of temperatures, which lie between 25°and 38°C. It does not rapidly adjust to downward shifts in temperature, but within its range several significant and interesting changes are seen in the lipid composition, which are probably related to the growth temperature rather than the rate of growth (Khuller and Goldfine, 1974).

As its growth temperature is lowered, the acyl chains of the total phospholipids become more unsaturated, as has been found in a wide variety of organisms (McElhaney, 1976). In addition, the proportion of 16:0, 16:1, and 17:cyc acyl chains increase relative to 18:0, 18:1, and 19:cyc resulting in a somewhat shorter average chain length at lower temperatures. An unexpected finding was that the alk-1-enyl chains were more saturated at lower growth temperatures (Table 1). This was especially true in the ethanolamine plus N-methylethanolamine plasmalogens, which represent about 30% of the total phosphatides. The alk-1-enyl and acyl chains of the glycerol phosphoglycerides did not show significant temperature-dependent changes in their degree of unsaturation, but the general trend to shorter chains at lower temperatures was also seen in this fraction (Khuller and Goldfine, 1974; Goldfine et al., 1977). The ethanolamine plus N-methylethanolamine diacylphosphatides, which represent 12 to 16% of the total phospholipids in these cells, are unusual in that they have a high proportion of unsaturated chains on C-1 and somewhat more saturated than unsaturated chains on C-2 (Hildebrand and Law, 1964). We found that this pattern was maintained at lower growth temperatures with little overall change in the degree of unsaturation, accompanied by some chain shortening (Khuller and Goldfine, 1974).

To summarize the results on chain composition, the sum of the fatty acyl plus alk-1-enyl chains becomes more unsaturated at lower temperatures, simply because the larger number of acyl chains more than compensate for the decreased unsaturation of the alk-1-enyl chains. In addition, both types of chains are somewhat shorter on average at lower temperatures.

Table I

Acyl and Alk-1-enyl Chain Composition of *C. butryicum*
Phospholipids as a Function of Growth Temperature

Fraction Examined	25°	30°	37°
	% Unsaturated + Cyclopropane		
Total Phospholipids			
Acyl Chains	51	48	43.5
Alk-1-enyl Chains	48	43	51
Ethanolamine plus N-methylethanolamine Plasmalogens			
Acyl Chains	45	47	41
Alk-1-enyl Chains	36	32	46
Glycerol Phosphoglycerides			
Acyl Chains	49	48	48
Alk-1-enyl Chains	54	52	50

The largest change observed in the phospholipid class composition was an increase in the total glycerol phosphoglycerides, largely the result of an increase in the plasmalogen form from 8.4% of the total phospholipids at 37° to 17% at 25°. This may represent an additional contribution to increased membrane fluidity, since phosphatidylglycerol of a given acyl chain composition has a lower melting point than the analogous phosphatidylethanolamine (Papahadjopoulos *et al.*, 1973).

THERMAL ADAPTATION IN *VEILLONELLA PARVULA*

All of these results must be viewed in the context of the overall lipid composition of *C. butyricum*. Other anaerobes with different polar head group, acyl and alk-1-enyl chain compositions may utilize other alterations if they are capable of changing their membrane fluidity at different growth temperatures. *Veillonella parvula* a small, gram-negative coccus, is one example. This organism has a relatively simple phospholipid composition consisting largely of phosphatidylserine (PS) and phosphatidylethanolamine (PE) in a ratio of 27:73, with the latter largely in the plasmalogen form (Van Golde et al., 1975). The major acyl chains are 17:1 > 15:0 > 18:1 > 17:0 > 13:0. The alk-1-enyl chains have more 15:0 than 17:0 and 17:1 (Verkley et al., 1975).

The range of growth temperatures is again somewhat limited, with 37° near the maximum and 25° near the minimum. Over this range we have seen little change in the ratio of PS to PE, or in the proportion of the plasmalogens in the total phospholipids. The major compositional changes are in the alk-1-enyl chains, which are balanced nearly equally between saturated and unsaturated species at 37°, but become 60% to 73% unsaturated at 25°. In this organism the acyl chains are highly unsaturated at 37° and there is little change at lower growth temperatures (Johnston and Goldfine, unpublished). Verkley et al. (1975) showed that the lipids of V. parvula grown at 37° are predominantly fluid at that temperature, and undergo a phase transition between 20° and 5°C. We would expect this transition temperature to be lowered by the increased content of unsaturated alk-1-enyl chains at 25° in view of the expected large increase in species with two unsaturated chains. We have not yet carried out physical studies with the lipids from this organism.

PHYSICAL STUDIES

The effects of the compositional changes in C. butyricum phospholipids on their thermotropic phase transitions or phase separations were studied with the spin probe 2,2,6,6-tetramethylpiperidine-1-oxyl (TEMPO) and the fluorescent probe 1-phenyl-6-phenylhexatriene (DPH). These results are summarized in Table 2 (Goldfine et al., 1977).

Both probes show that the bulk phospholipids must be nearly or completely melted at the growth temperatures. We believe that the deviations from linearity labelled t_2, seen in Arrhenius plots of TEMPO partitioning probably represent the completion of melting, since these breaks were within 1 to 3°C of the transitions from a flatter to a relatively steep curve in plots of DPH fluorescence

Table 2

Deviations from Linearity Observed with EPR and

Fluorescent Probes in C. butyricum Lipids

Growth Condition		t_1 TEMPO	t_1 DPH	t_2 TEMPO	t_2 DPH	t_3 TEMPO	t_3 DPH
Biotin	37°C	48.3°	n.s.[+]	34.5°	35.8°	n.s.	-5°
Biotin	25°C	45°	n.s.	29.4°	32°	n.s.	-5°
Oleate	37°C	44°	n.s.	28°	29°	n.s.	-2° *
Elaidate	37°C	46°	n.s.	24°	24°	12°	15°

* Polarization continued to increase below -2°

+ n.s. = not seen

polarization vs. temperature. This occurs at about 35° in the phospholipids from 37°-grown cells and at 29° to 32° in the phospholipids from 25°-grown cells. Since measurements of TEMPO partitioning are subject to large errors at lower temperatures, we were only able to detect the completion of these phase transitions in lipids from biotin-grown cells by measurement of DPH polarization. These occurred at -5° C in the lipids from cells grown at both 25° and 37°C.

C. butyricum is a biotin auxotroph and growth with suboptimal or no biotin requires the presence of unsaturated fatty acids (Broquist and Snell, 1951). We were thus able to study the effect of growth temperature on the utilization of exogenous fatty acids and the effects of their incorporation on membrane fluidity and on the properties of isolated lipids. A striking finding that emerged from these studies was that at 37°C in the presence of suboptimal biotin, incorporation of the exogenous fatty acids into phospholipid acyl chains was lower and more variable than their incorporation into the alk-1-enyl chains (Table 3). The greatest incorporation was seen with the highest melting fatty acid, elaidate, and the least with the lowest melting linoleate. These results are similar to those obtained previously with

Table 3
Phospholipid Chain Composition in C. butyricum grown
on Exogenous Fatty Acids with Suboptimal Biotin

Fatty Acid Supplement	Temp.	14:0	16:0	16:1 +17:cyc	18:1 +19:cyc	18:2
				Acyl Chains*		
Oleate	37°	7.5	35	18	39	-
Oleate	25°	4.6	14	9.1	72	-
Elaidate	37°	7.2	16	10.0	65	-
Linoleate	37°	10.0	40	15	6.2	28
				Alk-1-enyl Chains*		
Oleate	37°	—	7.9	11	81	—
Oleate	25°	—	5.6	6	88	-
Elaidate	37°	—	4.9	10	85	-
Linoleate	37°	—	7.0	19	12	63

* Data expressed as % by weight

unsaturated fatty acid auxotrophs of E. coli (Esfahani et al., 1969). When the growth temperature of cells growing in oleate plus suboptimal biotin medium was lowered to 25°C, the 18:1 plus 19:cyc in the acyl chains increased to 72%, with little increase in these chains in the alk-1-enyl groups (Table 3). Thus, as in biotin-grown cells, the increase in unsaturation at lower growth temperatures was largely in the acyl chains in oleate-grown cells.

In biotin-free media, the incorporation of oleate and elaidate into both acyl and alk-1-enyl chains was further enhanced at 37°C and both types of chains were ≫90% enriched in the fed fatty acids. Interestingly, C. butyricum did not grow on linoleate in biotin-free media. We can only speculate that the high levels of 18:2 incorporated under these conditions are incompatible with critical membrane functions.

A major change occurs in the phospholipid composition of oleate-grown cells. There is a decrease in the plasmalogen form of the ethanolamine plus N-methylethanolamine phosphatides concomitant with an increase in the glycerol acetals of these plasmalogens (Fig. 1b) from 29% to approximately 50% of the total phospholipids (Khuller and Goldfine, 1975). In these cells the plasmalogen form of phosphatidylglycerol also decreased from 8.4% to 2.4% of the phospholipids. The reverse was seen in elaidate-grown cells. There was a slight decrease in the glycerol acetal lipids accompanied by substantial increases of the plasmalogens in the glycerol phosphoglyceride fraction, as was seen in biotin-grown cells at 25°C (Goldfine *et al*., 1977).

Thermotropic phase transitions were readily seen in the phospholipids and membranes of elaidate-grown cells with both TEMPO and DPH (Table 2). These transitions began at 24°C with both probes and ended at 15° (DPH) or 12° (TEMPO) in the extractable lipids. Membranes isolated from these cells gave a similar, relatively narrow transition with DPH, but the midpoint was 5°C higher (Goldfine *et al*., 1977). The phospholipids from oleate-grown cells produced broad transitions beginning at 28 to 29° with both probes. The completion of the transition to the gel phase was not observed with either probe even though measurements were made to -17° C with DPH.

In recent studies on the thermotropic behavior of the pure lipids, we have worked with strain IFO 3852 of *C. butyricum*, which has a similar lipid composition to strain ATCC 6015, except that it has no N-methylethanolamine in its polar head groups. This base is replaced entirely by ethanolamine (Matsumoto *et al*., 1971), which greatly simplifies interpretation of the physical behavior of the lipids. In order to work with lipids with transition temperatures above 0° C we have concentrated on elaidate-grown biotin-free cultures. Under these conditions the total phospholipids of strain IFO 3852 have 99% 18:1 in the alk-1-enyl chains and 91% 18:1 in the acyl chains. The virtual absence of 19:cyc shows that most of the 18:1 is *trans* as was found in elaidate-grown strain ATCC 6015 (Khuller and Goldfine, 1975).

In these studies we have used the fluorescent probes *cis* and *trans*-parinaric acids (PnA). These long chain, polyunsaturated fatty acid probes have been shown to detect accurately the transition temperatures of pure lipids and the lateral phase separation of binary mixtures.

Trans-PnA partitions preferentially into the solid phase, whereas cis-PnA is found in both phases to an approximately equal extent (Sklar et al., 1977 a,b).

As with DPH and TEMPO, relatively narrow phase transitions were observed in the total phospholipids of strain IFO 3852 grown on elaidate, when the fluorescence intensity of cis or trans-PnA was was measured as a function of temperature. The midpoint of these transitions was 25° ± 1° C (Table 4).

The ethanolamine phosphatides, which constitute approximately 30% of the total, were isolated by sequential column chromatography on DEAE-cellulose and silicic acid. The diacyl and plasmalogen forms of PE were present in the ratio of 30:70. A narrow phase transition centered at 30°C was detected in the total PE with both cis

Table 4

Phase Transition Parameters for Parinaric Acids in Phospholipids from C. butyricum IFO 3852 Grown on Elaidate

	cis-Parinaric Acid			trans-Parinaric Acid		
	T_s* (°C)	T_c** (°C)	T_f* (°C)	T_s (°C)	T_c (°C)	T_f (°C)
Total Phospholipids	20(4.5)	24	28(5)	23(3)	25.5	28(4)
Total Phosphatidyl-ethanolamine	27(0.5)	29.5	32	28(4)	30	31.5(3)
Plasmalogen form of PE	29.5(4)	31	32(4)	30(3)	31	32(4)

* T_s and T_f as defined by Sklar et al. (1977b) and Tecoma et al. (1977) express the width of the transitions. Numbers in parenthesis represent the hysteresis between heating and cooling scans.

** T_c is the mid-point of the transition.

and *trans*-PnA (Table 4). The essentially pure plasmalogen form of PE was obtained by selective deacylation of the diacyl form with *Crotalus adamanteus* phospholipase A_2 (Waku and Nakazawa, 1972). This lipid, which had over 99% 18:1 in its alk-1-enyl chains and 91% 18:1 in its acyl chains, produced a narrow transition with both probes, with a midpoint at 31° (Table 4).

In collaboration with Dr. Michael C. Phillips of the Medical College of Pennsylvania, we have also determined the transition temperature of the elaidate-enriched PE plasmalogen by differential scanning calorimetry (DSC). The sample was hydrated in excess water and scanned at 10°/min. The lipid phase transition temperature, T_c, as defined by van Dijck *et al*. (1976) was 28°C. This can be compared to the transition temperature of synthetic dielaidoyl PE, which was found to be 35° by these authors. We believe that this is in large part a real difference between the diacyl and alk-1-enyl acyl lipids, and that the 9% of acyl chains other than *trans*-18:1 only serve to broaden the peak.

The thermal behavior of the unique glycerol acetal phospholipid of *C. butyricum* is as unusual as its structure. As shown in Fig. 2, studies with *cis* and *trans* PnA showed an exceptionally high degree of hysteresis; the average separation of the transitions between the heating and cooling scans was 12.5°C. These data were obtained at scan rates in the range of 0.5 to 1.0°/min. When the temperature was held at 26° for 30 min during a heating scan, the fluorescence intensity only decreased 15% and did not show the completion of the transition until the temperature was raised above 28°C. Similarly, during a cooling scan, the transition to the gel state did not occur for at least 30 min at 20.5°, but began once the temperature was lowered below 20°C.

DSC of this lipid has confirmed the hysteresis seen with fluorescent probes. In addition, it partially resolved both the melting transition and the transition to the gel state into two processes. The sum of the enthalpies is about twice that of PE. On heating at 10°/min the T_m values (the midpoints of the curves) were 30° and 34°, which corresponded well with the values obtained with the parinaric acids. On cooling at the same rate, two peaks were seen with T_m of 14° and 12.7°. When the sample was cooled at 0.62°/min, a rate close to that used with the fluorescent probes, the T_m values were

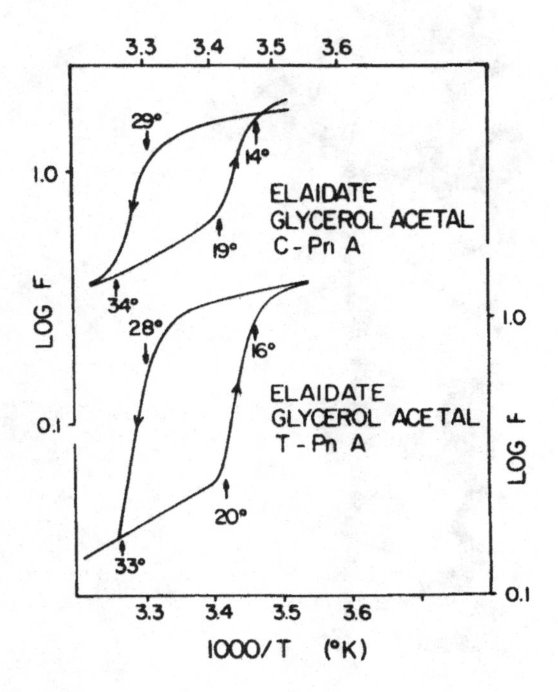

Fig. 2. Phase transitions detected in the glycerol acetal of phosphatidylethanolamine plasmalogen with <u>cis</u>-parinaric acid (top) and <u>trans</u>-parinaric acid (bottom).

19° and 15.5°, again in good agreement with the parinaric acid data.

A plausible, but unproven explanation of these results is based upon an examination of the space filling model of this structure (Fig. 3). The glycerol in acetal linkage is presumably at the surface of the bilayer and during cooling it may interfere with packing of the hydrocarbon chains, leading to supercooling. As noted above we observed that the transition temperatures of the cooling exothermal peaks were lower at faster scan speeds. The temperature of the gel to liquid-crystalline transition of the glycerol acetal is close to that observed for both the elaidate-enriched plasmalogen and dielaidoyl PE, but we have not determined which of the two peaks is due to

Fig. 3. Space filling model of the glycerol acetal of phosphatidyl-ethanolamine plasmalogen.

chain melting, nor do we know what the nature of the second process is.

CONCLUSIONS

Our results with two anaerobes, one a gram-positive spore-forming bacillus, the other a gram-negative coccus, show very different alterations in lipid composition in response to changes in growth temperatures. C. butyricum adjusts its polar head group composition and its acyl chains to provide a membrane with lower melting phospholipids. V. parvula, like E. coli adjusts its hydrocarbon chain composition. However, since the acyl chains are highly unsaturated at all temperatures studied, the adjustment is largely in the alk-1-enyl chain composition. These two sets of results clearly demonstrate that it is far too early to make generalizations. Other anaerobes, especially thermophiles and psychrophiles need to be studied to seek answers to the general question posed in this paper. Some work on the acyl chain compositions of thermophilic and psychrophilic anaerobes has been reported (Chan et al., 1971), but more work on the other components of the lipids needs to be done.

Another barrier to an understanding of these regulatory mechanisms, is our very incomplete understanding of the biosynthesis of plasmalogens in anaerobic bacteria. The mechanism has been shown to differ in a number of important ways from the mammalian pathway, and this has been discussed elsewhere (Goldfine and Hagen, 1972; Goldfine et al., 1976). Therefore work on the whole cell level will need to be supplemented and extended by work at the enzyme level, in order to begin to understand how these organisms adjust their membranes to a variety of conditions.

Acknowledgments: This work was supported by a research grant from the National Institute of Allergy and Infectious Diseases (AI-08903). We would like to express our appreciation to Dr. J.M. Vanderkooi for use of the fluorescence spectrophotometer and to Dr. A.F. Horwitz for helpful discussions and comments.

ABSTRACT

The presence of substantial amounts of alk-1-enyl acyl phospholipids (plasmalogens) in many anaerobic bacteria affords an opportunity to study the possible function of these lipids in the control of membrane fluidity. We describe work in our laboratory concerned with the thermal adaptation of *Clostridium butyricum* and *Veillonella parvula*. These two organisms show different changes in lipid composition as a function of decreasing growth temperatures. In *C. butyricum* the acyl chains but not the alk-1-enyl chains become more unsaturated, and changes in the proportions of the polar head groups and plasmalogens are seen at lower growth temperatures. In *V. parvula* the alk-1-enyl chains become more unsaturated at lower growth temperatures, whereas the acyl chains have a high proportion of monoenoic fatty acids at all temperatures studied. Changes in polar head groups were slight and no changes were seen in the plasmalogen content. The results of physical studies on the phospholipids from *C. butyricum* grown at different temperatures are reviewed. The plasmalogen form of phosphatidylethanolamine and its glycerol acetal were purified from cells highly enriched with *trans*-18:1. They were studied with the fluorescent probes, *cis*- and trans-parinaric acids, and by differential scanning calorimetry. An unusual degree of hysteresis was observed in the thermotropic phase transitions of the glycerol acetal. The phase transition temperature of the plasmalogen appeared to be several degrees lower than that of the corresponding diacyl form, dielaidoyl phosphatidylethanolamine.

REFERENCES

Baumann, N.A., P-O. Hagen, and H. Goldfine (1965). J. Biol. Chem. 240, 1559.

Broquist, H.P. and E.E. Snell (1951). J. Biol. Chem. 188, 431.

Chan, M., R.H. Himes and J.M. Akagi (1971). J. Bacteriol. 106, 876.

Clarke, N.G., G.P. Hazelwood, and R.M.C. Dawson (1976). Chem. Phys. Lipids 17, 222.

Debuch, H., and P. Seng (1972). In F. Snyder (ed.) Ether Lipids Academic Press, New York, p.1.

Esfahani, M., E.M. Barnes, Jr., and S.J. Wakil (1969). Proc. Natl. Acad. Sci. U.S.A. 64, 1057.

Goldfine, H. (1964). J. Biol. Chem. 239, 2130.

Goldfine, H. and P-O. Hagen (1972). In F. Snyder (ed.) Ether Lipids Academic Press, New York p. 329.

Goldfine, H. and C. Panos (1971). J. Lipid Res. 12, 214.

Goldfine, H., G.K. Khuller, and D.R. Lueking (1976). In R. Paoletti, G. Porcellati, and G. Jacini (eds.) Lipids Vol. 1 Biochemistry Raven Press, New York, p.11.

Goldfine, H., G.K. Khuller, R.P. Borie, B. Silverman, H. Selick, N.C. Johnston, J.M. Vanderkooi, and A.F. Horwitz (1977). Biochim. Biophys. Acta 488, 341.

Hagen, P-O. (1974). J. Bacteriol. 119, 643.

Hildebrand, J.C., and J.H. Law (1964). Biochemistry 3, 1304.

Kamio, Y., S. Kanegasaki, and H. Takahashi (1970). J. Gen. Appl. Microbiol. 16, 29.

Khuller, G.K., and H. Goldfine (1974). J. Lipid Res. 15, 500.

Khuller, G.K., and H. Goldfine (1975). Biochemistry 14, 3642.

Matsumoto, M., K. Tamiya, and K. Koizumi (1971). J. Biochem. (Tokyo) 69, 617.

McElhaney, R.N. (1976). In M.R. Heinrich (ed.) Extreme Environments Academic Press, New York, p. 255.

Papahadjopoulos, D., K. Jacobson, S. Nir, and T. Isac. (1973) Biochim. Biophys. Acta 311, 330.

Sklar, L.A., B.S. Hudson, M. Petersen, and J. Diamond (1977a). Biochemistry 16, 813.

Sklar, L.A., B.S. Hudson, and R.D. Simoni (1977b). Biochemistry 16, 819.

Tecoma, E., L.A. Sklar, R.D. Simoni, and B.S. Hudson (1977). Biochemistry 16, 829.

Van Dijck, P.W.M., B. DeKruijff, L.L.M. Van Deenen, J. DeGier, and R.A. Demel (1976). Biochim. Biophys. Acta 455, 576.
Van Golde, L.M.G., R.A. Prins, W. Franklin-Klein, and J. Akkermans-Kruyswijk (1973). Biochim. Biophys. Acta 326, 314.
Van Golde, L.M.G., J. Akkermans-Kruyswijk, W. Franklin-Klein, A. Lankhorst, and R.A. Prins (1975). FEBS Lett. 53, 57.
Verkley, A.J., P.H.J. Th. Ververgaert, R.A. Prins. and L.M.G. Van Golde (1975). J. Bacteriol. 124, 1522.
Waku, K., and Y. Nakazawa (1972). J. Biochem.(Tokyo) 72, 149.

REGULATION OF MEMBRANE FLUIDITY DURING TEMPERATURE ACCLIMATION BY *TETRAHYMENA PYRIFORMIS*

Guy A. Thompson, Jr.

Department of Botany

The University of Texas

Austin, Texas 78712

Nature seems to have chosen a variety of strategies for dealing with environmentally-induced perturbations of membrane fluidity. In recent years a number of investigators have examined the ability of the ciliated protozoan *Tetrahymena pyriformis* to offset the effects of extreme temperatures on membranes by altering its lipid composition. I should like to describe some of these observations, principally the ones made in my laboratory and the laboratory of Nozawa, which led us to believe that *Tetrahymena*, and perhaps many other organisms, respond to low temperature-induced suboptimal membrane fluidity by the same molecular sensing mechanism that they utilize for counteracting a variety of quite different fluidity-perturbing stimuli.

My presentation will cover the general properties of the *Tetrahymena* system and its response to temperature change. Dr. Nozawa (1979) will follow with a discussion of experiments done with membrane fluidity-perturbing agents other than temperature.

Tetrahymena is a typical eukaryotic cell with respect to the assortment of subcellular organelles that it possesses (Fig. 1). We chose it as a model system primarily for this reason, hoping that it would furnish answers to some of our many questions regarding structural and metabolic interrelationships among functionally distinct membranes within the same cell. I shall describe

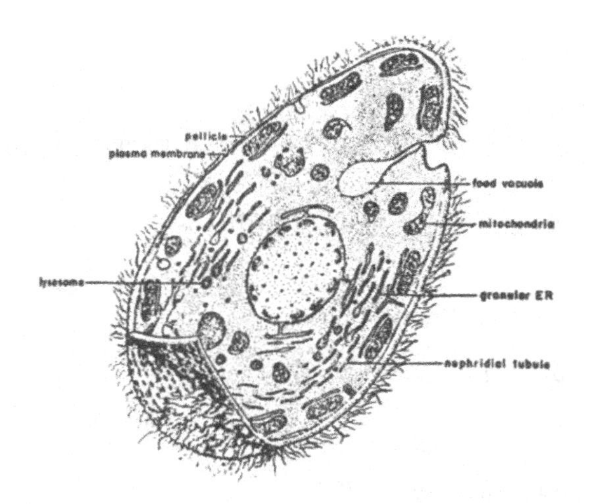

Figure 1. A diagrammatic cross-sectional view of *Tetrahymena pyriformis*.

---

shortly several interesting and important complexities that accompany temperature acclimation in the different membranes of this cell.

Although *Tetrahymena* resembles cells of higher animals and plants, its unicellular form and rapid growth rate make it convenient for laboratory manipulation. It is easily fractionated into its several principal membranous organelles (Nozawa and Thompson, 1971). While each different organelle has its own characteristic lipid composition, in all cases only 3 major phospholipid classes are involved (Fig. 2) (Nozawa and Thompson, 1971). These are 1) phosphatidylcholine, 2) phosphatidylethanolamine, and 3) the phosphonic acid analog of phosphatidylethanolamine, which I shall refer to as phosphonolipid. There is a gradual change in the relative proportions of these three classes of polar head groups in certain membranes as *Tetrahymena* cells are exposed to a temperature drop (Fukushima *et al.*, 1976), but the change requires several generations of growth and does not appear to be necessary for adaptation of the cells to altered temperature. All membranes also contain small quantities of a cholesterol analog known as tetrahymanol (Thompson *et al.*, 1971). The content of this does not change significantly during temperature acclimation. What changes most conspicuously during the

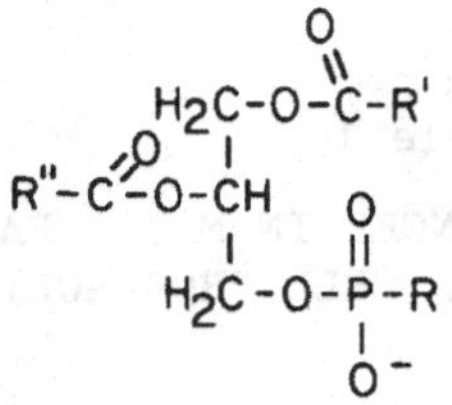

$$
\begin{array}{l}
H_2C-O-\overset{O}{\overset{\|}{C}}-R' \\
R''-\overset{O}{\overset{\|}{C}}-O-CH \\
H_2C-O-\overset{O}{\overset{\|}{P}}-R \\
\qquad\quad\;\; \mid \\
\qquad\quad\;\; O^-
\end{array}
$$

$$R = O-CH_2CH_2-N^+(CH_3)_3$$

$$O-CH_2CH_2-NH_2$$

$$CH_2CH_2-NH_2$$

Figure 2. The three major phospholipid classes found in Tetrahymena membranes. The R groups represent, from top to bottom, phosphatidylcholine, phosphatidylethanolamine, and diacyl-sn-glycero-2-aminoethylphosphonate (phosphonolipid).

---

acclimation process is the fatty acid composition of phospholipids. Analysis of whole cell phospholipids shows that, as in the case of most other cells that can adapt to chilling, Tetrahymena grown at low temperature have increased fatty acid unsaturation (Table I) (Fukushima et al., 1976). The major changes at 15° include an increase in 18:2 and 18:3 and a decrease in 16:0.

In order to determine what effects increased fatty acid unsaturation has on the structural properties of Tetrahymena membranes, our laboratories and that of Dr. Frank Wunderlich at Freiburg have conducted a fairly rigorous study using electron microscopy, electron spin resonance and fluorescence polarization spectrometry, X-ray diffraction, and other such techniques. Earlier papers in the symposium have discussed most of these physical chemical techniques. But before proceeding further, it might be advisable to mention the type of information that can be gained from freeze-fracture electron microscopy.

Figure 3 illustrates freeze-fracture replicas of a particular membrane of a 39.5°-grown cell fixed with glutaraldehyde a) after chilling quickly to 0.5°, b) after chilling quickly to 30° , and c) at 39.5°, the growth temperature. Such replicas and many others fixed at a variety of intermediate temperatures have been examined for

Table I

TEMPERATURE-INDUCED CHANGES IN MAJOR FATTY ACIDS OF _TETRAHYMENA_ WHOLE CELL PHOSPHOLIPIDS*

Fatty Acid	Growth at	
	15° (6)**	39.5° (5)
12:0	1***	1
14:0	7	6
_ai_-15:0	2	3
16:0	_9_	_13_
17:0 +16:2	2	4
17:1	1	1
16:1	9	9
18:1	10	11
18:2	_20_	_14_
γ-18:3	_31_	_24_

* data of Fukushima _et al_ (1976)

** figure in parentheses indicate the number of separate experiments

*** data are expressed as weight per cent of total recovered fatty acids

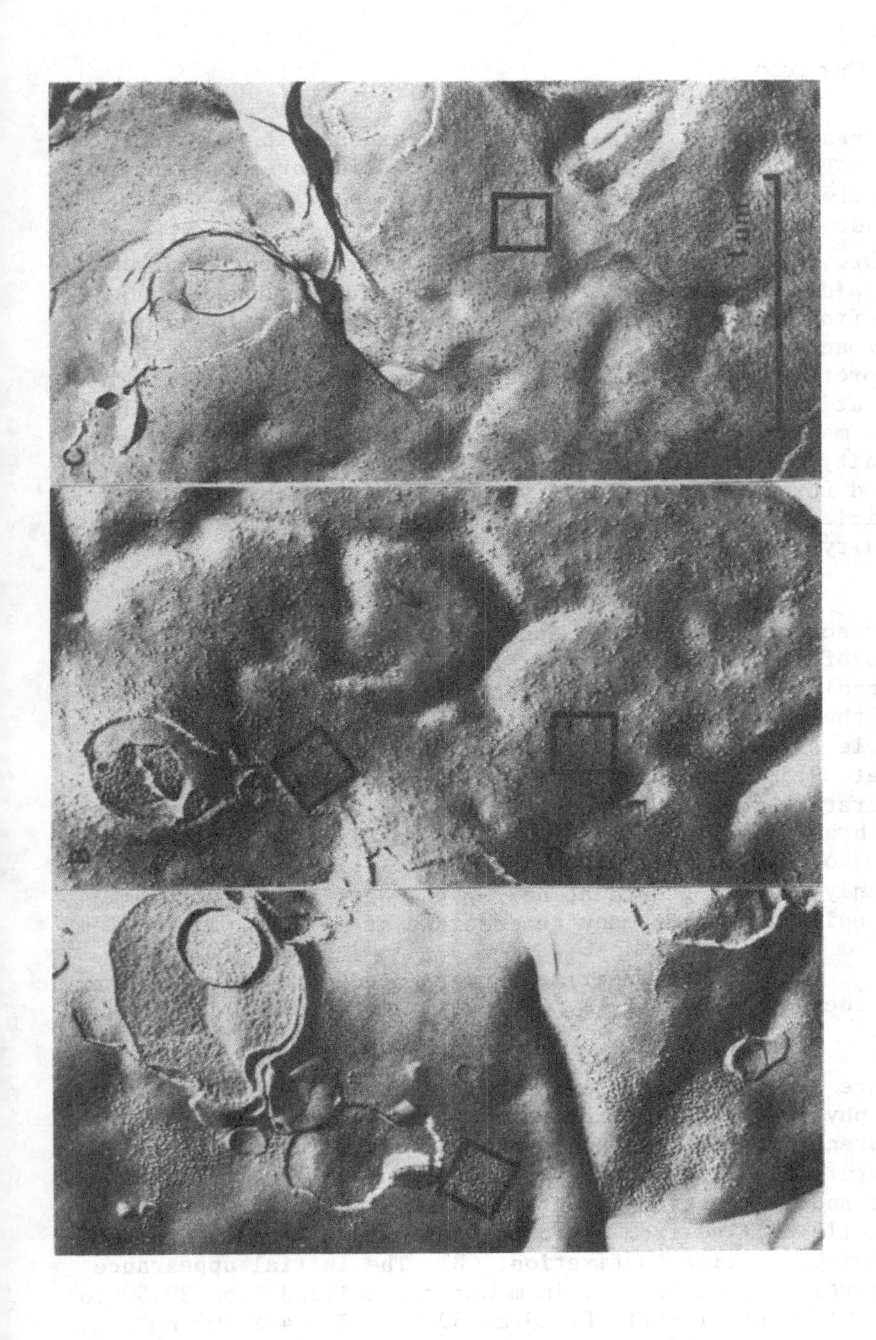

the arrangement of protein particles exposed at the fracture face. The intramembranous particles, which are apparently randomly distributed at 39.5°, begin to show small regions free of particles at about 34°. As the cells are chilled further, the particle-free areas, which represent domains of lipids that have undergone a temperature-induced transition from the liquid-crystalline to the gel phase, continue to expand until at 0.5° (Fig. 3) or even at 15°, most of the proteins are tightly aggregated together. The temperature at which particle-free lipid domains first appear in a given membrane type is a characteristic determined by the fluidity of the membrane and provides us with a sensitive method for inferring fluidity changes. The degree of particle aggregation can also furnish quantitative data on fluidity (see legend for Fig. 3).

Using freeze-fracture electron microscopy and fluorescence polarization spectrometry - two independent means of estimating membrane fluidity - we examined cells undergoing temperature acclimation and compared the changes with the concurrent modification in lipid composition. One example of this approach is illustrated below. Cells growing at 39.5° were shifted over a 30 min period to 15°. The temperature reduction caused an immediate cessation of cell growth and division, as illustrated in Fig. 4. After a period of growth inhibition lasting 10 - 12 hrs, the cells suddenly resumed growth at nearly the same rate observed with cells grown for many generations at 15°.

Only a few experiments were needed to make it clear that the key organelle in the acclimation process is the

---

Figure 3. The effects of rapidly decreasing temperature on the physical state of the Tetrahymena outer alveolar membrane (part of the pellicle), as visualized by freeze-fracture electron microscopy. A) pronounced lateral movement and aggregation of intramembranous protein particles in cells chilled from a growth temperature of 39.5° to 0.5°C immediately prior to fixation. B) The initial appearance of particle-free regions in membranes chilled from 39.5° to 30° C) Control cells fixed at 39.5°. The 4 X 10^4 nm^2 frames may be used for quantifying the degree of aggregation by computing a "particle density index" as described by Martin et al. (1976).

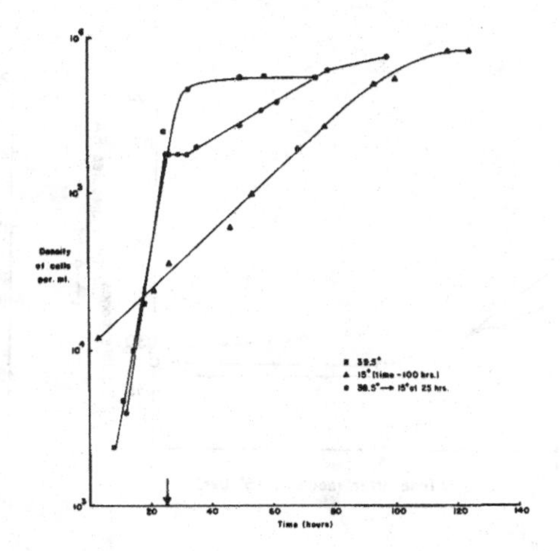

Figure 4. The growth rates of cells cultured at different temperatures. The curve depicting growth at 15° begins with a culture already grown for 100h to raise the cell density to a desirable value. The curve marked by filled circles represents cells shifted to 15° after growing at 39.5° for 25 h. The time of shifting is shown by an arrow. Data of Fukushima et al. (1976).

---

microsomal membrane. There were noticeable changes in the phospholipid fatty acid composition of microsomal membranes isolated within minutes after the temperature shift, whereas significant differences in whole cell phospholipids were not apparent until much later. The rapidity of change is illustrated in Figure 5. In this experiment we compared the extent of fatty acid unsaturation (dashed lines and Δ) with the increase in microsomal lipid fluidity, as inferred by measuring two parameters. The solid line and squares report the temperatures at which intramembranous particle-free areas could first be detected by freeze-fracture electron microscopy. In addition, decreasing fluorescence polarization of a probe added to the extracted lipids (solid lines and circles) confirmed that membrane fluidity, initially very much below the optimal level following the downward temperature shift, rapidly increased during the first 30 min at 15°.

Other membranes of the cell did not exhibit such a prompt response to low temperature. An example is the pellicle (the complex of surface membranes). Fluorescence polarization measurements (Fig. 6). show that lipids of the

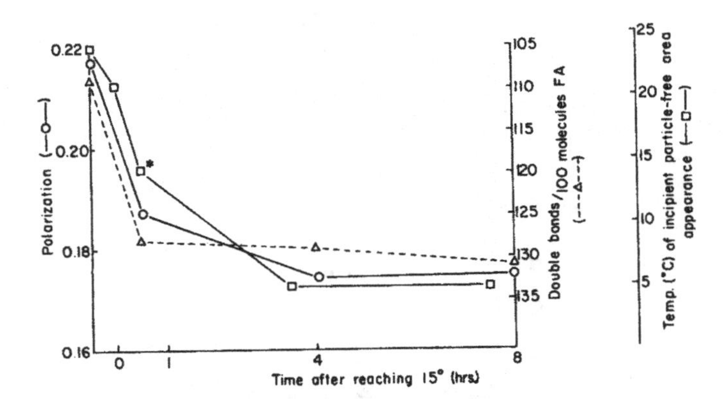

Figure 5. Comparison of the changes of three independently measured properties of microsomal membranes during temperature acclimation from 39.5° to 15°C. Cells were grown at 39.5°C, shifted to 15°C over a 30-min period, and then at the indicated times fixed for freeze-fracture analysis or harvested and fractionated for microsomal lipid isolation. Data from freeze-fracture observations of membrane particle redistribution (□), fluorescence polarization of 1,6-diphenyl-1,3,5-hexatriene in membrane lipids measured at 15°C (o), and the number of double bonds in phospholipid fatty acids (Δ) have been plotted to show the correspondence in rates of change of the three parameters. For further details see Martin and Thompson (1978).

---

pellicle underwent little polarization change during the early minutes following the drop in temperature. Only after a lag of 30 min did the fluidity of these membranes begin to increase. Lipid analysis (data not shown) indicated that the degree of unsaturation in phospholipid fatty acids also showed a sharp rise at precisely this time.

From experiments such as these, one can separate the phenomenon of temperature acclimation into two semi-independent processes. First there is an apparent increase in the activity of fatty acid desaturases, which exist as integral proteins of the microsomal membranes. Secondly, the phospholipids containing freshly desaturated fatty acids must be disseminated from the microsomal membranes to all other cellular membranes. A variety of data, including those in Figure 6, suggests that in Tetrahymena and in a

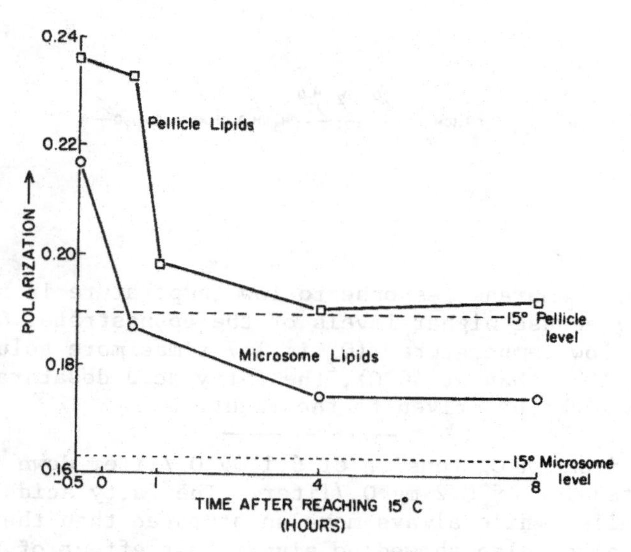

Figure 6. Time course of changes in diphenylhexatriene polarization in membrane lipids extracted from organelles of 39.5° C acclimated cells following a shift to 15°C. Cells were grown at 39.5° C, shifted to 15° as described in Fig. 5, and fractionated into microsomes and pellicles at the indicated times. Dashed lines show the polarization values in cells fully acclimated to 15°C. Data of Martin and Thompson (1978).

---

number of other cells, lipid dissemination is the rate limiting step in temperature acclimation.

After establishing the kinetics of the heightened fatty acid desaturation, we investigated the molecular basis for the increased desaturase activity. Several possibilities were considered. One of these was based on the observation of Harris and James (1969) (Figure 7) that increased availability of O_2, a cofactor in the fatty acid desaturation reaction, stimulates the reaction in non-photosynthetic plant tissues.

We proceeded to test the fatty acid composition of Tetrahymena cells grown isothermally but over a wide range of O_2 concentrations (Skriver and Thompson, 1976). Cells growing at 39.5° experienced little change in fatty acid

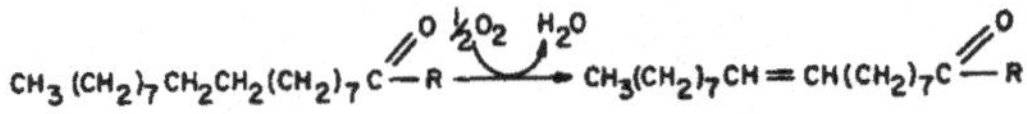

$$CH_3(CH_2)_7CH_2CH_2(CH_2)_7C(=O)-R \xrightarrow{\frac{1}{2}O_2 \quad H_2O} CH_3(CH_2)_7CH=CH(CH_2)_7C(=O)-R$$

Figure 7. Membrane response to low temperature in higher plants. Because higher levels of the cosubstrate, O_2, are found at low temperatures (O_2 is 1.7 times more soluble in water at 10°C than at 40°C), the fatty acid desaturation reaction might be driven to the right.

---

pattern from an O_2 tension of 6.1 mg O_2/liter down to a very low O_2 tension of 0.2 mg O_2/liter. The fatty acids of 15°-grown cells, while always more unsaturated than their 39.5° counterparts, also showed no significant effect of reduced O_2 tension.

A second possible mechanism for controlling desaturase activity calls for a change in the absolute amount of the enzyme. Rapid fluctuations in the quantity of fatty acid desaturase have been reported in diabetic and carbohydrate-fed rats (Prasad and Joshi, 1979).

Recent data from the laboratory of Nozawa (Fukushima *et al*., 1979) have indicated that palmitoyl-CoA desaturase does indeed increase in amount during the early part of the low-temperature acclimation process. There appears to be a 400% rise in activity of this enzyme, as compared with the basal level at 39.5°, during the first few min. at 15° (Table II). After that time palmitoyl-CoA desaturase activity returns to a normal level. This finding provides a logical explanation for the observed transient increase of palmitoleic acid *in vivo* during the first hour following a downward temperature shift.

However, an induction of desaturase synthesis does not seem to be essential for increased desaturase action. Cells whose capacity for protein synthesis was blocked with cycloheximide immediately prior to the 39.5° → 15° temperature shift could still undergo a nearly normal fatty acid

Table II

PALMITOYL COA DESATURASE ACTIVITY IN *TETRAHYMENA* MICROSOMES *

Cells were grown at 39.5° and slowly chilled to 15° over a 30 min. period. After isolation, microsomes were assayed at 39.5° and 15°C.

Pretreatment of cells	Assayed at 39.5°C		Assayed at 15°C	
	Absolute activity+	%**	Absolute activity+	%**
39.5°C-grown cells (before shift)	0.254	100	0.091	100
after shift to 15°C				
0 hr	1.066	420	0.407	447
2 hr	1.192	469	0.388	426
5 hr	0.475	187	0.204	224
15°C-grown cells	0.434	171		

* data of Fukushima, *et al.* (1979).

** % of value found in 39.5°C-grown cells

\+ nmol palmitoleate formed/min/mg protein.

alteration in the short period of time before secondary effects of cycloheximide began to be felt (Skriver and Thompson, 1979). No evidence has been found for the induction of desaturases responsible for the production of polyunsaturated fatty acids.

We have invested a lot of time and energy in testing yet another possible mechanism for controlling desaturase activity. According to this hypothesis, the activity of preexisting fatty acid desaturase molecules is regulated not by a direct effect of temperature on the enzyme, but by the physical state of the microsomal membrane lipid bilayer in which the desaturases are situated.

One method for distinguishing between membrane fluidity vs. temperature *per se* as a regulator of fatty acid desaturase activity is to vary the two factors independently. Thus membrane fluidity has been altered isothermally by treating *Tetrahymena* with a number of fluidizing compounds, such as polyunsaturated fatty acids, drugs, and general anesthetics. As Dr. Nozawa will describe in the following paper, increased fluidity in such experiments was almost invariably accompanied by decreased desaturase action and *vice versa*. In a sense, therefore, the fluidity of this membrane is truly "self-regulating."

Much evidence is available to support the notion that, in general, reduced membrane fluidity *decreases* the enzymatic activity of membrane-bound enzymes (Kimelberg, 1977). What we are proposing may appear to be the opposite: that decreased fluidity enhances desaturase action. Although a few instances of decreased membrane fluidity causing increased activity of membrane-bound enzymes are known (e.g. the presentation by Riordan (1979) in this symposium), no clear precedents exist to indicate how such an enhancement might be achieved. Two strictly speculative possibilities are shown in Figures 8 and 9. There are indeed some indications (Wunderlich *et al.*,1975), which I shall not describe here, that integral proteins of *Tetrahymena* microsomal membranes actually undergo the type of perpendicular movement illustrated in Figure 9.

The observed changes in fatty acid unsaturation and fluidity could more logically result from a condition in which fatty acid desaturase activity was actually

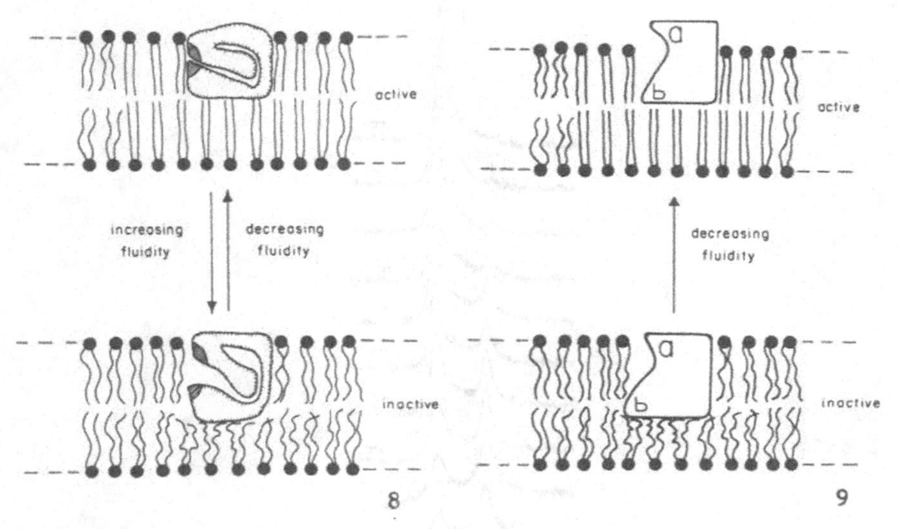

Figure 8. A possible mechanism for the enhancement of fatty acid desaturase activity due to a protein conformational change near the active site (hatched regions).
Figure 9. A possible mechanism for the enhancement of fatty acid desaturase activity due to a movement of the enzyme perpendicular to the plane of the membrane, perhaps orienting the active site (b) in a more favorable position relative to the phospholipid-bound fatty acid substrate.

---

reduced but to a much lesser extent than the activities of enzymes catalyzing fatty acid and phospholipid synthesis. This change in *relative* activities could quickly alter the degree of phospholipid fatty acid unsaturation in membranes. Unfortunately, the vast array of factors affecting the intracellular levels of lipid precursors and products and modulating the fluidity of selected membranes limits the effectiveness of *in vitro* experimentation in resolving the problems of functional desaturase activity levels.

The results described by me and those to be presented by Dr. Nozawa strongly implicate membrane fluidity as an important regulator of desaturase activity. Further insight into the regulatory process will require a fuller knowledge of the changing physical state of this key microsomal membrane. Accordingly, several laboratories have initiated detailed physical chemical studies of *Tetrahymena*

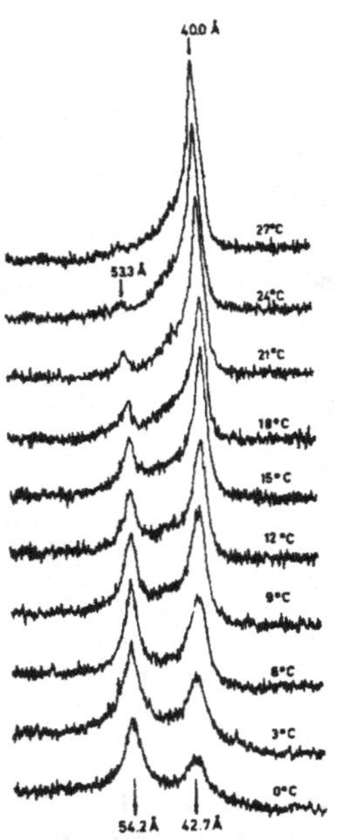

Figure 10. Densitometer tracings of a temperature series of small angle first-order X-ray diffractions of total microsomal lipids extracted from 18°-grown *Tetrahymena*. The emergence of reflections in the 53-54 Å range at 24°C and below indicate the presence of an ordered lipid phase as well as the liquid-crystalline phase represented by the 40-42 Å reflections. Data of Wunderlich *et al*., (1978).

microsomal membranes and membrane lipids.

Among the more useful analyses has been the X-ray diffraction examination carried out by Wunderlich et al., (1978). As illustrated in Fig. 10, the small angle X-ray diffraction patterns of microsomal total lipids from 18°-grown cells show the existence of two different microenvironments in the microsomal total lipids even above the growth temperature. This implies a possible lipid phase separation under physiologically normal conditions. Further evidence for the coexistence of a liquid-crystalline and a gel phase in biological membranes has recently emerged from the deuterium NMR studies of Smith (1979).

In my laboratory we have studied the same problem with fluorescence polarization measurements. The presence of very reproducible "break points" has been found in otherwise nearly straight lines obtained by plotting polarization vs. temperature of extracted membrane lipids (Fig. 11) or intact membranes. The characteristic temperatures delimiting these break points are fixed by the growth temperature of the cells. At the present time, the association of the break points with lipid phase changes has not yet been confirmed.

Current investigations in my laboratory and in several others are aimed at correlating more accurately any relationship between membrane fluidity and fatty acid desaturase activity. Such a means of control would afford a most efficient multipurpose mechanism capable of responding to the fluidity-perturbing actions of cations, drugs, pressure and other agents which alter membrane physical properties in basically the same way that it responds to temperature change.

ABSTRACT

The regulation of membrane fluidity in a eukaryotic cell involves two separate but coordinated phenomena: 1) the enzymatic alteration of membrane lipids, principally in the endoplasmic reticulum, and 2) dissemination of the newly altered lipids to the various functionally distinct membranes throughout the cell. The unicellular protozoan Tetrahymena pyriformis is a useful model system for examining both these processes. By analyzing several membrane types for their lipid composition, lipid physical

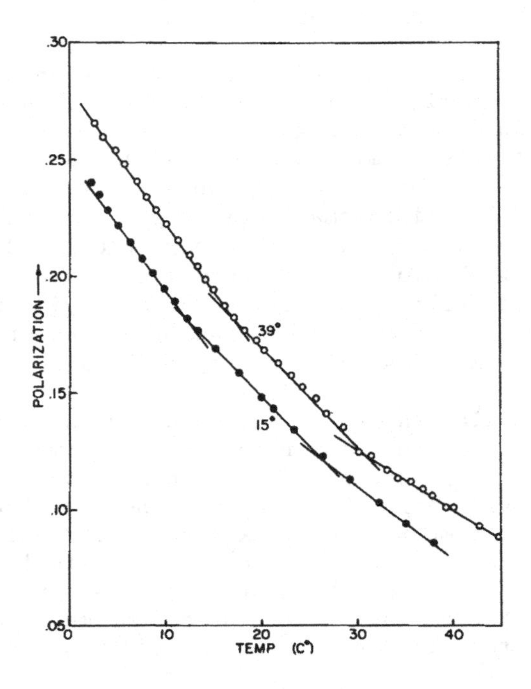

Figure 11. Plots of diphenylhexatriene polarization vs. temperature in multilamellar vesicles containing phospholipids isolated from microsomes of Tetrahymena grown at 39° or 15°. Unpublished findings of B. F. Dickens and G. A. Thompson, Jr.

properties and protein orientation in freeze-fracture electron microscopy replicas, it was possible to deduce the spatial and temporal pattern of intracellular lipid change during acclimation of cells chilled from 39°C to 15°C. Various possible mechanisms for regulating this temperature acclimation have been investigated. Evidence for an induction of palmitoyl-CoA desaturase at low temperature was found. There is also evidence that the physical state of the lipid bilayer environment near the several membrane-bound fatty acid desaturases is important in determining the activity of these enzymes.

ACKNOWLEDGMENTS

Work from the author's laboratory was supported by grants from the National Institute of General Medical Sciences, the National Cancer Institute, the Robert A. Welch Foundation, and the National Science Foundation.

REFERENCES

Fukushima, H., C. E. Martin, H. Iida, Y. Kitajima, G. A. Thompson, Jr. and Y. Nozawa (1976). Biochim. Biophys. Acta 431, 165.

Fukushima, H., S. Nagao, and Y. Nozawa (1979). Biochim. Biophys. Acta 572, 178.

Harris, P., and A. T. James (1969). Biochem. J. 112, 325.

Kimelberg, H. K. (1977) In G. Poste and G. L. Nicholson (eds.) Cell Surface Reviews, vol. 3, 205.

Martin, C. E., and G. A. Thompson, Jr. (1978). Biochemistry 17, 3581.

Martin, C. E., K. Hiramitsu, Y. Kitajima, Y. Nozawa, L. Skriver, and G. A. Thompson, Jr. (1976). Biochemistry 15, 5218.

Nozawa, Y. (1979). following article, this volume.

Nozawa, Y., and G. A. Thompson, Jr. (1971). J. Cell Biol. 49, 712.

Prasad, M. R., and V. C. Joshi (1979). J. Biol. Chem. 254, 997.

Riordan, J. R. Article in this volume.

Skriver, L., and G. A. Thompson, Jr. (1976). Biochim. Biophys. Acta 431, 180.

Skriver, L., and G. A. Thompson, Jr. (1979). Biochim. Bio-Acta 572, 376.

Smith, I. C. P. (1979) Abstracts of XIth Int. Congress of Biochem., 325.

Thompson, G. A. Jr., R. J. Bambery, and Y. Nozawa (1971). Biochemistry 10, 4441.

Wunderlich, F.,A. Ronai, V. Speth, J. Seelig, and A. Blume (1975). Biochemistry 14, 3730.

Wunderlich, F., W. Kreutz, P. Mahler, A. Ronai, and G. Heppeler (1978). Biochemistry 17, 2005.

MODIFICATION OF LIPID COMPOSITION AND MEMBRANE FLUIDITY IN *TETRAHYMENA*

Yoshinori Nozawa

Department of Biochemistry, Gifu University
School of Medicine
Tsukasamachi-40, Gifu, Japan

INTRODUCTION

It is now widely accepted that lipids play some important roles in a variety of functions taking place in the biological membrane. In order to shed light on the involvement of lipids in membrane functions, one useful approach has been to manipulate either quantitatively or qualitatively the membrane lipid composition of cells under the defined growth conditions, and to examine functions in the membranes with altered lipid composition. Although various methods for modifying membrane lipids have been available (Chapman and Quinn, 1976), one of the most commonly used methods is to supplement cells with different lipids or its precursors. The majority of studies of membrane lipid modification has largely involved alterations in the phospholipid acyl chain composition in bacteria and mycoplasma. Recently, such manipulation procedures have been applied to mammalian cells grown in tissue culture to alter not only fatty acid but also polar head group composition of membrane phospholipids.

Since the unicellular eukaryote, *Tetrahymena pyriformis* has a proven ability to alter its membrane lipid composition in response to growth conditions, this cell also is a promising model system for modifying the lipids and better understanding the interrelationship between lipid composition, fluidity and function in cell membranes.

This chapter summarizes mainly the results of our studies concerning the lipid modification and its effects on physical states and activities of some enzymes of *Tetrahymena* membranes

MODIFICATION OF MEMBRANE LIPID COMPOSITION

Several procedures have been employed to modify the lipid composition of membranes in *T. pyriformis* (Thompson and Nozawa, 1977; Nozawa and Thompson, 1979), which are listed as follows. The most commonly used method is to supplement cells with various substances which can be assembled into membranes.

A. Supplementation
   sterols, hexadecyl glycerol, fatty acids, isovalerate, sphingomyelin, choline analogs
B. Temperature acclimation
C. Starvation
D. Drug treatment
   phenethyl alcohol, ethanol, chlorophenoxy isobutyrate, triparanol, chlorpromazine
E. Enzyme treatment
   phospholipases
F. Age of culture

Supplementation

Ergosterol. Ferguson *et al.*(1977) have shown that ergosterol cannot be metabolized in *Tetrahymena* cells and its supplementation to the growth medium induces replacement of tetrahymanol, a sterol-like membrane component, with exogenous ergosterol. Such replacement was found to take place on a one-for-one molecule basis in the ciliary membrane. Moreover, since there is no significant difference in total phospholipid content between the native and replaced-membrane, it may be inferred that ergosterol molecules would be incorporated into the sites where the native tetrahymanol was located.

Ergosterol-replacement causes small significant quantitative changes in the fatty acid composition of the membrane phospholipids (Ferguson *et al.*, 1977; Nozawa *et al.*, 1975). A general trend demonstrates a decrease in ergosterol-replaced membranes both in the average fatty acyl chain length and the relative proportion of unsaturated acids. As shown in Table I, despite some variations between different membranes, the level of myristic (C14:0), palmitic

(C16:0), and palmitoleic (C16:1) acids increases at the expense of oleic (C18:1) and γ-linolenic (C18:3) acids. Similar alterations in the fatty acid composition are observed to occur in the cells supplemented with cholesterol, β-sitosterol and stigmasterol which are further metabolized to their derivatives. These marked changes are in common reflected in major phospholipids, phosphatidylcholine (PC), phosphatidylethanolamine (PE) and 2-aminoethylphosphonolipid (AEPL), but especially prominent in PE.

The polar head group composition also is appreciably modified by ergosterol substitution (Table II). In ergosterol-replaced membranes of pellicles and microsomes, there is a marked increase of PE with a concomitant decrease of PC. The level of AEPL shows no change in pellicles but a significant increase in microsomes. From these observations, phosphatidylethanolamine is assumed to play an important role in the surface membrane structure by compensating for the sterol change.

Hexadecyl glycerol. *Tetrahymena* cells contain a high level of glyceryl ether in phospholipids. When *T. pyriformis* NT-I cells are grown in the medium supplemented with 1-O-hexadecyl glycerol (HDG, chimyl alcohol) which is precursor of glyceryl ether-containing phospholipids, cells can incorporate large amounts of HDG into membranes, especially the pellicular membrane. By HDG-feeding both PC and AEPL become richer in glyceryl ether level, while no incorporation of supplemented HDG is observed in PE. For example, the molar content of glyceryl ether rises up to 71% of the total PC. Such marked incorporation of HDG induces alterations both in the phospholipid head group and fatty acyl chain composition (Fukushima *et al.*, 1976). Although no change in relative proportion of PC is noted in pellicles of HDG-fed cells, there is a large increase in AEPL with a compensating reduction in PE content. In contrast, small changes are seen in the microsomal phospholipid composition. As for the fatty acid composition of membrane phospholipids, a noticeable increase in palmitic acid together with a slight elevation of the content of linoleic and linolenic acids is found both in pellicles and microsomes. The percentage of oleic acid is much lower in HDG-fed microsomes than in control microsomes.

Fatty acids. Supplementation of fatty acids in *Tetrahymena* was first demonstrated by Lees and Korn (1966). Cells grown in the presence of 11,14-eicosadienoate, 8,11,

Table I Alterations induced by ergosterol replacement in phospholipid fatty acyl composition in various membranes from T. pyriformis WH-14 cells (values in %)

Fatty Acids	Cilia		Pellicles		Microsomes	
	Ergosterol-replaced	Control	Ergosterol-replaced	Control	Ergosterol-replaced	Control
C14:0	8.2	3.5	11.0	7.6	9.5	7.7
C16:0	10.1	10.1	15.7	12.9	12.3	12.2
C16:1 Δ^{9}	8.2	5.9	10.1	8.2	13.3	9.8
C18:0	3.8	4.5	2.1	2.9	1.9	2.0
C18:1 Δ^{9}	6.4	8.7	6.1	9.7	6.1	11.8
C18:2 $\Delta^{6,11}$	3.6	3.5	3.2	1.8	2.9	2.0
C18:2 $\Delta^{9,12}$	11.6	15.3	11.9	13.2	14.0	13.6
C18:3 $\Delta^{6,9,12}$	34.4	41.0	23.3	30.1	26.0	31.3

[Kasai et al., Maku (Membrane) 2, 301 (1977)]

Table II Alterations induced by ergosterol replacement in phospholipid composition of pellicles and microsomes from T. pyriformis WH-14 cells (values in %)

Phospholipids	Whole Cells		Pellicles		Microsomes	
	Ergosterol-replaced	Control	Ergosterol-replaced	Control	Ergosterol-replaced	Control
Phosphatidyl-choline	26.2	30.4	22.5	26.9	30.4	35.6
Phosphatidyl-ethanolamine	37.9	32.9	36.8	31.6	40.0	32.6
2-Aminoethyl-phosphonolipid	23.4	21.7	28.7	29.5	20.7	24.4
Cardiolipin	5.0	6.0	3.2	2.8	3.4	2.6

[Kasai et al., Maku (Membrane) 2, 301 (1977)]

Table III Alterations in phospholipid fatty acyl composition in various membranes from linoleic acid-supplemented cells of T. pyriformis NT-I (values in %)

Fatty Acids	Control Cells			C18:2-Fed Cells		
	Cilia	Pellicles	Microsomes	Cilia	Pellicles	Microsomes
C14:0	6.7	8.9	7.3	7.4	12.5	9.8
C16:0	16.8	16.4	13.2	19.5	18.8	15.8
C16:1 Δ^9	8.3	7.1	8.7	7.6	5.8	6.4
C18:0	6.1	3.0	2.2	5.6	2.8	2.3
C18:1 Δ^9	11.7	12.6	13.9	13.5	4.0	3.7
C18:2 $\Delta^{9,12}$	8.1	10.5	13.0	17.5	22.1	26.5
C18:3 $\Delta^{6,9,12}$	20.7	18.5	21.4	18.5	22.0	26.1

[Kasai et al., Biochemistry 15, 5228 (1976)]

14-eicosatrienoate, 11-eicosenoate, or 11-octadecenoate, were found to incorporate these fatty acids into neutral and phospholipids. We have fed T. pyriformis NT-I cells with large amounts of linolenic or linoleic plus linolenic acids, and investigated the effects of this supplementation upon membrane lipid composition (Martin et al., 1976; Kasai et al., 1976). As shown in Table III, a striking modification occurs in phospholipid fatty acid composition in various membranes. The supplemented linoleic acid is incorporated to a great extent into all membrane fractions examined. This drastic increase in linoleic acid level is accompanied by a compensating decrease in oleic acid in pellicular and microsomal membranes, but in linolenic acid in cilia. Other fatty acids such as palmitic, palmitoleic or cis -vaccenic acids also were found to be assembled into membrane phospholipids. Despite such profound modification of fatty acid profile of membrane phospholipids, little alteration occurs in the phospholipid polar head group composition.

Isovalerate. Conner and Reilly (1975) have shown that supplementation of isovalerate to T. pyriformis W culture causes quantitative but not qualitative alteration in fatty acid composition. The increase in odd-numbered iso-fatty acids, C13, C15, C17, C19, was compensated by a decrease in even-numbered normal acids. The saturated fatty acids in control cells consist predominantly of myristic and palmitic acid, while *iso*-C17:0 is most abundant in isovalerate -supplemented cells. The unsaturated fatty acid content is not changed. Recently, a novel *Tetrahymena* sp., as yet unclassified, was found to contain a strikingly large amount of *iso*-odd-numbered fatty acids, *i*-C15:0, *i*-C17:1, which are minor components in other classical *Tetrahymena* strains (Fukushima, 1978). Since it is expected that isovalerate can be utilized to synthesize *iso*-fatty acids in this strain, cells were incubated with 2 mM isovaleric acid. There was a marked increase in *i*-C15:0 and *i*-C17:1 in phospholipids of pellicles, mitochondria and microsomes (Kasai and Nozawa, unpublished).

Choline analogs and sphingomyelin. Choline analogs have been exploited to modify the polar head group composition in mammalian cells. Since no significant perturbation is induced in fatty acid profile, this technique is a useful system for studying the role of head groups in membranes.

When *T. pyriformis* NT-I culture was supplemented with choline analogs, ethanolamine, methylethanolamine, dimethylethanolamine, or choline, little or no modification was observed in phospholipid class composition except for methylethanolamine, in which case an unnatural phospholipid, phosphatidylmonomethylethanolamine, was accumulated up to one-third of the total phospholipid.

Sphingomyelin liposomes are ingested in food vacuoles and then incorporated as intact molecules into different membranes (Suezawa, Ohki and Nozawa, unpublished).

Temperature Acclimation

Tetrahymena is known to modify the fatty acid composition of membrane lipid in response to growth temperature changes (Wunderlich *et al*., 1973; Fukushima *et al*., 1976). In general, lowering the temperature induces an increase in the level of unsaturated fatty acids. *T. pyriformis* NT-I cells were grown isothermally at 15°C, 24°C, and 39.5°C. Profound changes were observed in the fatty acid composition of phospholipids from cilia, pellicles and microsomes. The principal effect of decreasing temperature is a marked increase in linoleic, linolenic and cilienic ($C18:2\Delta^{6,11}$) acids, accompanied with a decrease in palmitic acid. Temperature variation causes also alteration in the phospholipid distribution. There is a decrease in AEPL concentration and a corresponding increase in PE as the growth temperature increases, while the level of PC remains rather constant. The content of glyceryl ether-containing phospholipids drops markedly with increasing temperature, but little change is seen in the tetrahymanol level.

Drug Treatment

When T. pyriformis NT-I cells are grown in the presence of phenethyl alcohol (PEA), there are marked modifications in the relative proportions in phospholipid head, as well as fatty acyl group composition in pellicles, mitochondria, and microsomes (Nozawa et al., 1979). Compared with membranes from the control cells, the membranes from PEA-treated cells are found to contain a higher level of PC content with a compensating decrease in PE, while AEPL shows a slight

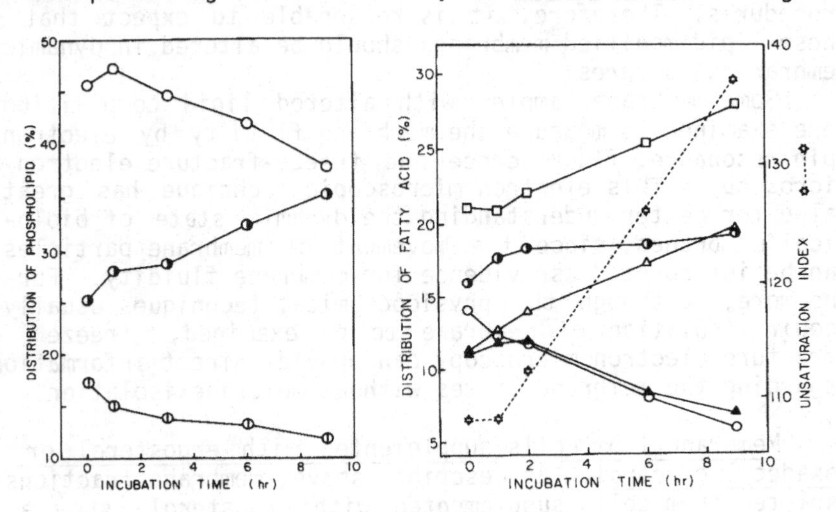

Fig. 1 Alterations induced by supplementation of hexadecyl glycerol in polar head group composition of T. pyriformis NT-I cells. (○) PE, (◑) PC, (⦶) AEPL.

Fig. 2 Alterations induced by HDG-supplementation in fatty acyl composition. (▲) C14:0, (◑)C16:0, (○)C16:1, (△) C18:2, (□) C18:3, (✡) unsaturation index. [Nozawa et al., Biochim. Biophys. Acta 552, 38 (1979)]

decrease (Fig. 1). Moreover, the acyl group profile of membrane phospholipids is modified so that there is a high elevation of the content of linoleic and linolenic acids. In contrast, there is a decrease in palmitoleic acid.

Ethanol treatment also causes perturbation in membrane lipid composition (Nandini-Kishore et al., 1979). The phospholipid fatty acid is altered gradually, with the principal changes being a decrease in palmitoleate and an increase in linoleate (Fig. 2). A decrease in the relative proportion of AEPL is offset by an increase in PE.

PHYSICAL STATES OF LIPID-MODIFIED MEMBRANES

Several lines of evidence have demonstrated that the physical properties of membranes are largely dependent on their lipid composition. As described above, the lipid composition of *Tetrahymena* membranes can be modified by various procedures. Therefore, it is reasonable to expect that these lipid-modified membranes should be altered in dynamic membrane structures.

Some membrane samples with altered lipid composition were examined to measure the membrane fluidity by electron spin resonance, fluorescence and freeze-fracture electron microscopy. This electron microscopic technique has great value for better understanding the dynamic state of biological membranes, since the movement of membrane particles can be interpreted as evidence for membrane fluidity. Furthermore, although the physicochemical techniques usually require isolation of membrane to be examined, freeze-fracture electron microscopy can provide direct information regarding the membrane states without membrane isolation.

Membrane from cells supplemented with ergosterol or hexadecyl glycerol. As described above, membrane fractions isolated from cells supplemented with ergosterol, show a marked alteration both in polar head group and fatty acid composition. The most profound changes are an increase of phosphatidylethanolamine and a decrease in the average length of fatty acids. The surface membrane, pellicle was examined by freeze-fracture electron microscopy. When cells are cooled to lower temperatures, the distribution pattern of membrane particles changes from the random to aggregated state in varying degrees. Fig. 3 illustrates a remarkable difference in particle distribution in the outer alveolar

Fig. 3 Freeze-fracture patterns of outer alveolar membranes of control (A) and ergosterol-replaced (B) T.pyriformis WH-14 cells grown at 34°C. Cells were cooled to 22°C over 4 min and then fixed.

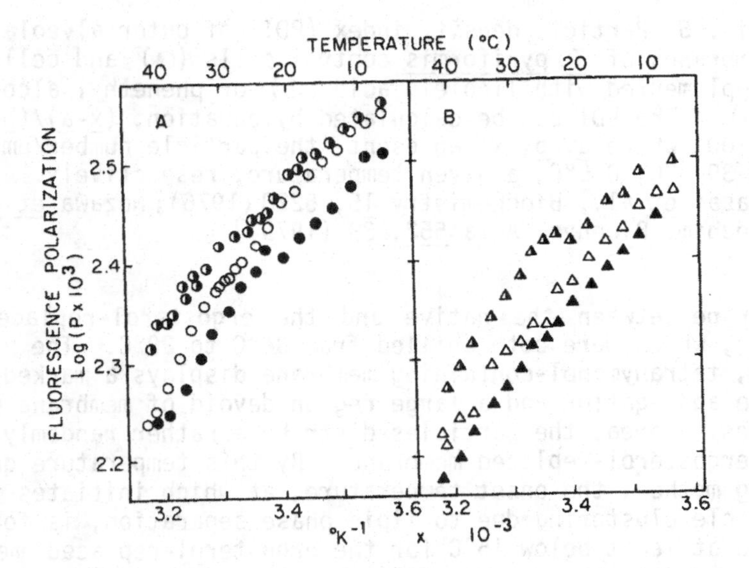

Fig. 4 Fluorescence polarization of the pellicular (A) and microsomal (B) membranes from control cells (○△) and cells supplemented with ergosterol (◑▲) or hexadecyl glycerol (●▲). [Shimonaka et al., Experientia 34, 586 (1978)]

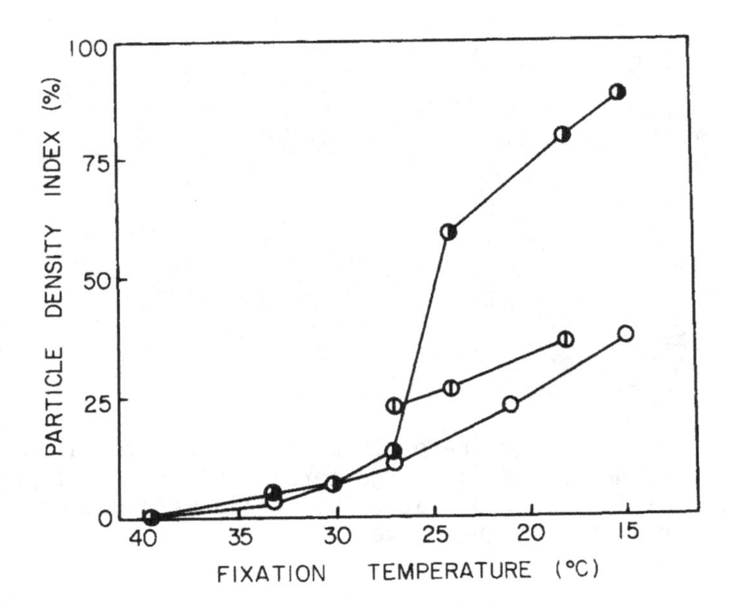

Fig. 5 Particle density index (PDI) of outer alveolar membranes of T. pyriformis control cells (◑) and cells supplemented with linoleic acid (⦶) or phenethyl alcohol (○). The PDI can be calculated by equation, $(x-a)/(b-a) \times 100$, where a, b, x represent the particle number/μm^2 at 39.5°C, 0.5°C, a given temperature, respectively. [Kasai et al., Biochemistry 15, 5228 (1976); Nozawa et al., Biochim. Biophys. Acta 552, 38 (1979)]

membrane between the native and the ergosterol-replaced cells, which were both chilled from 34°C to 22°C. The native, tetrahymanol-containing membrane displays a marked particle aggregation and a large region devoid of membrane particles, whereas the particles distribute rather randomly in the ergosterol-replaced membrane. By this temperature quenching method, the onset temperature, at which initiates the particle clustering due to lipid phase separation, is found to be at least below 15°C for the ergosterol-replaced membrane, as compared with 22°C for the native membrane (Thompson and Nozawa, 1977). Cilia, pellicles and microsomes isolated from cells grown in the presence and absence of ergosterol were compared for membrane fluidity by electron spin resonance (ESR). The values of order parameter were measured at various temperatures for membranes labeled with a spin probe, 5'-nitroxystearate. ESR data indicate that all membrane fractions from ergosterol-replaced cells are lower in fluidity than membranes from control cells. The decrease in membrane fluidity of these ergosterol-replaced membranes was also revealed by the fluorescence polarization using a probe, 1,6-diphenyl-1,3,5-hexatriene(DPH). Fig. 4 shows the Arrhenius plots of fluorescence polarization of lipid-modified pellicles and microsomes (Shimonaka _et al_., 1978). Pellicular and microsomal membranes from ergosterol -replaced cells become less fluid than those from the control cells. Similar trends were also observed with dispersions of extracted lipids.

On the other hand, membranes from hexadecyl glycerol -supplemented cells contain more polyunsaturated fatty acids and a higher level of AEPL. As seen in Fig. 4, supplementation of hexadecyl glycerol exerts fluidizing effects for both pellicles and microsomes. Especially this effect is predominant below about 20°C.

Membranes of fatty acid-supplemented cells. Cells grown in the medium supplemented with linoleic acid incorporate large amounts of the fatty acid into the membrane phospholipids. The physical states of membranes with a markedly altered fatty acid composition were examined by freeze-fracture electron microscopy. For quantitative analysis of changes in lipid-modified membranes, a parameter, particle density index (PDI), has been developed (Martin _et al_., 1976). The PF face of the outer alveolar membrane(OAM) is used to determine PDI, since like a sensor this membrane is extremely sensitive to various environmental factors, particularly temperature, thus undergoing rapid rearrangement of membrane particles. This index can be calculated on the basis of particle number/μm^2 of membranes fixed at 39.5°C, 0°C and a given temperature.

The PDI of OAM of C18:2-fed cells is plotted vs. temperature (Fig. 5), which depicts a significant difference in the degree of particle aggregation from the control membrane (Kasai _et al_., 1976). This implies that membranes of C18:2-fed cells are more fluid than the control membranes. Similar fluidizing effects of membranes were observed in the case of supplementation of C18:2 plus C18:3.

Membranes from cells grown at different temperatures. The membranes from 39.5°C-grown cells are strikingly different in lipid composition from those of 15°C-grown cells, in that unsaturation of phospholipid fatty acids and

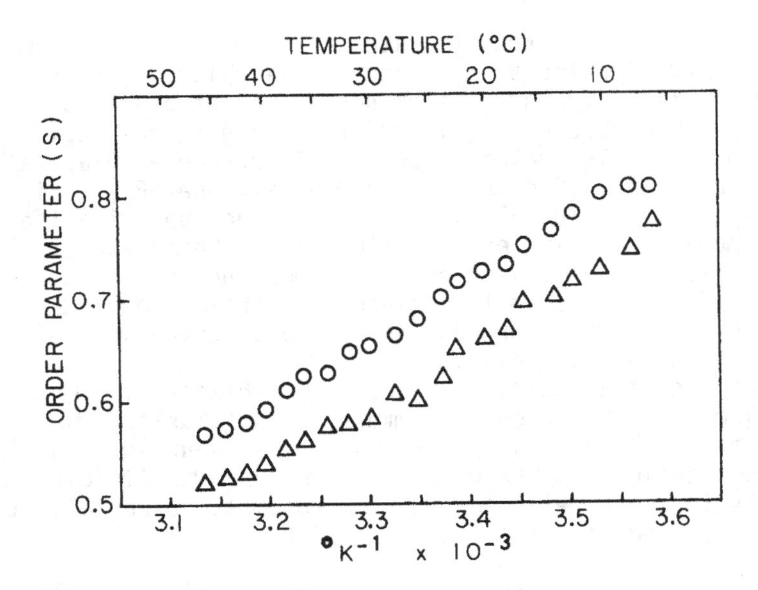

Fig. 6 Order parameters measured by electron spin resonance of the pellicular membranes from 39.5°C-grown (○) or 15°C-grown (△) cells of T. pyriformis NT-I. [Iida et al., Biochim. Biophys. Acta 508, 55 (1978)]

2-aminoethylphosphonolipid content are higher in the cold cells. The fluidities of major membrane fractions isolated from cells grown at two extreme temperatures were measured by electron spin resonance as a function of temperature(Iida et al., 1978). The membranes of cilia, pellicles and microsomes prepared from cells grown at 15°C have greater fluidities than membranes from 39.5°C-grown cells, when compared at the same temperature. Fig. 6 displaying an example of pellicles indicates that order parameters are lower in membranes from 15°C-cells at all temperatures examined. Such increased fluidity of the cold cells' membranes can be mainly interpreted by an increased level of unsaturated fatty acids in membrane phospholipids.

Membranes of phenethyl alcohol-treated cells. The principal perturbation in membrane lipid composition of phenethyl alcohol(PEA)-treated cells is an increase in phosphatidylcholine and polyunsaturated fatty acids. Compared with the control membranes, the PEA-membranes reveal less aggregation of intramembranous particles. The slope of the PDI curve of the PEA-membranes is much lower below about 27°C than the control membranes (Nozawa et al., 1979). Since such an abrupt increase of PDI in the control membranes is thought to be presumably due to a thermal lipid phase transition, the membranes of PEA-treated cells do not undergo a sharp phase transition and are rather more fluid than those of the untreated cells. Nandini-Kishore et al.(1979) have shown that the long-term growth in the presence of ethanol give rises to membranes with altered lipid composition: increase in phosphatidylcholine and linoleic acid with an accompanying decrease in palmitoleic acid. The microsomal phospholipids from ethanol-treated cells are greater in the fluidity as measured by fluorescence polarization using the probe diphenylhexatriene.

EFFECTS OF LIPID MODIFICATION ON MEMBRANE FUNCTIONS

A large body of evidence is accumulating to support the concept that the physical properties of membrane lipids are important in determining the biological activity of membrane functions. It was shown that exposure of Tetrahymena cells to various substances or different temperatures induces modification of membrane lipid composition, thus resulting in changes in the membrane fluidity. Therefore, some of these lipid modifications would be expected to affect biological functions taking place in membranes.

Membrane-Bound Enzymes

Adenylate cyclase. Tetrahymena has both adenylate and guanylate cyclase with high activities and they are found to be entirely localized in and tightly associated with the surface membrane, pellicle (Shimonaka and Nozawa, 1977). Rather surprisingly, the adenylate cyclase is hormone-sensitive, though this cell is not exposed to hormones under its normal growth conditions. The membrane-bound cyclase is partially solubilized by washing with 0.25 M sucrose, and this extracted enzyme still responds to epinephrine (Kassis and Kindler, 1975). However, the cyclase treated with Triton X-100 does no longer show stimulation by the hormone. On the other hand, recent studies have demonstrated the dependence

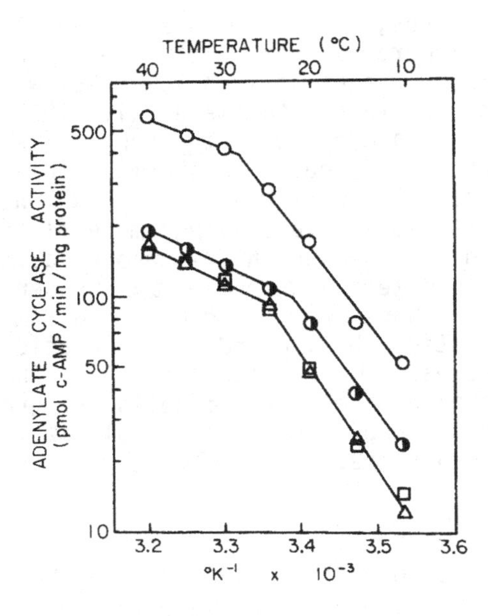

Fig. 7 Arrhenius plots of adenylate cyclase activity in the pellicular membranes from control (○) and ergosterol -replaced (◑) cells of T. pyriformis NT-I. The ergosterol -replaced membranes were treated with filipin (△) or amphotericin B (□).

of adenylate cyclase activity upon the physical properties of membrane lipids (Brivio-Hangland et al., 1976). We have examined the influence of ergosterol-supplementation upon the properties of membrane-bound enzyme. Fig. 7 illustrates the Arrhenius plots of adenylate cyclase activity of pellicles isolated from 39.5°C-grown cells. The specific activities of the ergosterol-replaced membrane are much lower than that of the control membrane, particularly above 22°C. The sharp discontinuity is seen at 28°C for the control pellicle, whereas there is an abrupt change at 22°C of the cyclase activity for the ergosterol-replaced pellicle. These two temperature are coincident with the initiation temperatures of membrane particle aggregation of two types of the pellicle. Further treatment of ergosterol-replaced pellicles by polyene antibiotics, amphotericin B and filipin, may affect to some extent the profile of the cyclase activities.

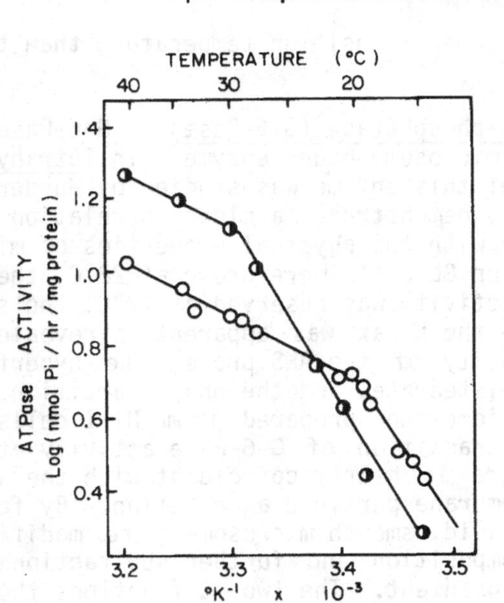

Fig. 8 Arrhenius plots of ATPase activity in the pellicular membranes from control (◑) or hexadecyl glycerol -supplemented (○) cells of T. pyriformis NT-I.

Adenosine triphosphatase (ATPase). The pellicular membrane has high activity of ATPase which has not yet been characterized. In Arrhenius plots of ATPase activity of pellicles from 39.5°C-grown cells, there is a discontinuity at 28°C. However, the lipid-modified pellicles prepared from hexadecyl glycerol (HDG)-supplemented cells exhibit a temperature-dependent transition at 19°C, and also have much lower specific activities above 28°C (Fig. 8). This drastic change in the transition temperature may be explained by differences in the membrane fluidity between the control and HDG-pellicles. The latter membranes are more fluid than the former as demonstrated in Fig. 4. The results of other recent studies support the concept that the activity of this membrane-bound ATPase might be influenced by the membrane fluidity. For example, treatment of the pellicles with the local anesthetic, dibucaine, or the inhalation anesthetic methoxyflurane, was observed to inhibit ATPase activity and lower its transition temperature to 21°C (Saeki et al., 1979). Also the pellicle from 15°C-grown cells has a lower transition temperature than that from 39.5°C-cells.

Glucose-6-phosphatase (G-6-Pase). G-6-Pase is well known to be a microsome-bound enzyme. In *Tetrahymena* cells, the activity of this enzyme was studied by Wunderlich *et al*. (1975) and they demonstrated a clear correlation of the G-6-Pase activity with the physical properties of microsomal membranes. When GL cells were grown at 28°C, the discontinuity of its activity was observed at 17°C, the same temperature at which the break was apparent as revealed by fluorescence intensity of the ANS probe, the hyperfine splitting of 5-doxylstearate, and the phase partition of 4-doxyldecane. The microsomes prepared from NT-I cells grown at 39.5°C show a transition of G-6-Pase activity at 25°C, and this temperature is nearly coincident with the onset temperature of membrane particle aggregation. By feeding cells with palmitic acid, smooth microsomes are modified in fatty acid composition and further subfractionated by sucrose density gradient. The two subfractions thus prepared

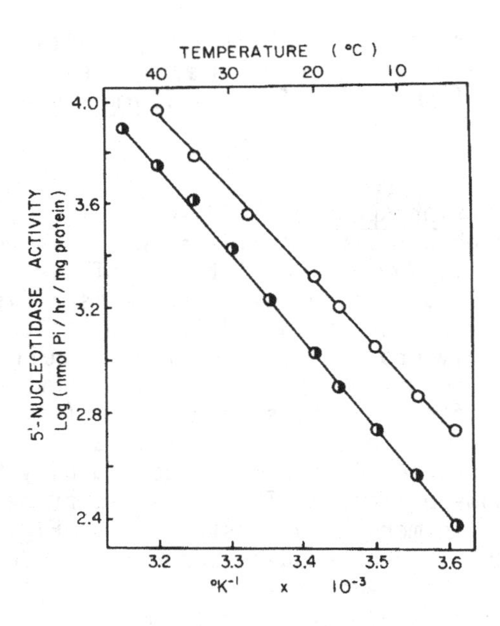

Fig. 9 Arrhenius plots of 5'-nucleotidase activity in the pellicular membranes from 39.5°C-grown (○) or 15°C-grown (◑) cells of *T. pyriformis* NT-I.

have different fluidities measured by electron spin resonance. In Arrhenius plots, the membrane subfraction with the smaller order parameter exhibits a higher specific activity and a transition at a higher temperature, as compared with the membrane with the greater order parameter (Kameyama et al., submitted).

5'-Nucleotidase. In contrast to the above-mentioned enzymes, 5'-nucleotidase which has been called an ecto -enzyme does not seem to respond to changes in the physical properties of membranes. No transition is observed in the Arrhenius plots of this enzyme activity in the pellicular membranes from either the 39.5°C- or 15°C-grown cells, despite the marked difference in their lipid composition (Fig. 9). The specific activity is higher in the 39.5°C-membrane than in the 15°C-membrane (Kawai and Nozawa, unpublished).

Phospholipid Transfer (Exchange) between Membranes

It is well known that phospholipid exchange contributes to membrane formation or dynamic equilibrium in the membrane constituents of cell organelles. We have shown using the radioisotope technique that there is a rapid movement of phospholipid molecules between different membranes in Tetrahymena (Nozawa and Thompson, 1971). This exchange process was found to occur even in the non-growing cells in which occurs active net synthesis of phospholipids.

The kinetics of this exchange observed in Tetrahymena membranes have been analyzed by electron spin resonance using spin-labeled phosphatidylcholine (Iida et al., 1978). The microsomes labeled with this spin probe were incubated with unlabeled cilia, pellicles and microsomes. The transfer rate is estimated from the increase in the central peak height which is caused by decrease in the exchange broadening in the electron spin resonance spectra due to dilution of the probe. The transfer rate from microsomes to other target membranes increases in the order, cilia < pellicles < microsomes. It should be noted that the fluidity of these membranes also becomes greater in the same order, implying that the incorporation of phospholipid is highly dependent upon the membrane fluidity at the acceptor sites. The fluidity dependence of lipid transfer is supported by other evidence which demonstrates marked differences in the transfer rate from microsomes between two types of membranes prepared from 39.5°C- and 15°C-grown cells. Phos-

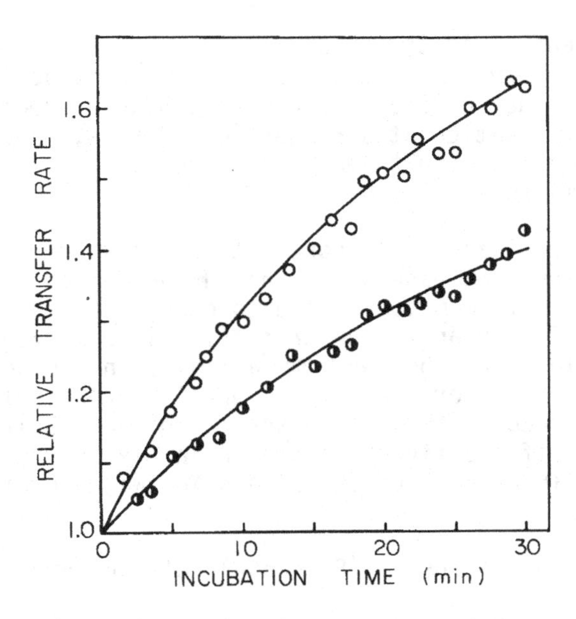

Fig. 10 Transfer of phosphatidylcholine from microsomes to pellicular membranes from 39.5°C-grown(◑) or 15°C -grown(○) cells of T. pyriformis NT-I. The donor membrane, microsomes were prepared from cells grown at 24°C. [Iida et al., Biochim. Biophys. Acta 508, 55 (1978)]

phatidylcholine molecules are assembled at a higher rate into the target membranes (cilia, pellicles or microsomes) from 15°C-grown cells than into membranes from 39.5°C-grown cells. The membranes from cells grown at 15°C have greater fluidity than those from cells grown at 39.5°C, as shown previously in Fig. 6. Fig. 10 illustrates an example of the transfer from microsomes to the acceptor membrane, pellicular membrane.

Electrophysiological Responses

The electrical response, one of the most typical biological events taking place in cell membranes, would be speculated to be somehow related to the physical state of membranes. In order to investigate correlation between electrical properties and lipid composition of membranes, a convenient organism is required which is suited to both electrophysiological techniques and modification of membrane lipid composition. Recently, we have disclosed that Tetrahymena is nearly as useful as Paramecia with which extensive studies on membrane excitability have been performed. Therefore, some electrical parameters were examined in the ergosterol-replaced Tetrahymena cells of GL strain (Onimaru et al., submitted). The preliminary results obtained suggest that ergosterol-replacement might induce alteration in membrane action potential. By further experiments with other membrane lipid-modified cells as described in the earlier section, it is hoped to gain a clue toward better understanding the lipids' role in various electrophysiological phenomena in biological membranes.

ABSTRACT

Since the membrane lipid composition of a lower eukaryote Tetrahymena pyriformis can be modified with ease by changing growth conditions, this cell is a highly useful model system for better understanding the interrelationship between lipid composition, physical states and function of biological membranes. In the present study we modified the lipid composition of the ciliary, pellicular, and microsomal membranes by supplementation, temperature acclimation and drug treatment, and found that the phospholipid polar head as well as fatty acyl group composition was appreciably altered. It was also shown that such lipid modification resulted in changes in the membrane fluidity and some membrane-bound enzymes in this cell.

ACKNOWLEDGEMENTS

This work was in part supported by grants from the Ministry of Education and the Japan Society for Promotion of Science under the Japan-U.S. Co-operative Science Research Program (5R070). The author is grateful to Dr. G.A. Thompson for stimulating discussion, and also to Miss Reiko Kasai for assistance in the preparation of this manuscript.

REFERENCES

Brivio-Hangland, R.P., S.L. Louis, K. Musch, N. Waldech, and M.A. Williams (1976). Biochim. Biophys. Acta 433, 150.
Chapman, D., and P.J. Quinn (1976). Chem. Phys. Lipids 17, 363.
Conner, R.L., and A.E. Reilly (1975). Biochim. Biophys. Acta 398, 209.
Ferguson, K.A., R.L. Conner, and F.B. Mallory (1971). Lipids 144, 448.
Fukushima, H., C.E. Martin, H. Iida, Y. Kitajima, G.A. Thompson, and Y. Nozawa (1976). Biochim. Biophys. Acta 431, 165.
Fukushima, H., T. Watanabe, and Y. Nozawa (1976). Biochim. Biophys. Acta 436, 249.
Fukushima, H., R. Kasai, N. Akimori, and Y. Nozawa (1978). Japan J. Exp. Med. 48, 373.
Iida, H., T. Maeda, K. Ohki, Y. Nozawa, and S. Ohnishi (1978) Biochim. Biophys. Acta 508, 55.
Kameyama, Y., K. Ohki, and Y. Nozawa, submitted.
Kasai, R., Y. Kitajima, C.E. Martin, Y. Nozawa, L. Skriver, and G.A. Thompson (1976). Biochemistry 15, 5228.
Kassis, S., and S.H. Kindler (1975). Biochim. Biophys. Acta 391, 513.
Lees, A.M., and E.D. Korn (1966). Biochemistry 5, 1475.
Martin, C.E., K. Hiramitsu, Y. Kitajima, Y. Nozawa, L. Skriver, and G.A. Thompson (1976). Biochemistry 15, 5218.
Nandini-Kishore, S.G., S.M. Mattox, C.E. Martin, and G.A. Thompson (1979). Biochim. Biophys. Acta 551, 315
Nozawa, Y., and G.A. Thompson (1971). J. Cell Biol. 49, 722.
Nozawa, Y., and G.A. Thompson (1979). In S.H. Hutner (ed.), Biochemistry and Physiology of Protozoa 2. Academic Press, New York. 275 pp.
Nozawa, Y., H. Fukushima, and H. Iida (1975). Biochim. Biophys. Acta 406, 248.
Nozawa, Y., R. Kasai, and T. Sekiya (1979). Biochim. Biophys. Acta 552, 38.
Onimaru, T., Y. Naitoh, K. Ohki, and Y. Nozawa, submitted.
Saeki, T., Y. Nozawa, H. Shimonaka, K. Kawai, M. Ito, and M. Yamamoto (1979). Biochem. Pharmacol. 28, 1095.
Shimonaka, H., and Y. Nozawa (1977). Cell Struct. Funct. 2, 81.
Shimonaka, H., H. Fukushima, K. Kawai, S. Nagao, Y. Okano, and Y. Nozawa (1978). Experientia 34, 586.
Thompson, G.A., and Y. Nozawa (1977). Biochim. Biophys. Acta 472, 55.
Wunderlich, F., V. Speth, W. Batz, and H. Kleinig (1973). Biochim. Biophys. Acta 298, 39.
Wunderlich, F., A. Ronai, V. Speth, J. Seelig, and A. Blume (1975). Biochemistry 14, 3730.

REGULATION OF PHOSPHOLIPID N-METHYLATION IN THE HEPATOCYTE

Björn Åkesson

Department of Physiological Chemistry,

University of Lund, P.O.Box 750, Lund,Sweden

ABSTRACT

The evidence for control of phosphatidylethanolamine N-methylation to phosphatidylcholine by substrate availability is reviewed. In cultured hepatocytes the amount of methionine in the medium markedly influences the rate of methylation. Furthermore, in vitamin B_{12}-deficient rats the methylation is reduced, most probably due to a diminished biosynthesis of methionine. The regulatory role of the amount of phosphatidylethanolamine has been unclear, but our recent experiments show that manipulation of its concentration in membranes or in intact hepatocytes greatly influences the methylation rate. This provides a mechanism for an autoregulation of the relative proportions of the two major phospholipids in liver membranes. Also methionine deprivation and phenobarbital treatment of hepatocytes increases phospholipid methylation due to an increased amount of phosphatidylethanolamine. Previous controversies on dietary induced changes in phospholipid methylation can be reconciled with the interpretation given here on the properties of different assays.

INTRODUCTION

The step-wise N-methylation of phosphatidylethanolamine (PE) to phosphatidylcholine (PC) in a liver microsomal fraction was described by Bremer and Greenberg (1960). The methylation occurs in several organs but is most active in the endoplasmic reticulum of the liver (Bjørnstad and Bremer, 1966; Van Golde et al., 1974). There is evidence that the first of the three methylation steps is rate-limiting but studies on this point have been hampered by the fact that exogenous PE is not methylated in vitro (Rehbinder and Greenberg, 1965). Experiments with mutants of Neurospora provided evidence that the first methylation is catalyzed by one enzyme and the two other methylations by another enzyme (Scarborough and Nyc, 1967). Direct experimental support for the existence of two enzymes in adrenal medulla was obtained recently by Hirata et al.(1978). The enzyme catalyzing the methylation of PE (PEMT) had a low Km for S-adenosyl-methionine (AdoMet) and was stimulated by Mg^{+2}. The second methylating enzyme (PDMEMT) had a higher Km for AdoMet and was not Mg^{+2}-dependent.

The activity of methyl tranfer has been reported to change in choline deficiency, after alcohol consumption and after treatment with drugs (Fallon et al., 1969; Glenn and Austin, 1971; Feuer et al., 1973), but it is not known whether this applies to one or several enzymes.

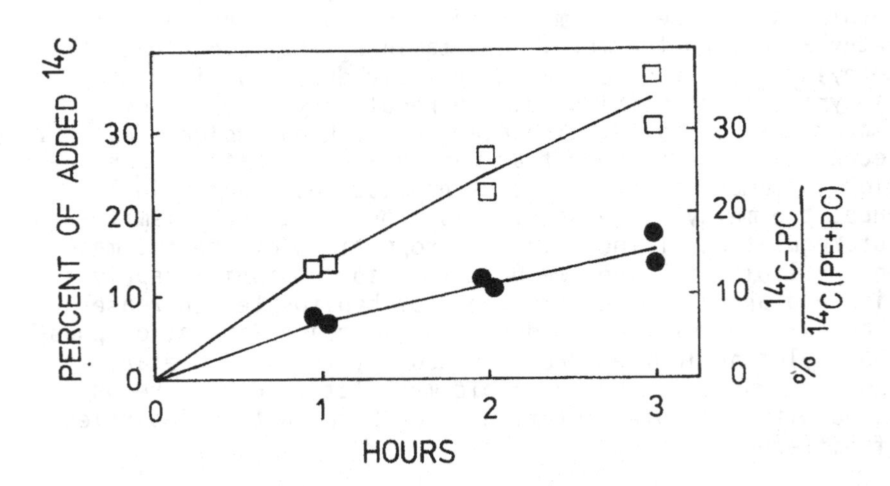

Figure 1. Assay of phospholipid methylation in cultured hepatocytes incubated with (^{14}C)ethanolamine.□, % of added ^{14}C in lipids;●, ^{14}C-PC/^{14}C(PE+PC).

REGULATORY INFLUENCE OF METHIONINE/ADO-MET

Phospholipid methylation can be assayed in several ways. In isolated hepatocytes the methyl incorporation from (Me-^{14}C)methionine into phospholipids reaches a saturation level at 0.4 mM (Sundler and Åkesson, 1975). Alternatively, the conversion of PE to methylated phospholipids can be measured after labeling PE by incubation of hepatocytes with labeled ethanolamine (Fig. 1). This conversion is stimulated markedly by methionine. The two measures correlate very well in most conditions, but can diverge e.g. after addition of N-methylethanolamines (Sundler and Åkesson, 1975; Åkesson, 1977a).

The concentration range of methionine in the hepatocyte medium which affected PE methylation is close to the physiological range in the plasma (Sundler and Åkesson, 1975). An increased load of methionine increases the liver concentration of AdoMet in vivo (Baldessarini, 1966). Consequently, agents which change the cellular uptake of methionine and/or AdoMet synthesis will probably affect PE methylation.

Recently, Hoffman et al.(1979) determined the activities of phospholipid methylation and the concentration of AdoMet in liver during development. AdoMet did not change much but the ratio AdoMet/AdoHcy and also phospholipid methylation in vitro decreased with age. The AdoMet/AdoHcy ratio may therefore be an alternative regulatory parameter to the concentration of AdoMet for PE methylation in vivo. However, this can probably not explain changes in phospholipid methylation during development determined in vitro.

IMPORTANCE OF PHOSPHOLIPID SUBSTRATES

As mentioned above the natural substrate of phospholipid methylation, PE, is not methylated when added to subcellular fractions in vitro. Instead N-methylated PE's, above all PDME (phosphatidyldimethylethanolamine), have been used by many workers to saturate the phospholipid requirement in the assay. It is important to note, especially in view of the results of Hirata et al.(1978), that assays with added PDME may not give a measure of the rate-limiting enzyme activity in phospholipid methylation. Since PE is the second major phospholipid in the liver, one could visualize that PEMT always would have access to large amounts of substrate, but the influence of the cellular amount of PE on the methylation rate has not been studied, although it is known that its fatty acid composition affects the methylation (Lyman et al., 1968; Arvidson, 1968). We therefore used primary cultures of rat hepatocytes, where the PE content was modified. In hepatocytes maintained in a medium lacking ethanolamine for several days, the proportion of PE decreased (Åkesson, 1977b). This decrease could be prevented by 1 mM ethanolamine. The rate of PE methylation followed a similar course as the amount of PE, both when measured as incorporation of ^{14}C from (Me-^{14}C)methionine into phospholipids in intact cells or as synthesis from AdoMet in cell homogenates (Åkesson, 1978). This suggests that also the amount of phospholipid substrate in the methylation has a marked influence on the reaction rate. The maintenance of PEMT during incubation of hepatocytes with ethanolamine was not due to increased synthesis of new enzyme protein, since the difference to controls persisted, when incubation was carried out in the presence of actinomycin D or cycloheximide (Åkesson, 1979).

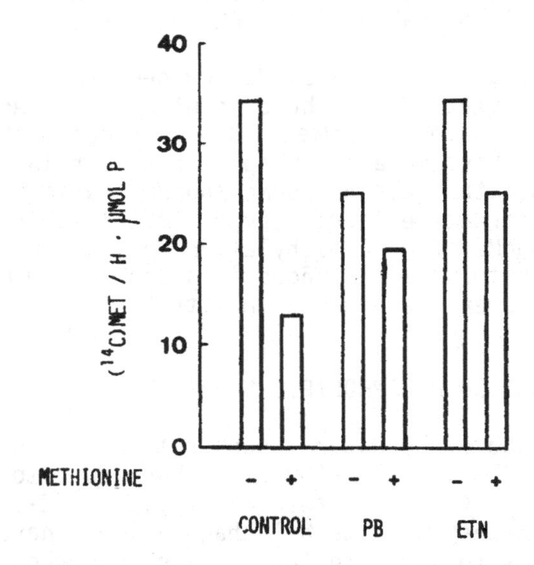

Figure 2. Effect of preincubation with methionine, ethanolamine and phenobarbital on phospholipid methylation from (Me-^{14}C)methionine in cultured hepatocytes. Data are expressed as nmol methyl per hour per µmol cell phospholipids.

Also maintenance of hepatocytes in the absence of methionine for 2 days resulted in an increase in PE. In these cells PE methylation was enhanced, reflecting the increased availability of PE substrate. Addition of ethanolamine in the absence of methionine did not stimulate methylation further (Fig. 2), suggesting that an optimal enrichment of PE was attained already by omission of methionine.

PE methylation was usually expressed as nmol methyl per hour per μmol phospholipid, the latter as a measure of hepatocyte cell mass. If the methylation rate only reflects the amount of PE, the rate would be constant if expressed per μmol PE. This was not the case (Fig. 3), which points to the possibility that the amount of PE in a specific subcellular pool may be the regulating parameter.

A more direct way to study the influence of the PE substrate was to modify the amount of membrane-bound PE in subcellular fractions in vitro. Therefore a rat liver microsomal fraction was treated with increasing amounts of methylacetimidate, an amino-group-blocking agent. In the reisolated

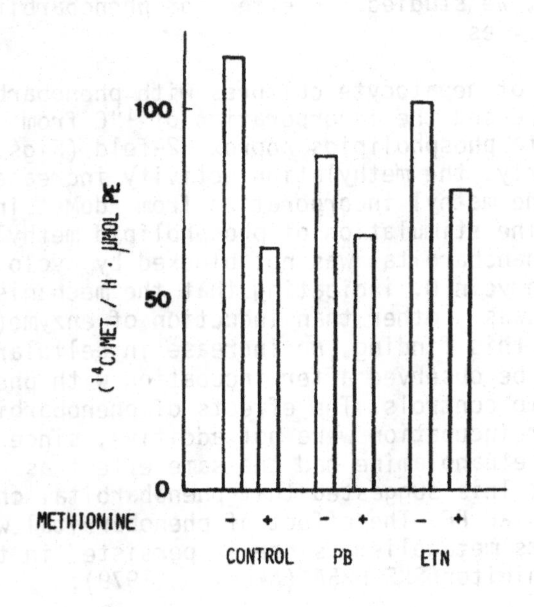

Figure 3. Phospholipid methylation and cellular phosphatidylethanolamine. The data in Fig. 2 were expressed as nmol per hour per μmol phosphatidylethanolamine.

microsomes, there was a parallel decrease in the amount of PE and activity of PEMT, again indicating the importance of the PE amount (Åkesson, 1979). It could not be excluded that PEMT was inactivated by the methylacetimidate treatment, but the methylation of added PDME remained constant.

EFFECT OF BARBITURATES

Administration of phenobarbital to intact animals for several days increases the amount of enzyme proteins and phospholipids of the endoplasmic reticulum. Since the accumulation is most marked for PC, a stimulation of phospholipid methylation has been suspected. This has also been observed by some investigators (Young *et al*., 1971; Feuer *et al*., 1973), but others found a more complicated response (Davison and Wills, 1974). On the other hand, a decrease in the conversion of PE to PC in rat liver after phenobarbital treatment has been reported (Davison and Wills, 1974). Also a phenobarbital-induced decrease in phospholipid catabolism has been suggested (Stein and Stein, 1969). To evaluate these alternatives, we studied the effect of phenobarbital in cultured hepatocytes.

Preincubation of hepatocyte cultures with phenobarbital for 2 days increased the incorporation of ^{14}C from (^{14}C)methionine into phospholipids approx. 2-fold (Figs. 2 and 3). Concomitantly, the methylation activity increased, as measured by the methyl incorporation from AdoMet in cell homogenates. The stimulation of phospholipid methylation induced by phenobarbital was not blocked by cycloheximide and actinomycin D, indicating that the mechanism behind this effect was other than induction of enzyme(s). In accordance with this finding, no increase in cellular phospholipid could be observed after incubation with phenobarbital compared to controls. The effects of phenobarbital and ethanolamine preincubation were not additive, since phenobarbital plus ethanolamine had the same effect as ethanolamine alone. This suggested that phenobarbital changed the amount of cellular PE. The effect of phenobarbital was not dependent on its metabolism, since it persisted in the presence of the inhibitor SKF-525A (Åkesson, 1979).

When phenobarbital was added together with (^{14}C)methionine and (^{3}H)ethanolamine in the assay, it inhibited phospholipid methylation. The degree of inhibition with both assays was the same and 2 mM phenobarbital inhibited phospholipid methylation approximately 50 %. This immediate effect of phenobarbital on phospholipid methylation was also observed in hepatocyte cultures. The similar inhibition in both assays suggested that the rapid effect of phenobarbital was due to a decreased availability of methionine and thereby of AdoMet. As described above the amount of methionine in the incubation medium has a pronounced effect on the methylation of PE

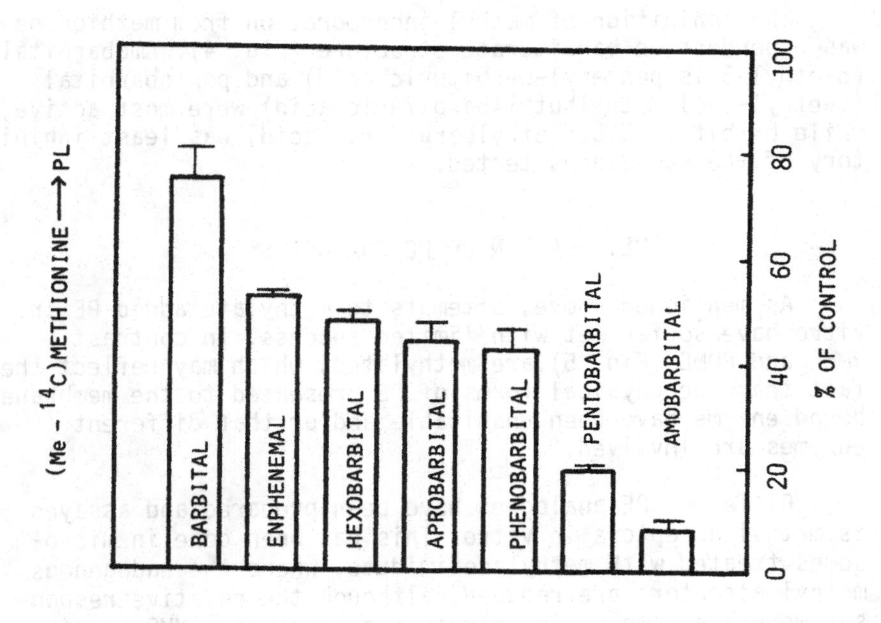

Figure 4. Immediate effects of different barbiturates (2 mM) on phospholipid methylation from (Me-^{14}C)methionine in suspended hepatocytes.

in hepatocytes (Sundler and Åkesson, 1975). Therefore, the long-term effect of phenobarbital preincubation is probably another example of methionine deprivation, which increases the amount of cellular PE and thereby the methylation rate.

To elucidate the mechanism of action of phenobarbital, several types of experiments were performed. Phenobarbital did not inhibit PEMT activity in vitro. Similarly, no inhibition was observed for the synthesis of AdoMet in rat liver supernatant. Since phenobarbital inhibited also the incorporation of (^{14}C)methionine into hepatocyte protein, the target of phenobarbital action may be the cellular uptake of methionine. Methionine is taken up by several transport systems in hepatocytes (Fehlmann et al., 1979), and it remains to be studied whether phenobarbital influences any of these systems.

The inhibition of methyl incorporation from methionine was dependent on barbiturate structure (Fig. 4). Amobarbital (5-ethyl-5-isopentenyl-barbituric acid) and pentobarbital (5-ethyl-5-(1-methylbutyl)barbituric acid) were most active, while barbital (5,5-diethylbarbituric acid) was least inhibitory of the substances tested.

METHYLATION OF PE ANALOGUES

As mentioned above, attempts to methylate added PE in vitro have so far met with limited success. In contrast, PMME and PDME (Fig. 5) are methylated, which may reflect the fact that the physical forms of PE presented to the membrane-bound enzyme have been unsuitable and/or that different enzymes are involved.

Different PE analogues have been prepared and assayed as methyl acceptors in vitro. This has been done in microsomes treated with methylacetimidate, where the endogenous methyl acceptors are reduced, although the relative responses were the same as in untreated microsomes. PMME was the best acceptor followed by PDME and PMEE. The relative efficiency of the three acceptors was maintained over a broad concentration range. Added PE was not methylated, nor were phosphatidylpropanolamine or PDEE. The results with the latter compound indicate that phospholipids with too bulky alkyl substituents do not act as substrates. Neither N-acetimidoyl-PE nor N-acetyl-PE was methylated.

Although the experiments with microsomal systems gave valuable data on the substrate specificity, it is possible that the added phospholipids do not reflect the behaviour of membrane-bound substrates. To circumvent this problem, the methylation reaction can be studied after modifying the hepatocyte phospholipid composition by incubation with different phospholipid base precursors (Åkesson, 1977b). As discussed above, preincubation with ethanolamine gave approx. twofold higher methylation rate than in control incubations. The efficiency of the phospholipid substrate in the hepatocyte is probably influenced both by the extent of phospholipid accumulation and the reactivity of the membrane-bound substrate. Further experiments to correlate the phospholipid accumulation to the rate of phospholipid methylation are necessary to clarify this point.

PHOSPHATIDYLETHANOLAMINE ANALOGUES

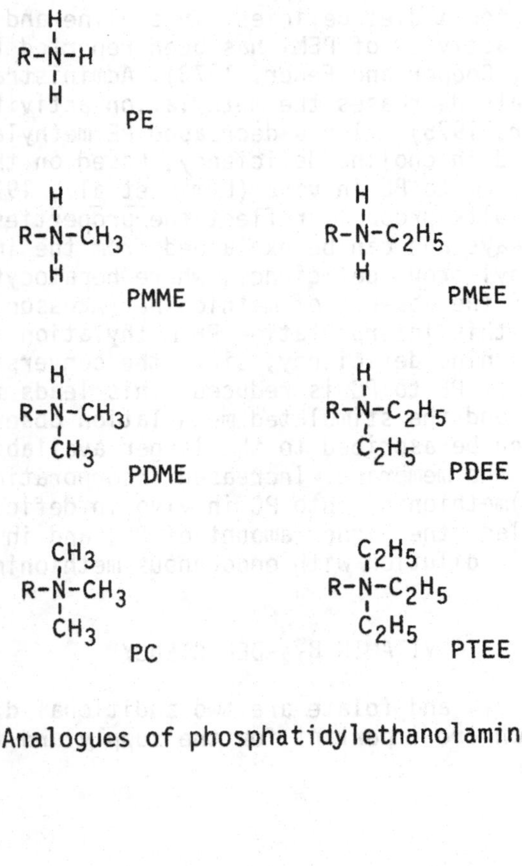

H
R-N-H
H
PE

H
$R-N-CH_3$
H
PMME

H
$R-N-C_2H_5$
H
PMEE

H
$R-N-CH_3$
CH_3
PDME

H
$R-N-C_2H_5$
C_2H_5
PDEE

CH_3
$R-N-CH_3$
CH_3
PC

C_2H_5
$R-N-C_2H_5$
C_2H_5
PTEE

Figure 5. Analogues of phosphatidylethanolamine.

So far, only the total incorporation of methyl groups into phospholipids has been considered, but evidently different phospholipid substrates can accept between 1 and 3 methyl groups. In hepatocytes preincubated with ethanolamine and in control incubations most of the labeled methyl groups was in PC with little accumulation in partially methylated PE's. The increase in phospholipid methylation after preincubation with monoethylethanolamine was mainly accounted for by formation of phospholipids with higher chromatographic mobility than PC. This probably represents phosphatidylethylmethylethanolamine since another possible product, phosphatidyldimethylethylethanolamine would migrate very close to PC. Significant formation of non-quartenary phospholipids occurred also after preincubation with 2-aminobutanol. A related compound, monomethylated phosphatidyl-N-isopropylethanolamine, has been isolated by Moore *et al*. (1978).

PHOSPHOLIPID METHYLATION IN DIFFERENT NUTRITIONAL STATES

In rats fed a diet deficient in choline and methionine, an increased activity of PEMT has been reported (Glenn and Austin, 1971; Cooper and Feuer, 1973). Administration of choline to rats decreases the methylation activity (Skurdal and Cornatzer, 1975). Also a decreased PE methylation has been suggested in choline deficiency, based on the slower conversion of PE to PC in vivo (Lyman *et al*., 1973). The divergent results probably reflect the properties of the different assays and can be explained from the in vitro model of methyl-group deficiency, where hepatocytes are maintained in the absence of methionine (Åkesson, 1978). According to this interpretation PE methylation is low in choline-methionine deficiency, since the conversion in intact cells of PE to PC is reduced. This leads to an increase of PE and the stimulated methylation observed in vitro can then be ascribed to the larger availability of PE substrate in the membrane. Increased incorporation of ^{14}C from (Me-^{14}C)methionine into PC in vivo in deficient animals may also reflect the higher amount of PE, and in addition differences in dilution with endogenous methionine.

VITAMIN B_{12}-DEFICIENCY

Vitamin B_{12} and folate are two additional dietary components, which are important for the supply and turnover of one-carbon fragments in the body. The effect of these nutrients on phospholipid methylation has not previously been studied as extensively as those of methionine and choline. We have used a rat model where pronounced vitamin B_{12}-deficiency was produced (Fehling et al., 1978). The diet contained adequate amounts of choline, methionine and folate. The incorporation of injected (^{14}C)formaldehyde into liver PC was reduced to 50 % in deficient rats, which is consistent with a lower conversion to methionine due to reduced activity of tetrahydropteroylglutamate methyltransferase (Åkesson et al., 1978). PE methylation, determined after the injection of labeled ethanolamine, was significantly lower in both male and female rats (Åkesson et al., 1978) and also in rat sucklings (Åkesson et al., 1979) compared to supplemented controls (Fig. 6). The reduced methylation in vitamin B_{12}-deficiency led to a higher mass ratio PE/PC in the liver.

Also these results can be interpreted according to the model of methyl-group deficiency described above. Accordingly, the availability of methionine and AdoMet in the liver decreased during vitamin B_{12}-deficiency, due to inhibited endogenous synthesis, although the dietary supply of methio-

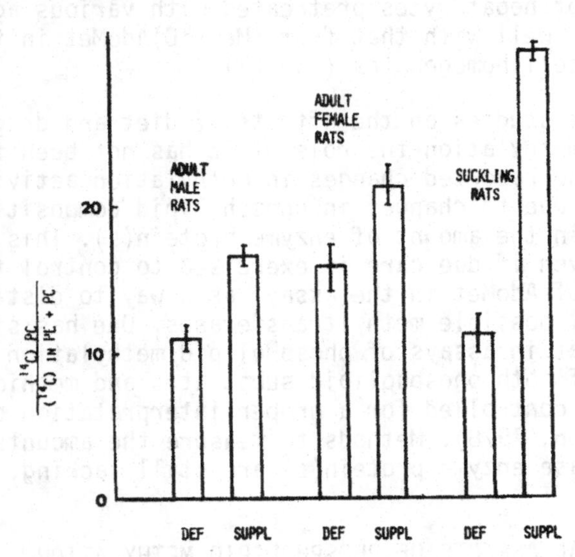

Figure 6. Methylation of liver phosphatidylethanolamine to phosphatidylcholine after injection of (^{14}C)ethanolamine to vitamin B_{12}-deficient rats and their supplemented controls. Drawn from (Åkesson et al., 1978; Åkesson et al., 1979).

nine was kept at a level that is normally adequate. The concentration of AdoMet in liver decreased in vitamin B_{12}-deficiency, although it could be normalized by including methionine in the diet (Vidal and Stokstad, 1974). These data were, however, obtained in rats with less prolonged deficiency than ours (Fehling et al., 1978). At present it is not known whether the activities of phospholipid-methylating enzyme(s) are also changed in vitamin B_{12}-deficiency.

ASSAY OF PHOSPHOLIPID METHYLATION

As pointed out several times in this paper, the assay of phospholipid methylation may give divergent results depending on the principle of the assay. In vivo measurement of the conversion of PE to PC reflects the availability of methionine/AdoMet to a large degree and this also applies when this parameter is measured in isolated cells. The incorporation of methyl groups from methionine or AdoMet in vivo or in vitro will be influenced by the amount of available phospholipid in the system tested. Accordingly, the incorporation of methyl groups from (Me-^{14}C)methionine into phospholipids of hepatocytes pretreated with various agents correlated very well with that from (Me-^{14}C)AdoMet in the corresponding cell homogenates (Fig. 7).

In several studies on the effects of diet and drugs on phospholipid methylation the role of PE has not been fully realized and the reported changes in methylation activity may as well be due to changes in phospholipid composition as in changes in the amount of enzyme protein(s). This will be a problem even if due care is exercised to control the concentration of AdoMet in the assay, as a way to distinguish between several possible methyltransferases. One has simply to conclude that in assays of phospholipid methylation the availability of both phospholipid substrates and methionine/AdoMet must be controlled for a proper interpretation of the results (Åkesson, 1978). Methods to measure the amounts of methyltransferase enzyme protein(s) are still lacking.

FUNCTIONAL ASPECTS OF PHOSPHOLIPID METHYLATION

PE and PC are the major lipids in the hepatocyte and much work has been done to establish their configuration in model systems. Also partially N-methylated phospholipids have been studied (Vaughan and Keough, 1974), and it was found that the temperature for the gel-to-liquid-crystalline transition decreased with increasing number of N-methyl groups. Therefore changes in the mass ratio PE/PC may influence membrane fluidity.

The ratio PE/PC is controlled via a variety of synthetic and degradative pathways and since PE methylation accounts for a sizeable part of PC synthesis (Åkesson and Sundler, 1977), it has a major role for the control of the PE/PC ratio. It also provides a mechanism for autoregulation of this ratio since in conditions with low PE methylation due to limited amounts of methionine/AdoMet, the amount of membrane-bound PE will increase. In turn, this will increase PE methylation, at least when AdoMet is available, which will counteract the increase in the PE/PC ratio.

PE methylation is also the pathway for endogenous synthesis of choline and forms a part of a complex metabolic cycle of one-carbon fragments (Bremer et al., 1960). In this

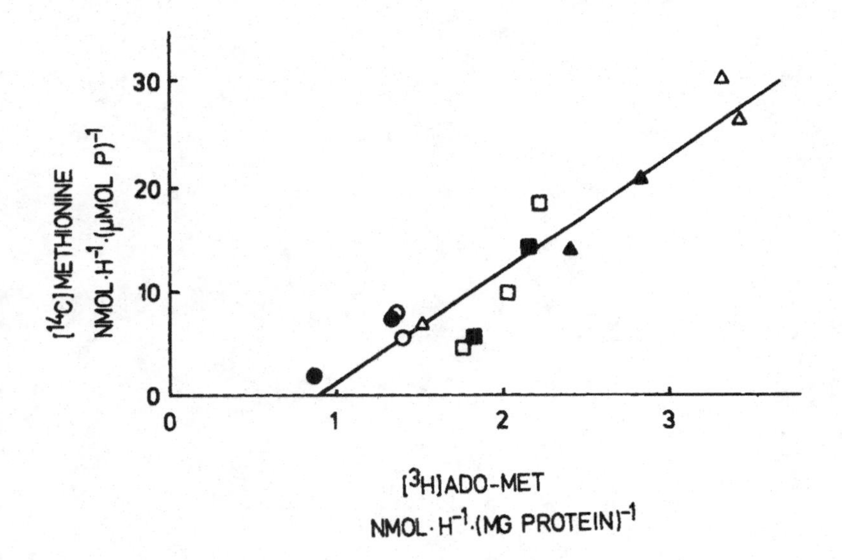

Figure 7. Hepatocytes were cultured in the presence of different agents. In intact cells or homogenates the methylation of phospholipid was assayed from (Me-^{14}C)methionine or (^{3}H)AdoMet, respectively.

context PC can be regarded as a large reservoir of choline and one-carbon fragments. Dietary choline deficiency produces several symptoms in animals, but the molecular mechanisms behind them are not known. This is partially due to the complex metabolic relationship of choline to other dietary components, such as methionine, folate and vitamin B_{12} (Krebs _et al_., 1976).

(The studies from the author´s laboratory were supported by the Swedish Medical Research Council (project 3968) and A. Påhlsson´s Foundation.)

REFERENCES

Åkesson, B. (1977a). Biochem. J. 168, 401.
Åkesson, B. (1977b). Biochem. Biophys. Res. Commun. 76, 93.
Åkesson, B. (1978). FEBS Lett. 92, 177.
Åkesson, B. (1979). (to be published).
Åkesson, B. and R. Sundler (1977). Biochem. Soc. Trans. 5, 43.
Åkesson, B., C. Fehling, and M. Jägerstad (1978). Br. J. Nutr. 40, 521.
Åkesson, B., C. Fehling, and M. Jägerstad (1979). Br. J. Nutr. 41, 263.
Arvidson, G.A.E. (1968). Eur. J. Biochem. 5, 415.
Baldessarini, R.J. (1966). Biochem. Pharmacol. 15, 741.
Bjørnstad, P. and J. Bremer (1966). J. Lipid Res. 7, 38.
Bremer, J. and D.M. Greenberg (1960). Biochim. Biophys. Acta 37, 173.
Bremer, J,. P.H. Figard, and D.M. Greenberg (1960). Biochim. Biophys. Acta 43, 477.
Cooper, S.D. and G. Feuer (1973). Toxicol. Appl. Pharmacol. 25, 7.
Davison, S.C. and E.D. Wills (1974). Biochem. J. 142, 19.
Fallon, H.J., P.M. Gertman, and E.L. Kemp (1969). Biochim. Biophys. Acta 187, 94.
Fehling, C., M. Jägerstad, B. Åkesson, J. Axelsson, and A. Brun (1978). Br. J. Nutr. 39, 501.
Fehlmann, M., A. Le Cam, P. Kitabgi, J.-F. Rey, and P. Freychet (1979). J. Biol. Chem. 254, 401.
Feuer, G., D.R. Miller, S.D. Cooper, F.A. de la Iglesia, and G. Lumb (1973). Int. J. Clin. Pharmacol. 7, 13.
Glenn, J.L. and W. Austin (1971). Biochim. Biophys. Acta 231, 153.
Hirata, F., O.H. Viveros, E.M. Diliberto, Jr., and J. Axelrod (1978). Proc. Natl. Acad. Sci. USA 75, 1718.
Hoffman, D.R., W.E. Cornatzer, and J.A. Duerre (1979). Can. J. Biochem. 57, 56.
Krebs, H.A., R. Hems, and B. Tyler (1976). Biochem. J. 158, 341.
Lyman, R.L., S.M. Hopkins, G. Sheehan, and J. Tinoco (1968). Biochim. Biophys. Acta 152, 197.
Lyman, R.L., G. Sheehan, and J. Tinoco (1973). Lipids 8, 71.
Moore, C., M.L. Blank, T.-C. Lee, B. Benjamin, C. Piantadosi, and F. Snyder (1978). Chem. Phys. Lip. 21, 23.
Rehbinder, D. and D.M. Greenberg (1965). Arch. Biochem. Biophys. 108, 110.
Scarborough, G.A. and J.F. Nyc (1967). J. Biol. Chem. 242, 238.

Skurdal, D.N. and W.E. Cornatzer (1975). Int. J. Biochem. 6, 579.
Stein, Y. and O. Stein (1969). Israel J. Med. Sci. 5, 985.
Sundler, R. and B. Åkesson (1975). J. Biol. Chem. 250, 3359.
Van Golde, L.M.G., J. Raben, J.J. Batenburg, B. Fleischer, F. Zambrano, and S. Fleischer (1974). Biochim. Biophys. Acta 360, 179.
Vaughan, D.J. and K.M. Keough (1974). FEBS Lett. 47, 158.
Vidal, A.J. and E.L.R. Stokstad (1974). Biochim. Biophys. Acta 362, 245.
Young, D.L., G. Powell, and W.O. McMillan (1971). J. Lipid Res. 12, 1.

AUTHOR INDEX

SUBJECT INDEX